Plant Improvement: Molecular Breeding and Genetic Perspectives

Plant Improvement: Molecular Breeding and Genetic Perspectives

Edited by Kiara Woods

SYRAWOOD
PUBLISHING HOUSE

New York

Published by Syrawood Publishing House,
750 Third Avenue, 9th Floor,
New York, NY 10017, USA
www.syrawoodpublishinghouse.com

Plant Improvement: Molecular Breeding and Genetic Perspectives
Edited by Kiara Woods

International Standard Book Number: 978-1-64740-412-3 (Hardback)

Trademark Notice: Registered trademark of products or corporate names are used only for explanation and identification without intent to infringe.

Cataloging-in-publication Data

Plant improvement : molecular breeding and genetic perspectives / edited by Kiara Woods.
    p. cm.
Includes bibliographical references and index.
ISBN 978-1-64740-412-3
1. Plant molecular genetics. 2. Plant breeding. 3. Plants--Molecular aspects. I. Woods, Kiara.
QK981.4 .M65 2023
572.865 2--dc23

TABLE OF CONTENTS

Permissions

List of Contributors

Index

PREFACE

This book aims to highlight the current researches and provides a platform to further the scope of innovations in this area. This book is a product of the combined efforts of many researchers and scientists, after going through thorough studies and analysis from different parts of the world. The objective of this book is to provide the readers with the latest information of the field.

The study of genes, genetic variation and heredity in plants is referred to as plant genetics. It is a specialized field of biology and botany. Molecular breeding refers to the application of molecular biology tools in breeding crop plants. It is used for plant improvement in terms of starch composition, grain quality, flower color, leaf color, flower development and plant architecture. The aim of molecular plant breeding is to create, select and fix superior plant phenotypes that are suitable for the needs of farmers and consumers. Marker-assisted backcrossing (MABC), marker-assisted selection (MAS), marker-assisted recurrent selection (MARS), and genome-wide selection (GWS) are a few important molecular breeding techniques. These techniques are applied to analyze the different agronomic traits in order to increase the production, reduce the cost of producing plants with biotic and abiotic stress tolerance, and increase the quality of plants. The book aims to shed light on some of the unexplored aspects of plant genetics and molecular breeding. A number of latest researches have been included to keep the readers updated with the global concepts in this area of study. The book will serve as a valuable source of reference for graduate and postgraduate students.

I would like to express my sincere thanks to the authors for their dedicated efforts in the completion of this book. I acknowledge the efforts of the publisher for providing constant support. Lastly, I would like to thank my family for their support in all academic endeavors.

Editor

Conventional and Molecular Techniques from Simple Breeding to Speed Breeding in Crop Plants: Recent Advances and Future Outlook

Sunny Ahmar [1,†], **Rafaqat Ali Gill** [2,†], **Ki-Hong Jung** [3,*], **Aroosha Faheem** [4],
Muhammad Uzair Qasim [1], **Mustansar Mubeen** [5] and **Weijun Zhou** [6,*]

[1] National Key Laboratory of Crop Genetic Improvement, College of Plant Science and Technology, Huazhong Agricultural University, Wuhan 430070, Hubei, China; sunny.ahmar@yahoo.com (S.A.); uzairqasim1149@yahoo.com (M.U.Q.)

[2] Oil Crops Research Institute, Chinese Academy of Agriculture Sciences, Wuhan 430070, China; drragill@caas.cn

[3] Graduate School of Biotechnology & Crop Biotech Institute, Kyung Hee University, Yongin 17104, Korea

[4] State Key Laboratory of Agricultural Microbiology and State Key Laboratory of Microbial Biosensor, College of Life Sciences Huazhong Agriculture University, Wuhan 430070, China; arushafaheem@hotmail.com

[5] State Key Laboratory of Agricultural Microbiology and Provincial Key Laboratory of Plant Pathology of Hubei Province, College of Plant Science and Technology, Huazhong Agricultural University, Wuhan 430070, China; mustansar01@yahoo.com

[6] Institute of Crop Science and Zhejiang Key Laboratory of Crop Germplasm, Zhejiang University, Hangzhou 310058, China

* Correspondence: khjung2010@khu.ac.kr (K.-H.J.); wjzhou@zju.edu.cn (W.Z.)

† These authors contributed equally to this work.

Abstract: In most crop breeding programs, the rate of yield increment is insufficient to cope with the increased food demand caused by a rapidly expanding global population. In plant breeding, the development of improved crop varieties is limited by the very long crop duration. Given the many phases of crossing, selection, and testing involved in the production of new plant varieties, it can take one or two decades to create a new cultivar. One possible way of alleviating food scarcity problems and increasing food security is to develop improved plant varieties rapidly. Traditional farming methods practiced since quite some time have decreased the genetic variability of crops. To improve agronomic traits associated with yield, quality, and resistance to biotic and abiotic stresses in crop plants, several conventional and molecular approaches have been used, including genetic selection, mutagenic breeding, somaclonal variations, whole-genome sequence-based approaches, physical maps, and functional genomic tools. However, recent advances in genome editing technology using programmable nucleases, clustered regularly interspaced short palindromic repeats (CRISPR), and CRISPR-associated (Cas) proteins have opened the door to a new plant breeding era. Therefore, to increase the efficiency of crop breeding, plant breeders and researchers around the world are using novel strategies such as speed breeding, genome editing tools, and high-throughput phenotyping. In this review, we summarize recent findings on several aspects of crop breeding to describe the evolution of plant breeding practices, from traditional to modern speed breeding combined with genome editing tools, which aim to produce crop generations with desired traits annually.

Keywords: food security; food scarcity; conventional breeding; CRISPR/Cas9; CRISPR/Cpf1; high-throughput phenotyping; speed breeding

1. Introduction

Since the early 1900s, plant breeding has played a fundamental role in ensuring food security and safety and has had a profound impact on food production all over the world [1,2]. In recent years, however, problems related to food quality and quantity globally have arisen as a consequence of the excessive food requirement for the rapidly increasing human population. Furthermore, radical changes in weather conditions caused by global climate change are causing heat and drought stress; consequently, farmers around the world are facing significant yield losses [3]. Global epidemics, such as the Irish potato blight of the 1840s and the Southern corn leaf blight in the United States in the 1970s, were disastrous events leading to the deaths of millions of people due to food shortage [4,5]. In recent years, the ratio of food production to consumption has decreased considerably, while both urbanization rates and demographic growth have increased globally. In this era of fast development and rapid growth, people prefer to consume processed foods, where nutritional quality is compromised. The world is expected to reach 10 billion by 2050, but no satisfactory strategies are in place to feed this massive population [6,7]. Developed countries have increased their agricultural productivity, partially meeting their food requirements, but this has resulted in increased stress on food manufacturing departments [8].

Plant breeding can be used to develop plants with desired traits [9]. Artificial plant selection has been used by humans for the past 10,000 years, selecting and breeding plants with higher nutritional values [10] (Figure 1). Traditional agricultural methods aimed to improve the nutritional status of different food plants. Recent scientific developments provide a wide range of possibilities and innovations in plant breeding [11]. To satisfy the continuously increasing demand for plant-based products, the current level of annual yield enhancement in major crop species (varying from 0.8–1.2%) must be doubled [12].

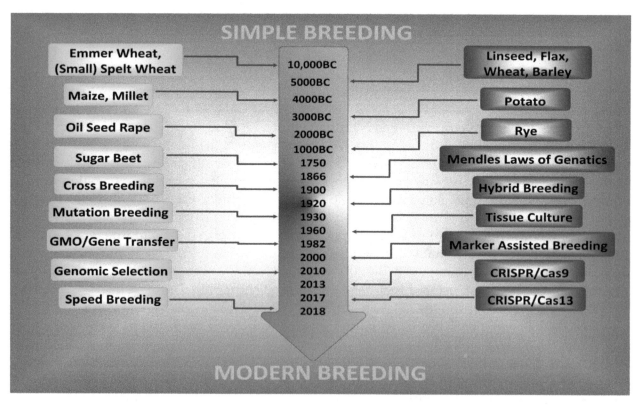

Figure 1. Historical milestones in plant breeding. For 10,000 years, farmers and breeders have been developing and improving crops. Presently, farmers feed 10 times more people using the same amount of land as 100 years ago.

The introduction of Mendelian laws revolutionized the field of crop breeding. Over the last 150 years, crop development has been altered to a great extent as a consequence of contemporary cutting-edge genomics [13]. Different approaches have been used to shorten the duration of plant reproductive cycles. Novel techniques developed in this decade, such as genomic selection, high-throughput phenotyping (HTP), and modern speed breeding, have been shown to accelerate plant breeding. Genetic engineering and molecular methods have also played a role in developing crops with desirable characteristics using gene transformation [14–17]. Other techniques like large-scale sequencing, genomics, rapid gene isolation, and high-throughput molecular markers have also been proposed to improve the breeding of commercially important crop species, such as cisgenesis, intragenesis, polyploidy breeding, and mutation breeding [18–21].

Conventional breeding techniques are inadequate for plant genome enhancement to develop new plant varieties. To overcome this obstacle in plant breeding practices, molecular markers have been used since the 1990s for the selection of superior hybrid lines [22]. Improving plant phenotype for a specific desirable trait involves the artificial selection and breeding of this given trait by the plant breeder. Generally, breeders tend to focus on traits of diploid or diploid-like crops (e.g., maize and tomatoes) rather than polyploid crops (e.g., alfalfa and potatoes), which have more complex genetics. Breeders hence prefer to use crops with shorter reproductive cycles, which allow the production of several generations in a single year and, leading to faster production of the desired phenotypes by artificial breeding compared to crops that only reproduce annually or perennial plants that only reproduce every few years [23–25]. Plant breeding, combined with genome studies, enhances the accuracy of breeding practices and saves time [26]. Compared to other kingdoms, plants are more easily genetically manipulated to obtain desired genetic combinations by selfing, crossbreeding (or both) given their short generation time, and large population size available for analyses [27]. In the early 1980s, NASA partnered with Utah State University to explore the possibility of growing rapid cycling wheat under constant light in space stations. This joint effort resulted in the development of "USU-Apogee", a dwarf wheat line bred for rapid cycling [28,29]. Recently, Lee Hickey and colleagues solved this issue by presenting the idea of "speed breeding", a non-GMO path enabling researcher to turn over many generations and select plants for desired traits between many variations [8,16]. This method uses regulated environmental conditions and prolonged photoperiods to achieve between four and six generations per year of long duration crops (i.e., wheat, barley, and canola) [16,30,31].

Researchers outlined the evolving EU regulatory framework for GMOs and discussed potential ways of regulating plant varieties developed using precision breeding approaches such as clustered regularly interspaced short palindromic repeats (CRISPR), and CRISPR-associated (Cas) proteins CRISPR/Cas9 [32]. Research interest in genetically engineered crops (and more precisely "biotech crops") has been increasing, given the urgent need to ensure food security for the growing human population [33].

Genome editing involves inserting, deleting, or substituting a foreign gene in the organism's DNA. Upon successful transformation, this new sequence is integrated into the host genome [34,35]. Several processes are involved in the fixation of specific DNA sequences, cut with the help of nucleases. Plant breeding alone cannot achieve the required traits, but using the CRISPR-associated (Cas) enzymes (CRISPR/Cas and CRISPR/Cpf1) can help meet the needs for efficient crop research [36,37]. In this review, we discuss the use of conventional and non-conventional plant breeding techniques for different crops, as well as the use of genome editing techniques to change and improve desired phenotypes. Moreover, the potential correlations between these approaches used to develop future strategies for crop improvement will also be explored.

2. Mutation through Traditional or Conventional Breeding

The advantage of conventional plant breeding consists of increasing the availability of genetic resources for crop improvement through introgression of the desired traits. However, some plants are at risk of becoming susceptible to environmental stress and losing genetic diversity [38]. Thus, traditional cultivation methods are not sufficient to resolve global food security issues. Combining multiple phenotypic characters within a single plant variety would successfully increase yield and has been widely used, however, new breeding techniques are less expensive and will enable faster production of genetically improved crops [39].

In recent times, improvements in traditional plant breeding have been introduced, such as wide crosses, introgression of traits from wild relatives by hybrid breeding, mutagenesis, double haploid technology, and some tissue culture-based approaches such as embryo and ovule rescue (to achieve maximum plant regeneration) and protoplast fusion [40–42]. Food and feed crops developed by conventional plant breeding have specific natural phenotypic and agronomic properties. To improve crop quality, researchers have introgressed many beneficial traits through plant breeding with wild relatives, such as higher yield, abiotic and biotic stress resistance, and increased nutritional value [39,43,44]. The identification and combination of traits in familiar genotypes and the selection of high-performing varieties can establish a crop lineage with the desired properties. That being said, this approach can have potentially adverse impacts on food and environmental safety as it occasionally gives rise to safety concerns through unpredictable effects [9,45].

A trait (e.g., stress tolerance) can be improved by selecting the best hybrid progeny with the desired trait using cross breeding [46] (Figure 2a). Desired traits can also be introduced into a chosen 'best' recipient line through backcrossing of the selected progeny with the recipient line for several generations to reduce unwanted phenotype combinations [47]. Genetic variability can be reduced by the use of long-term traditional breeding methods; thus, the introduction of new genes is required for the improvement of desired traits by speed breeding, mutation breeding, and rapid generation advance (RGA) [16,31,48]. From this point of view, mutations could be useful in plant breeding programs and all these precision breeding tools can contribute to the improvement of specific features during the breeding cycle. Plant breeding is always approached holistically by analyzing all applicable agricultural functionality (Figure 3).

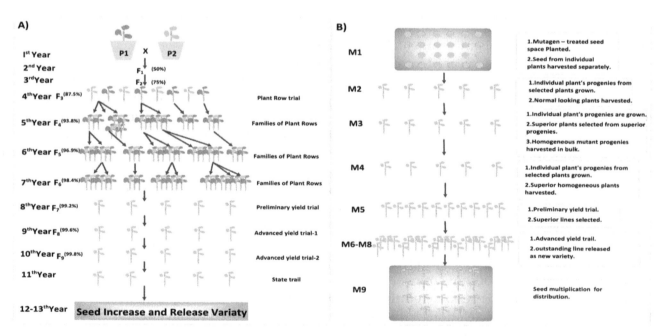

Figure 2. Improvement of agronomic traits using traditional breeding and chemical or physical mutagenic approaches. (**A**) Improving a trait (e.g., disease resistance) by the traditional breeding and for the introduction of the desired donor trait into the 'chosen' recipient line by selecting the progeny with the desired traits from the recipient line and crossing it with the donor line. (**B**) This process uses chemical or physical mutagens to generate mutants via random mutagenesis.

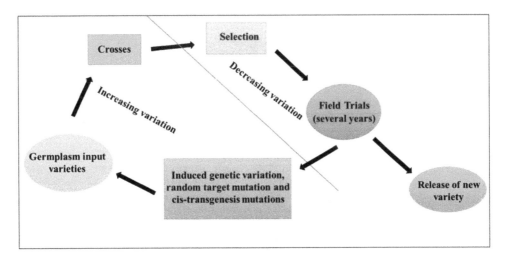

Figure 3. The plant breeding innovation cycle.

Identifying plants with desirable traits among existing plant varieties (or developing new phenotypes if these are not found naturally) is the initial and most important step in plant breeding. It would be impossible to develop new varieties or improve existing ones without natural genetic variation determined by spontaneous mutations. Ossowski et al. [49] concluded that the de novo spontaneous mutation rate was 7×10^{-9} base replacements per site per generation in all the nuclear genomes of five *Arabidopsis thaliana* accumulation lines sustained by single seed descent (SSD) over 30 generations. [50]. This is expected to be true for the genomes of most other plant species: for example, about 20 billion mutations occur each year in a one-hectare wheat field (personal communication with Professor Detlef Weigel, Max Planck Institute for Developmental Biology, Germany).

Another technique to improve plant varieties by conventional breeding is through mutation breeding. Mutagenesis is the phenomenon in which sudden heritable changes occur in the genetic material of an organism. It can occur spontaneously in nature or can be a result of exposure to different chemical, physical, or biological agents [51]. Mutation breeding is classified based on the three known types of mutagenesis. The first is radiation-induced mutagenesis in which mutations occur as a result of exposure to radiation (gamma rays, X-rays, or ion beams.); second is chemically induced mutagenesis; while the third is insertional mutagenesis, a consequence of DNA insertions either through the genetic transformation and insertion of T-DNA or the activation of transposable elements (i.e., site-directed mutagenesis; Table 1) [50,52]. According to Van Harten (Professor Agricultural University, Wageningen, The Netherlands), the history of plant mutation spans back to 300 BC, while the term mutation was first used in 1901 by Hugo De Vries, who reported during the final year of his studies that heredity might be changed by another mechanism, different from recombination and segregation [53]. He examined genomic variations and described them as heritable changes arising from this unique mechanism [54]. Numerous steps are required in any mutation breeding strategy: first, reducing the number of potential variants among the mutagenized seeds or other propagules for close evaluation of the first (M1) plant generation [51,54]. The benefits of mutation breeding over other breeding methods rely on the ability to select useful variant mutants in the second (M2) or third (M3) generations (Figure 2b).

Artificial mutation-causing agents are called mutagens; they are generally classified into two categories: physical and chemical mutagens [55]. They can induce mutations in almost any planting materials, including in vitro cultured cells, seedlings, and whole plants. Seeds are the most frequently used plant material for this specific purpose, but recently, various forms of plant propagules, such as tubers, bulbs, rhizomes, and mutation-induced vegetative propagated plants, are being used more frequently, as scientists take advantage of totipotency in single cells [56]. For example, with the use of ethyl methanesulfonate (EMS) and fast neutrons, collections of M82 tomato mutants were produced and more than 3000 phenotype alterations were classified [57]. An EMS-induced mutation library for the miniature dwarf tomato cultivar Micro-Tom has also been created, creating another resource for tomato genetic studies [58].

Table 1. Examples of commonly used physical and chemical mutagens, their characteristics, and hazard impacts.

Types	Mutagens	Characteristics (Sources and Description)	Hazards	References
Physical Mutagens	X-rays	Electromagnetic radiation; penetrates tissues from just a few millimeters to many centimeters.	Dangerous, penetrating	[59]
	Gamma rays	60Co (Cobalt-60) and 137Cs (Caesium-137); electric magnet radiation generated with radiation isotope and nuclear reactors.	Dangerous, penetrating	[59,60]
	Neutron	235U; there are fast, slow, thermal types; formed in nuclear reactors; unloaded particles; penetrate tissues up to large numbers centimeter;	Very dangerous	[59,60]
	Beta particles	32P and 14C; reduced particle accelerators or radioisotopes; electrons; ionizing and penetrating tissues shallowly	Maybe dangerous	[60]
	Alpha particles	Sources originating from radiological isotopes; helium nucleus able to penetrate tissues heavily	Very dangerous	[59]
	Proton	Present in nuclear reactors and accelerators; derived from the nucleus of hydrogen; penetrate tissues up to several inches.	Very dangerous	[59,60]
	Ion beam	Positively charged ions are accelerated at a high speed and used to irradiate living materials, including plant seeds and tissue culture.	Dangerous	[60]
Chemical Mutagens	Alkylating agents	The alkylated base can then degrade with bases to create a primary site which is mutagenic or recombinogenic or mispairs in DNA replication mutations, depending on the atom concerned.	Dangerous	[59]
	Azide	Just like alkylating agents.	Dangerous	[59]
	Hydroxylamine	Just like alkylating agents.	Dangerous	[56,59]
	Nitrous acid	Acts through deamination, replacing cytosine with uracil, which can pair with adenine and thus result in transitions via subsequent replication cycles.	Very Hazard	[56]
	Acridines	Interspersing between the DNA bases, thus distorting the DNA double helix and the DNA polymerase, recognizes the new basis for this expanded (intercalated) molecule and inserts a frameshift in front of it.	Dangerous	[56]
	Base analog	Comprises the transformations (purine to purine and pyrimidine to pyrimidine) into DNA in place of the regular bases during DNA replication and tautomerizing (existent in two forms, which interconvert into one another such that guanine may be present in keto and enol forms).	Some may be dangerous	[56]

Despite considerable success during the last century, the advances in yields of major crops (e.g., wheat) stabilized or even declined in many regions of the world [61,62]. Restrictions on phenotyping efficiency are increasingly being perceived as key constraints to genetic enhancements in breeding practices [63,64]. Specifically, HTP may cause a bottleneck in traditional breeding, marker-assisted selection (MAS), or genomic selection, where phenotyping is important to establish the accuracy of statistical models [63,65]. Accurate phenotyping is also required to replicate the outcomes of mutagenesis (i.e., GMOs) [66]. Deery et al. [67] and White and Conley [68] reviewed in great detail the benefits and challenges of potential phenotyping platforms, such as HTP.

Furthermore, SSD can be accelerated through the use of HTP [69,70]. SSD is most suitable for handling large segregating populations; while HTP tools are used in breeding programs [71]. Without undermining genetic variability and genetic development, SSD optimizes resource distribution, reducing the time spent growing crops and lowering costs associated with earlier generations' progress [72]. SSD has been successfully used in the groundnut breeding program, with the implementation of an inbreeding cycle producing multiple generations annually to advance fixed lines to multisite evaluation tests [73]. This speed breeding approach is ideal for SSD programs, particularly in cereal crops, allowing for the rapid cycling of multiple lines with healthy plants and viable seeds [30].

3. Mutagens for Molecular Breeding

One of the principal goals in the field of molecular biology is to identify and manipulate genes involved in human, animal, and plant disorders. Genomic tools used in such studies include restriction enzymes, biomarkers, molecular glue (ligases), as well as transcription and post-translational modification machinery [74]. Furthermore, molecular biological approaches are widely used to develop biofortified crops and plant varieties with high yield, new traits, and resistance to insect pests and diseases [75,76]. Globally, about 40 million hectares have been assigned to transgenic cultivars, which were commercialized after testing their biosafety level in 1999 [75]. Plant breeding was then reformed when researchers started to combine traditional practices with molecular tools to address phenotypic changes concerning the genotype of plant traits [77]. Accurate genome sequencing is essential before molecular tools can be used, and next-generation sequencing (NGS) allows researchers to decipher entire genomes and produce vast gene libraries for bioinformatics studies [78]. NGS opens new possibilities in phylogenetic and evolutionary studies, enabling the discovery of novel regulatory sequences and molecular markers [79]. Molecular biology is also facilitating the identification of diverse cytoplasmic male sterility sources in hybrid breeding. Some fertility restorer genes have been cloned in maize, rice, and sorghum [80]. Mutations in the target gene can be screened using target-induced local lesions in the genome (TILLING) and Eco-TILLING, which can directly identify allelic variations in the genome [81]. The most recent studies have determined the structure of plant germplasm using bulked segregant analysis [82], association mapping, genome resequencing [83,84], and fine gene mapping. This allows for the identification of single base-pair polymorphisms based on single sequence repeats, single nucleotide polymorphisms (SNPs), and unique biomarkers linked to quantitative trait loci (QTL) for genome manipulation, germplasm enhancement, and creating high-density gene libraries [85]. Traditional mutagenesis has certain limitations, as it can produce undesirable knockout mutations. It is also time-consuming and requires large-scale screening [86]. However, MAS is a direct approach for tracking mutations that improve backcrossing efficiency (or "breeding by design") [87] and determining the homogeneity of the progeny phenotypes.

In principle, all genome cleavage techniques produce double-stranded breaks (DSBs), blunt ends, or overhangs of the target nucleotide fragment, whether by homologous recombination, site-directed insertion/substitution of genes, or knockout mutations [88]. These DSBs, produced as a result of the action of sequence-specific nucleases (SSNs), are repaired by the non-homologous end joining (NHEJ) mechanism, which adds or removes nucleotides by the homology-directed repair pathway, directing DNA substitutions at target sites [89]. Various literature reviews report three primary SSN systems for genome editing. The first involves zinc finger nucleases (ZFNs), which form the basis for DNA manipulation. The second system involves transcription activator-like effector nucleases (TALENs), while the third system, the most important revolution in cutting-edge genomics, is a clustered regularly interspaced short palindromic repeats/associated protein 9 (CRISPR/Cas9) system [88,90–92]. The use of ZFNs has certain limitations: the constructs are not easy to design and transform, even in plants, and it is an expensive approach. Moreover, some researchers have reported non-specific nucleotide recognition because of their origin from eukaryotic transcription motifs, making this approach less reliable for genome editing [93,94]. Most restriction nucleases are derived from bacteria and TALENs were isolated from the prokaryotic plant pathogen *Xanthomonas* [95]. However, TALENs comprise large and repetitive constructs that require a lot of time and precision to edit the target sequence [96]. Soon after the discovery of TALENs, another promising nuclease (CRISPR/Cas9) was found in a bacterial immune system [97]. This system has been widely used in recent plant genome editing studies and has started replacing the TALEN and ZFN systems due to its high efficiency and accuracy in inducing site-directed breaks in double-stranded DNA [98]. Recently, a CRISPR-associated endonuclease from *Prevotella* and *Francisella* (Cpf1) has emerged as a replacement tool for precise genome editing, including DNA-free dissection of plant material, with higher potency, specificity, and enormous possibilities of wider application [99,100]. The base-editing approach using CRISPR/nCas9 (Cas9 nickase) or dCas9 (deactivated Cas9) fused with cytidine deaminase is a powerful tool to create point mutations. In this study, we point out the remarkable *G. hirsutum*-base editor 3 (GhBE3) base enhancing system developed to create single base mutations in the allotetraploid genome of cotton (*Gossypium hirsutum*) [101].

4. CRISPR/Cas9 and CRISPR/Cpf1 as Genetic Dissection Tools

The CRISPR/Cas9 system is a high-throughput discovery system in cutting-edge genomics, with recent studies reporting extensive use of Cas9 in gene transformation, drug delivery, and knockout mutations based on NHEJ-mediated DSBs [102]. Several studies investigated the mode of action of this potent nuclease and discovered the presence of a CRISPR loci, a cluster of repeating nucleotides in bacterial and archaeal immune systems [103]. These loci have a unique sequence, comprising of Cas9-encoding operons, transcription machinery, and consecutive repeats originating from various viral genomes separated by spacer sequences. These repeats were incorporated into the bacterial genome either by a virus or another foreign invader following an immune reaction [17].

Yin et al. (2017) found that Cas9 can be 'tricked' by supplementing any foreign nucleotide sequence that is digested and inserted in the bacterial genome. To knock a gene out, the CRISPR/Cas9 system is designed accordingly and transformed to explants via *Agrobacterium*, electroporation or the biolistic method. The regenerated plantlets' grown from the transformed callus are then transferred to planting soil [104]. The CRISPR/Cas9 gene knockout system has four significant features: (a) synthetic guide RNA (about 18–20 nucleotides) binding to target DNA, (b) Cas9 cleavage at 3–4 nucleotides after the adjacent proto spacer motif (PAM) (generally, 50 NGG identifies the PAM sequence) [105], (c) selection of a suitable binary vector and sgRNA cloning, and (d) transforming the construct in explants via *Agrobacterium* or microprojectile gene bombardment (Figure 4). *Agrobacterium*-mediated transformation is preferred in most studies given its efficiency and secure delivery [106]. The transformants are raised in growth chambers and examined for mutation studies using PCR, western blotting, ELISA, genotyping, sequencing, and other molecular techniques.

With recent advances in molecular biology and the discovery of sequences in the microbial immune system, biotechnologists can manipulate the organism's genome in a specific and precise way with the aid of CRISPR and its associated Cas proteins. This remarkable genome editing technique is categorized into two broad classes and six types: class 1 with types I, III, and IV and class 2 with types II, type V, and type VI [107]. Type II is the most widely used system in genome editing, while CRISPR/Cas9 from the *Streptococcus* pyogenes is the most commonly used method in the genome editing process. CRISPR class II has a type V effector named Cpf1, which can be designed with highly specific CRISPR RNA to cleave corresponding DNA sequences [108,109]. Cpf1 was recently developed as a substitute to Cas9, because of its unique ability to target T-rich motives through staggered DSBs without the need to trans-activate crRNA. Cpf1 can also process RNA and the DNA nuclease operation. Studies have been conducted to examine the Cpf1 mechanism, aiming to achieve more precise DNA editing and to address it the crystal structure of Cas12b homologous [110]. Another study reported that small molecular compounds can enhance Cpf1 efficiency as they are directly involved in activating or suppressing signaling pathways for cellular repair. Thus, small-molecule–mediated DNA repair aids in useful CRISPR mediated knock-outs [111].

Unfortunately, the development of new crop varieties by genome editing has been delayed in many countries by strict GMO regulations across the globe. This is particularly true for areas obeying a process rather than a regulatory framework based on the product, like in the EU, where authorizations for new varieties developed by genome editing techniques are subject to time- and cost-intensive verification procedures [112]. A recent decision by the European Court of Justice announced the enforcement of strict GMO legislation on target genome editing tools, even if the product is entirely free of transgenes [113]. Process-based regulations were also introduced in at least 15 countries, such as Brazil, India, China, and Australia, while 14 countries, including Canada, Argentina, and the Philippines adopted a product-based regulation. Several countries still have no specific regulatory system in operation, including Paraguay, Myanmar, Chile, and Vietnam. One of the most interesting aspects of regulation is Argentina's adoption, which is more versatile as it allows recent developments in genome editing to be taken into account. In the EU, genome-edited plants are typically listed as GMOs in compliance with the current legislation [114]. The control of genetically modified (GM) crops in the United States is authorized on a case-by-case basis, as set out in the structured framework for the control of biotechnology [115].

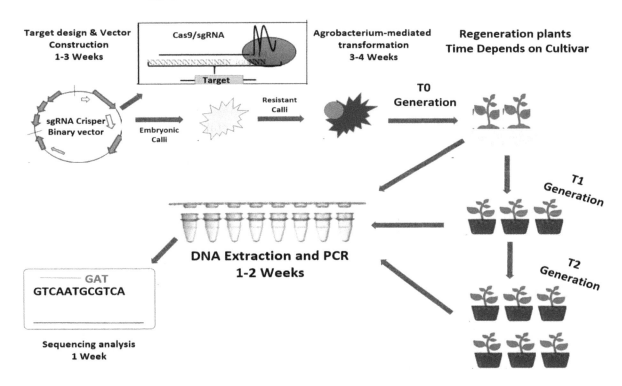

Figure 4. CRISPR–Cas9-based genome editing. CRISPR/Cas9 system uses Cas9 and sgRNA to cleave foreign DNA. It works in three steps: (1) the expression of the nuclear-localized Cas9 protein, (2) the generation of gRNA containing first 20-nt complementary to the target gene, and (3) the NGG PAM site recognition located nearly at the 3′ end of the target site. This process is followed by three additional steps: (1) design target and construction of a gene-specific sgRNA (vector), (2) CRISPR–Cas9 sgRNA can be transfected into the plant protoplast through *Agrobacterium*-mediated transformation, and (3) regenerated plants are screened for mutation via PCR-assay and sequencing. The estimated time needed is indicated for most steps.

5. Speed Breeding (Time-Saving Tools) for Accelerating Plant Breeding

Most plant species create a bottleneck in their applied research and breeding programs, generating the need for technologies to accelerate up plant growth and generation turnover. In the early 1980s, NASA's work was an inspiration for all plant scientists. In 2003, researchers at the University of Queensland coined the term "speed breeding" as a combination of methods developed to accelerate the speed of wheat breeding. Speed breeding protocols are currently being developed for several crops [16,30]. Speed breeding is suitable for diverse germplasm and does not require specific equipment for in vitro culturing, unlike doubled haploid (DH) technology, in which haploid embryos are produced to yield completely homozygous lines [116]. The principle behind speed breeding is to use optimum light intensity, temperature, and daytime length control (22 h light, 22 °C day/17 °C night, and high light intensity) to increase the rate of photosynthesis, which directly stimulates early flowering, coupled with annual seed harvesting to shorten the generation time [16,117]. Light intensity and wavelength plays a key in the regulation of flowering [118,119]. Croser et al. [120] developed early- and late-flowering genotypes for peas, chickpeas, faba beans, and lupins under controlled conditions using various parts of the light spectrum (blue and far red-improved LED lights and metal halide). These species showed a positive correlation to the diminishing red:far red-red proportion (R:FR). Accordingly, light with the most elevated power in the FR area is the most inductive [121,122]. In general, light with high R:FR (e.g., from fluorescent lamps) reduces stem enlargement and increases lateral branching, whereas light with a low R:FR (e.g., from incandescent lamps) strongly enhances stem elongation but inhibits lateral branching and flowering. This process is regulated by FR, while blue light mediates phytochrome FR (Pfr). Furthermore, the effect of R light on flowering repression is mediated by phytochrome R (Pr) [122,123].

Species-specific protocols to induce early flowering using certain environmental signals have been developed, such as short days or vernalization like RGA [48]. Greenhouse strategies under controlled conditions were compared with in vitro plus in vivo strategies and fast generation cycling by extended photoperiod [124–126]. The cost and space requirements associated with developing a large number of inbred lines can be reduced by implementing these practices in the breeding of small grain cereals grown at high densities (e.g., 1000 plants/m^2) [127].

Until recently, speed breeding had been reported to shorten generation time by extending photoperiods (Figure 5), while certain crop species, such as radish (*Raphanus sativus*), pepper (*Capsicum annum*), and leafy vegetables such as Amaranth (*Amaranthus* spp.) and sunflower (*Helianthus annuus*) responded positively to increased day length [27,30,117,128]. Speed breeding of short-day crops has been limited because of their flowering requirements. Nevertheless, recently, Lee Hickey and his research team worked on developing protocols for short-day crop like sorghum, millet and pigeon pea with the International Crop Research Institute for the Semi-Arid Tropics (ICRISAT) as part of a project funded by the Bill and Melinda Gates Foundation (https://geneticliteracyproject.org/2020/03/02/how-speed-breeding-will-help-us-expand-crop-diversity-to-feed-10-billion-people/). Sorghum, millet, and pigeon peas are important plants for many smallholder farmers in Africa and Asia, refining protocols targeted for these types of users has significant implications for global subsistence agriculture. This goal involves improving the protocols and conditions required for the induction of early flowering and rapid crop development [117]. O'Connor et al. (2014) already reported successful results in the speed breeding of peanuts (*Arachis hypogaea*). Increased day length helped amaranth (*Amaranthus* spp.) to achieve more generations annually [129]. In staple food crops requiring shorter photoperiods to initiate the reproductive phases, such as rice (*Oryza sativa*) and maize (*Zea mays*), speed breeding can accelerate vegetative growth [130]. Using speed breeding, it is possible to develop successive generations of improved crops for field examination via SSD, which is cheaper compared to the production of DHs. Speed breeding is also favorable to gene insertion (common haplotypes) of distinct phenotypes followed by MAS of elite hybrid lines [31,131].

In conclusion, recent advances in plant breeding and genomics have contributed to the development of qualitatively and quantitatively improved cultivars. Innovative agronomic strategies, in addition to the usual practices, have led to remarkable agricultural outcomes. However, sustainable crop development to ensure global food security can only be achieved with the combined investments of private firms, extension workers, and the public sector.

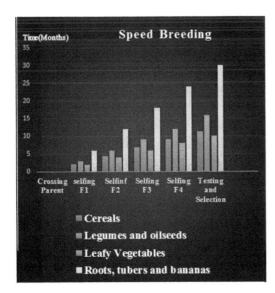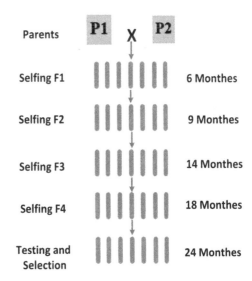

Figure 5. Graphical presentation of the elite line development procedure. Comparison of time (in months) required to develop elite lines from selected parents of some crops. Extended photoperiods induced earlier flowering and created 4 generations annually. The optimal temperature regime (maximum and minimum temperatures) should be applied for each crop. A higher temperature should be maintained during the photoperiod, whereas a fall in temperature during the dark period can aid in stress recovery. At the University of Queensland; (UQ), a 12-h 22 °C/17 °C temperature cycling regime with 2 h of darkness occurring within 12 h of 17 °C has proven successful. The figure is briefly modified from Watson et al. (2018).

6. Contribution of Plant Breeding to Crop Improvement

Molecular plant breeding was revolutionized in the 21st century, leading to crop improvement based on genomics, molecular marker selection, and conventional plant breeding practices [10,39]. For instance, the average yield of wheat (*Triticum spp.*), maize (*Zea mays*), and soybean (*Glycine max*), all significant crops in the United States, showed a positive linear increase from 1930 to 2012 [132,133]. The introduction of recessive genes in off-season nurseries was commercialized by pioneering plant breeder Norman Borlaug (among others), which helped to reduce the time needed to develop new cultivars. For example, the time for developing a new wheat cultivar was reduced from 10–12 years to only 5–6 years [134].

In hybrid and pure line crop breeding, developing similar and homozygous lines is a time-consuming process. Cycle time has been reduced from five to two generations by producing homogeneous and homozygous lines using DHs in diverse crops [135,136]. The maize DH system is one of the most common, it uses the R1-NJ color marker. However, the DH system has various genotypic and biological limitations [136,137]. Different crop species show dependence on the genotype for haploid induction [138], adapting tissue culture (e.g., in case of anther culture), and chromosome doubling by colchicine [139]. Breeders using the DH system unintentionally practice many selections for loci, increasing the success rate of this approach [140], but this might limit genetic variation in the breeding populations in responsive genome regions. Another approach is the RNAi suppression of plant genes (for instance, the MutS HOMOLOG1 (*MSH1*) gene) in multiple plant species, which produces a variety of developmental modifications accompanied by adjustments in plant defense, phytohormones, and abiotic stress response pathways combined with methylome repatterning [141].

Although the evaluation and production of GM crops is an active area of research, this technology is currently restricted because of political and ethical concerns. Nevertheless, GMO technologies make use of the variations that are present in deliberately mutated or naturally occurring populations [39,142,143]. GMOs have a variety of practical applications, for instance, they can be used to produce plant proteins that are toxic to insect pests, create herbicide tolerance genes for weed control, and create "golden rice" biofortified with vitamin A [144]. The characterization and discovery of genes and promoters can offer precise and effective temporal and spatial control of the expression of different genes, which is crucial for the future use of GM crops [145].

The availability of published genome sequences for different crops is increasing every year, facilitated by the use of sequencing technologies that improve sequencing speed and cost [146,147]. Current sequencing technologies, such as the NGS technique, can sequence multiple cultivars with both small and large genomes at a reasonable cost [148]. Although various published genomes are considered to be incomplete, they remain a valuable tool to evaluate important crop traits such as grain traits, fruit ripening, and flowering time adaptation [83,149].

Modern plant breeding programs have engaged interdisciplinary teams with expertise in the fields of statistics, biochemistry, physiology, bioinformatics, molecular biology, agronomy, and economics [150]. Crop breeding has been revolutionized and research on the advancement of DNA sequencing technologies has started the "genomics era" of crop improvement [151]. The genomes of most of the essential crops have been sequenced, creating a much cheaper genotyping platform for DNA fingerprinting. SNPs are ubiquitous DNA markers in crop genomes, they are also cost-efficient and easy to handle. Therefore, in today's crop improvement practices, genotyping large populations with a large number of markers is standard practice. Even whole-genome resequencing data are becoming easily available, giving unprecedented access to the structural diversity of crop genomes [65,83,152].

Currently, researchers are also using molecular genetic mapping of QTL of many complex traits vital in plant breeding. The detection and molecular cloning of genes underlying QTL enable the investigation of naturally occurring allelic variations for specific complex traits [85,153]. Plant productivity can be improved by identifying novel alleles through functional genomics or haplotype analysis. Advances in cereal genomics research in recent years have enabled scientists to improve the prediction of phenotypes from genotypes in cereal breeding [11,46].

Recently developed DNA-free CRISPR/Cas9 system delivery methods, different Cas9 variants, and RNA-guided nucleases offer new possibilities for crop genomic engineering [154]. The need to increase food security makes boosting crop production the primary objective of gene editing (Table 2). Crop yield is a complex trait that depends on several factors. The required phenotypes were found in plants with the loss of function mutations in yield related genes, highlighting the usefulness of CRISPR/Cas9 in improving yield-related traits by knocking out negative regulators affecting yield-determination factors, such as OsGS3 for grain size, OsGn1a for grain number; OsGW5, TaGW2, TaGASR7, and OsGLW2 for grain weight; TaDEP1 and OsDEP1 for panicle size, and OsAAP3 for tiller numbers. [155,156]. Similarly, three rice weight-related genes (GW5, GW2, and TGW6) were knocked out, causing pyramiding and increased weight [157]. The knockout of the *Waxy* gene using CRISPR/Cas9 resulted in the development of rice cultivars with higher nutritional quality [158]. DuPont Pioneer introduced a CRISPR/Cas9 knockout waxy corn line with high yields, ideal for commercial use [155]. A knockout of the *MLO* gene in tomato using CRISPR/Cas9 resulted in resistance to powdery mildew [159]. CRISPR has also been used to mutate the *OsERF922* transcription factor, resulting in resistance to rice blast, a destructive fungal disease [160].

By adopting a 22 h photoperiod and a temperature-controlled regime, generation times were considerably reduced in durum wheat (*T. durum*), spring bread wheat (*T. aestivum*), chickpea (*Cicer arietinum*), pea (*Pisum sativum*), barley (*Hordeum vulgare*), stiff brome (*Brachypodium distachyon*), canola (*Brassica napus*), and barrel clover (*Medicago truncatula*), compared with plants grown in a greenhouse with no supplementary light or those grown in the field. Under rapid growth conditions, plant development was normal, plants (such as wheat and barley) could be crossed easily, and seed germination rates were high [31,161–163].

Table 2. Application of breeding techniques toward crop improvement.

Sr.no.	Species	Method	Traits	References
1	Rice	Cross Breeding	Increased spikelet number per panicle	[164]
2	Rice	Cross Breeding	Yield Increases	[165]
3	Wheat	Cross Breeding	Increase Grain Yield	[166]
4	Tomato	Mutation Breeding	Resistance to bacterial wilt (*Ralstonia solanacearum*)	[167]
5	Rapeseed	Mutation Breeding	Resistance to stem rot (*Sclerotinia sclerotiorum*)	[168]
6	Cotton	Mutation Breeding	Resistance to bacterial blight, cotton leaf curl virus	[169]
7	Barley	Mutation Breeding	Salinity tolerance	[170]
8	Sunflower	Mutation Breeding	Semi-dwarf cultivar/dwarf	
9	Cassava	Mutation Breeding	High-amylose content preferred by diabetes patients because it lowers the insulin level, which prevents quick spikes in glucose contents.	[171]
10	Groundnut	Mutation Breeding	Dark green, obovate leaf pod; increased seed size, higher yield, moderately resistant to diseases, increased oil and protein content	[172]
11	Maize	Transgenic Breeding	increased vitamin content (vitamins C, E, or provitamin A)	[173]
12	Tomato	Transgenic Breeding	Dry Matter Increases	[174]
13	Soybean	Transgenic Breeding	Altered carbohydrates metabolism	[174]
14	Barley	Molecular Marker	Adult resistance to stripe rust	[175]
15	Maize	Molecular Marker	Development of quality protein maize	[22]
16	Watermelon	Marker-Assisted Selection	Early Flowering	[176]
17	Canola	QTL	Dynamic growth QTL	[153]
18	Alfalfa	Intragenesis	Lignin content	[129]
19	Apple	Cisgenesis, Intragenesis	Scab resistance	[177,178]
20	Barley	Cisgenesis	Grain phytase activity	[179]
21	Durum wheat	Cisgenesis	Baking quality	[180]
22	Perennial ryegrass	Intragenesis	Drought tolerance	[181]
23	Poplar	Cisgenesis	Plant growth and stature, wood properties	[181]
24	Potato	Cisgenesis	Late blight resistance	[182]
25	Strawberry	Intragenesis	Gray mold resistance	[183]
26	Tomato	Gene editing/ZFN	Reduction of cholesterol and steroidal glycoalkaloids, such as toxic α-solanine and α- chaconine	[184]
27	Wheat	Gene editing/TALEN	Heritable Modification	[185]
28	Rice	Gene knockout/ CRISPR/Cas9	Fragrance	[186]
29	Bread Wheat and Maize	Gene knockout/ CRISPR/Cas9	Leaf development; Male fertility, Herbicide resistance	[187]
30	Poplar	Gene knockout/ CRISPR/Cas9	Lignin content; Condensed tannin content	[188]
31	Tomato	Gene editing/ CRISPR/Cas9	Leaf development	[189]
32	Soybean	Gene replacement/ CRISPR/Cas9	Herbicide resistance	[190]
33	Maize	Gene replacement/ CRISPR/Cas9	Herbicide resistance	[187]
34	Cotton	Genome Editing/ CRISPR/Cas9	Produce transgenic seeds without regeneration	[191]
35	Soybean	Genome Editing/ CRISPR/Cas9	Early Flowering	[192]
36	Rice	Genome Editing/ CRISPR/Cas9	Increased grain weight	[157]
37	Tomato	Genome Editing/ CRISPR/Cas9	Resistance to powdery mildew	[159]
38	Wheat	Gene knockout/ CRISPR/Cas9	low-gluten foodstuff	[193]
39	Rice	Gene knockout/ CRISPR/Cas9	Generate mutant plants which is sensitive to salt stress	[194]
40	Rapeseed	Gene knockout/ CRISPR/Cas9	Controlling pod shattering resistance in oilseed rape	[195]
41	Tomato, Potato	CRISPR/Cas9 Cytidine Base Editor	Transgene-free plants in the first generation in tomato and potato	[196]
42	Tobacco	Genome Editing /CRISPR/Cpf1	Plants harboring	[197]
43	Rice	Genome Editing /CRISPR/Cpf1	Regulate the stomatal density in leaf	[198]
44	Rice	Genome Editing /CRISPR/Cpf1	Stable mRNA equal	[100,199]
45	Maize	Genome Editing /CRISPR/Cpf1	Mutation frequencies doubled	[199]

Table 2. *Cont.*

Sr.no.	Species	Method	Traits	References
46	Chickpea	Rapid generation advance (RGA)	Seven generations per year and enable speed breeding	[48]
47	Pea	Greenhouse strategy	6 Generation/year	[124]
48	Chickpea	Speed Breeding	4-6 Generation/year	[200]
49	Barley	Speed Breeding	Resistance to Leaf Rust	[16]
50	Spring wheat	Speed Breeding	Resistance to Stem Rust	[201]
51	Spring wheat	Speed Breeding	4-6 Generation/year	[16]
52	Barley	Speed Breeding	4-6 Generation/year	[16]
53	Peanut	Speed Breeding	2-3 Generation/year	[200]
54	Canola	Speed Breeding	4-6 Generation/year	[16]
55	Wheat	High-throughput phenotyping (HTP)	Development of improved, high-yielding crop varieties	[202]
56	Tomato	High-throughput phenotyping (HTP)	Using biostimulants to increase the plant capacity of using water	[203]

7. Future Outlook

Although modern plant breeding relies on traditional techniques, the emergence of new approaches will undoubtedly increase its efficiency and effectiveness. In the future, we can expect a wide range of techniques to be developed using interdisciplinary principles to increase their benefits. Strategies for crop production, breeding methods, approaches to field testing, genotyping technologies, even equipment and facilities need to be implemented across crop species to keep our food, fiber and biobased economy diverse. The discovery of CRISPR/Cas9, CRISPR/Cpf1, base-editing, and RGA has revolutionized molecular biology and its innovative applications in agriculture, setting a turning point in plant breeding and cultivation. GMOs can positively disseminate a selectable gene across wild populations in a gene drive process.

Altogether, CRISPR-based gene drive systems will prove—in time—to be beneficial for mankind. They will, for instance, prevent epidemics, improve agricultural practices, and control the spread of invasive species as plant cultivars resistant to insect pests and pathogens and tolerant to herbicides are developed. Existing genome editing techniques can be improved with the help of speed breeding (e.g., genes responsible for late flowering could be knocked out using CRISPR/Cas9). After the successful transfer of Cas9 into the plant, the transgenic plant can then be grown under speed breeding conditions rather than the usual glasshouse conditions to obtain transgenic seeds as early as possible. Using this method, it is possible to obtain stable homozygous phenotypes in less than a year. Furthermore, this method also decreases generation time, as it normally takes several years to develop a GMO crop. However, more efficient breeding strategies combining these technologies could lead to a step-change in the rate of genetic gain. Therefore, CRISPR/Cas9, primarily based on genome editing and speed breeding, will likely gain in popularity. It will be a crucial technique to obtain plants with specific desirable traits and contribute to reaching our objectives for zero-hunger globally. The development of innovations is often applied to a few important economic crops, which require specific adaptation to the reproduction and propagation method and the "process" of the new line development for the various crops of interest. Likewise, the transition of technology usually originates in developed countries, mostly in the private sector. It should be transferred to the public sector and into the developing world, given the significant financial investment required for groundbreaking work.

8. Conclusions

The primary methods for crop improvement in modern agriculture are cross breeding, mutation breeding, and transgenic breeding. Such time-consuming, laborious, and untargeted breeding programs cannot satisfy the increasing global food demand. To deal with this challenge and to enhance crop selection efficiency, marker-assisted breeding, and transgenic approaches have been adopted, generating desired traits via exogenous transformation into elite varieties. These genome editing systems are excellent tools that provide rapid, targeted mutagenesis and can identify the specific plant molecular mechanisms for crop improvement. Crop breeding was revolutionized by the development of next-generation breeding techniques. Genome editing technologies have many advantages over traditional agricultural methods, given their simplicity, efficiency, high specificity, and amenability to multiplexing. We conclude that speed breeding, combined with genetic tools and resources, enable plant biologists to scale up their research in the field of crop improvement.

Author Contributions: S.A., R.A.G., K.-H.J., and W.Z. designed the study; S.A., R.A.G., and M.U.Q. wrote the MS; S.A., A.F., M.M., R.A.G., W.Z., and K.H.J. revised the MS. In this study, S.A. and R.A.G. contributed equally. All authors have read and agreed to the published version of the manuscript.

Acknowledgments: Authors are thankful to an anonymous reviewer for their comments and critical reading of the manuscript.

Abbreviations

CRISPR	Clustered Regularly Interspaced Short Palindromic Repeats (CRISPR);
Cas9	CRISPR-associated Proteins;
Cpf1	CRISPR-associated endonuclease in Prevotella and Francisella;
DSB	Double Strand Breaks;
HTP	High-Throughput Phenotyping;
NASA	National Aeronautics and Space Administration;
USU	Utah State University;
GMO	Genetically Modified Organism;
EU	European Union;
Non-GMO	Non-Genetically modified Organism;
SSD	Single Seed Descent;
SB	Speed Breeding;
RGA	Rapid Generation Advance;
UQ	University of Queensland;
DH	Double Haploid;
ICRISAT	International Crop Research Institute for the Semi-Arid Tropics.

References

1. Tester, M.; Langridge, P. Breeding technologies to increase crop production in a changing world. *Science (80-.).* **2010**, *327*, 818–822. [CrossRef] [PubMed]
2. Shiferaw, B.; Smale, M.; Braun, H.-J.; Duveiller, E.; Reynolds, M.; Muricho, G. Crops that feed the world 10. Past successes and future challenges to the role played by wheat in global food security. *Food Secur.* **2013**, *5*, 291–317. [CrossRef]
3. Von Braun, J.; Rosegrant, M.W.; Pandya-Lorch, R.; Cohen, M.J.; Cline, S.A.; Brown, M.A.; Bos, M.S. *New Risks and Opportunities for Food Security Scenario Analyses for 2015 and 2050*; IFPRI: Washington, DC, USA, 2005.
4. Ristaino, J.B. Tracking historic migrations of the Irish potato famine pathogen, Phytophthora infestans. *Microbes Infect.* **2002**, *4*, 1369–1377. [CrossRef]
5. Tatum, L.A. The Southern Corn Leaf Blight Epidemic. *Science* **1971**, *171*, 1113–1116. [CrossRef]
6. UN World Population Projected to Reach 9.8 Billion in 2050, and 11.2 Billion in 2100. Available online: https://www.un.org/development/desa/en/news/population/world-population-prospects-2017.html.
7. FAO How to Feed the World in 2050. In *Insights from an Expert Meet*; FAO: Roma, Italy, 2009; Volume 2050, pp. 1–35.

8. Voss-Fels, K.P.; Stahl, A.; Hickey, L.T. Q&A: Modern crop breeding for future food security. *BMC Biol.* **2019**, *17*, 18.

9. Cheema, K.S. K. Plant Breeding its Applications and Future Prospects. *Int. J. Eng. Technol. Sci. Res.* **2018**, *5*, 88–94.

10. Moose, S.P.; Mumm, R.H. Molecular Plant Breeding as the Foundation for 21st Century Crop Improvement. *Plant Physiol.* **2008**, *147*, 969–977. [CrossRef]

11. Varshney, R.K.; Hoisington, D.A.; Tyagi, A.K. Advances in cereal genomics and applications in crop breeding. *Trends Biotechnol.* **2006**, *24*, 490–499. [CrossRef]

12. Li, H.; Rasheed, A.; Hickey, L.T.; He, Z. Fast-forwarding genetic gain. *Trends Plant Sci.* **2018**, *23*, 184–186. [CrossRef] [PubMed]

13. Collins, F.S.; Green, E.D.; Guttmacher, A.E.; Guyer, M.S. A vision for the future of genomics research. *Nature* **2003**, *431*, 835–847.

14. Majid, A.; Parray, G.A.; Wani, S.H.; Kordostami, M.; Sofi, N.R.; Waza, S.A.; Shikari, A.B.; Gulzar, S. Genome Editing and its Necessity in Agriculture. *Int. J. Curr. Microbiol. Appl. Sci.* **2017**, *6*, 5435–5443. [CrossRef]

15. Araus, J.L.; Kefauver, S.C.; Zaman-Allah, M.; Olsen, M.S.; Cairns, J.E. Translating High-Throughput Phenotyping into Genetic Gain. *Trends Plant Sci.* **2018**, *23*, 451–466. [CrossRef] [PubMed]

16. Watson, A.; Ghosh, S.; Williams, M.J.; Cuddy, W.S.; Simmonds, J.; Rey, M.D.; Asyraf Md Hatta, M.; Hinchliffe, A.; Steed, A.; Reynolds, D.; et al. Speed breeding is a powerful tool to accelerate crop research and breeding. *Nat. Plants* **2018**, *4*, 23–29. [CrossRef] [PubMed]

17. Zhang, F.; Wen, Y.; Guo, X. CRISPR/Cas9 for genome editing: Progress, implications and challenges. *Hum. Mol. Genet.* **2014**, *23*, R40–R46. [CrossRef] [PubMed]

18. Murovec, J.; Pirc, Ž.; Yang, B. New variants of CRISPR RNA-guided genome editing enzymes. *Plant Biotechnol. J.* **2017**, *15*, 917–926. [CrossRef]

19. Acquaah, G. Polyploidy in Plant Breeding. In *Principles of Plant Genetics and Breeding*; John Wiley & Sons: Hoboken, NJ, USA, 2012; pp. 452–469.

20. Muth, J.; Hartje, S.; Twyman, R.M.; Hofferbert, H.R.; Tacke, E.; Prüfer, D. Precision breeding for novel starch variants in potato. *Plant Biotechnol. J.* **2008**, *6*, 576–584. [CrossRef]

21. Mujjassim, N.E.; Mallik, M.; Rathod, N.K.K.; Nitesh, S.D. Cisgenesis and intragenesis a new tool for conventional plant breeding: A review. *J. Pharmacogn. Phytochem.* **2019**, *8*, 2485–2489.

22. Dreher, K.; Morris, M.; Khairallah, M.; Ribaut, J.M.; Shivaji, P.; Ganesan, S. Is marker-assisted selection cost-effective compared with conventional plant breeding methods? The case of quality protein Maize. *Econ. Soc. Issues Agric. Biotechnol.* **2009**, 203–236.

23. Abreu, G.B.; Ramalho, M.A.P.; Toledo, F.H.R.B.; De Souza, J.C. Strategies to improve mass selection in maize. *Maydica* **2010**, *55*, 219–225.

24. Kandemir, N.; Saygili, İ. Apomixis: New horizons in plant breeding. *Turkish J. Agric. For.* **2015**, *39*, 549–556. [CrossRef]

25. Leifert, C.; Tamm, L.; Lammerts van Bueren, E.T.; Jones, S.S.; Murphy, K.M.; Myers, J.R.; Messmer, M.M. The need to breed crop varieties suitable for organic farming, using wheat, tomato and broccoli as examples: A review. *NJAS - Wageningen J. Life Sci.* **2011**, *58*, 193–205.

26. Doust, A.; Diao, X. Plant Genetics and Genomics: Crops and Models Volume 19. *Genet. Genom. Setaria Ser.* **2017**, *19*, 377.

27. Stetter, M.G.; Zeitler, L.; Steinhaus, A.; Kroener, K.; Biljecki, M.; Schmid, K.J. Crossing Methods and Cultivation Conditions for Rapid Production of Segregating Populations in Three Grain Amaranth Species. *Front. Plant Sci.* **2016**, *7*, 816. [CrossRef] [PubMed]

28. Bugbee, B.; Koerner, G. Yield comparisons and unique characteristics of the dwarf wheat cultivar "USU-Apogee". *Adv. Sp. Res.* **1997**, *20*, 1891–1894. [CrossRef]

29. Bula, R.J.; Morrow, R.C.; Tibbitts, T.W.; Barta, D.J.; Ignatius, R.W.; Martin, T.S. Light-emitting diodes as a radiation source for plants. *HortScience* **1991**, *26*, 203–205. [CrossRef]

30. Ghosh, S.; Watson, A.; Gonzalez-Navarro, O.E.; Ramirez-Gonzalez, R.H.; Yanes, L.; Mendoza-Suárez, M.; Simmonds, J.; Wells, R.; Rayner, T.; Green, P.; et al. Speed breeding in growth chambers and glasshouses for crop breeding and model plant research. *Nat. Protoc.* **2018**, *13*, 2944–2963. [CrossRef]

31. Hickey, L.T.; Germa, S.E.; Diaz, J.E.; Ziems, L.A.; Fowler, R.A.; Platz, G.J.; Franckowiak, J.D.; Dieters, M.J. *Speed Breeding for Multiple Disease Resistance in Barley*; Springer: New York, NY, USA, 2017.

32. Chen, K.; Wang, Y.; Zhang, R.; Zhang, H.; Gao, C. CRISPR/Cas Genome Editing and Precision Plant Breeding in Agriculture. *Annu. Rev. Plant Biol.* **2019**, *70*, annurev. [CrossRef]
33. Godwin, I.D.; Rutkoski, J.; Varshney, R.K.; Hickey, L.T. Technological perspectives for plant breeding. *Theor. Appl. Genet.* **2019**, *132*, 555–557. [CrossRef]
34. Lee, J.; Chung, J.H.; Kim, H.M.; Kim, D.W.; Kim, H. Designed nucleases for targeted genome editing. *Plant Biotechnol. J.* **2016**, *14*, 448–462. [CrossRef] [PubMed]
35. Zhang, H.; Zhang, J.; Lang, Z.; Botella, J.R.; Zhu, J.K. Genome Editing—Principles and Applications for Functional Genomics Research and Crop Improvement. *Crit. Rev. Plant Sci.* **2017**, *36*, 291–309. [CrossRef]
36. Hsu, P.D.; Scott, D.A.; Weinstein, J.A.; Ran, F.A.; Konermann, S.; Agarwala, V.; Li, Y.; Fine, E.J.; Wu, X.; Shalem, O.; et al. DNA targeting specificity of RNA-guided Cas9 nucleases. *Nat. Biotechnol.* **2013**, *3*, 827. [CrossRef] [PubMed]
37. Zetsche, B.; Gootenberg, J.S.; Abudayyeh, O.O.; Slaymaker, I.M.; Makarova, K.S.; Essletzbichler, P.; Volz, S.E.; Joung, J.; Van Der Oost, J.; Regev, A.; et al. Cpf1 Is a Single RNA-Guided Endonuclease of a Class 2 CRISPR-Cas System. *Cell* **2015**. [CrossRef]
38. Basey, A.C.; Fant, J.B.; Kramer, A.T. Producing native plant materials for restoration: 10 rules to collect and maintain genetic diversity. *Nativ. Plants J.* **2015**, *16*, 37–53. [CrossRef]
39. Krimsky, S. Traditional Plant Breeding. In *GMOs Decoded*; MIT Press: Cambridge, MA, USA, 2019.
40. Shepard, J.F.; Bidney, D.; Barsby, T.; Kemble, R. Fusion of Protoplasts. *Biotechnol. Biol. Front.* **2019**.
41. Marthe, F. Tissue culture approaches in relation to medicinal plant improvement. In *Biotechnologies of Crop Improvement*; Research Gate: Berlin, Germany, 2018; Volume 1, pp. 487–497. ISBN 9783319782836.
42. Germana, M.A. Anther culture for haploid and doubled haploid production. *Plant Cell Tissue Organ Cult.* **2011**, *104*, 283–300. [CrossRef]
43. Hajjar, R.; Hodgkin, T. The use of wild relatives in crop improvement: A survey of developments over the last 20 years. *Euphytica* **2007**, *156*, 1–13. [CrossRef]
44. Ceccarelli, S.; Guimaraes, E.P.; Weltzien, E. *Plant breeding and farmer participation*; NHBS: Devon, UK, 2009; ISBN 9789251063828.
45. Cellini, F.; Chesson, A.; Colquhoun, I.; Constable, A.; Davies, H.V.; Engel, K.H.; Gatehouse, A.M.R.; Kärenlampi, S.; Kok, E.J.; Leguay, J.-J. Unintended effects and their detection in genetically modified crops. *Food Chem. Toxicol.* **2004**, *42*, 1089–1125. [CrossRef]
46. Dolferus, R.; Ji, X.; Richards, R.A. Abiotic stress and control of grain number in cereals. *Plant Sci.* **2011**, *181*, 331–341. [CrossRef]
47. Caligari, P.D.S.; Brown, J. Plant Breeding, Practice. In *Encyclopedia of Applied Plant Sciences*; Academic Press: Cambridge, MA, USA, 2016; Volume 2, pp. 229–235. ISBN 9780123948083.
48. Samineni, S.; Sen, M.; Sajja, S.B.; Gaur, P.M. Rapid generation advance (RGA) in chickpea to produce up to seven generations per year and enable speed breeding. *Crop J.* **2019**. [CrossRef]
49. Ossowski, S.; Schneeberger, K.; Lucas-Lledó, J.I.; Warthmann, N.; Clark, R.M.; Shaw, R.G.; Weigel, D.; Lynch, M. The rate and molecular spectrum of spontaneous mutations in Arabidopsis thaliana. *Science (80-.).* **2010**, *327*, 92–94. [CrossRef] [PubMed]
50. Oladosu, Y.; Rafii, M.Y.; Abdullah, N.; Hussin, G.; Ramli, A.; Rahim, H.A.; Miah, G.; Usman, M. Principle and application of plant mutagenesis in crop improvement: A review. *Biotechnol. Biotechnol. Equip.* **2016**, *30*, 1–16. [CrossRef]
51. Roychowdhury, R.; Tah, J. Mutagenesis—A potential approach for crop improvement. In *Crop Improvement*; Springer: New York, NY, USA, 2013; pp. 149–187.
52. Forster, B.P.; Shu, Q.Y.; Nakagawa, H. Plant mutagenesis in crop improvement: Basic terms and applications. *Plant Mutat. Breed. Biotechnol.* **2012**, 9–20.
53. Van Harten, A.M. *Mutation Breeding: Theory and Practical Applications*; Cambridge University Press: Cambridge, MA, USA, 1998; ISBN 0521470749.
54. Kharkwal, M.C. A brief history of plant mutagenesis. *Plant Mutat. Breed. Biotechnol.* **2012**, 21–30.
55. Mba, C.; Afza, R.; Bado, S.; Jain, S.M. Induced Mutagenesis in Plants. *Plant Cell Cult. Essent. Methods* **2010**, 111–130.
56. Mba, C. Induced Mutations Unleash the Potentials of Plant Genetic Resources for Food and Agriculture. *Agronomy* **2013**, *3*, 200–231. [CrossRef]

57. Menda, N.; Semel, Y.; Peled, D.; Eshed, Y.; Zamir, D. In silico screening of a saturated mutation library of tomato. *Plant J.* **2004**, *38*, 861–872. [CrossRef]

58. Watanabe, S.; Mizoguchi, T.; Aoki, K.; Kubo, Y.; Mori, H.; Imanishi, S.; Yamazaki, Y.; Shibata, D.; Ezura, H. Ethylmethanesulfonate (EMS) mutagenesis of Solanum lycopersicum cv. Micro-Tom for large-scale mutant screens. *Plant Biotechnol.* **2007**, *24*, 33–38. [CrossRef]

59. Wani, M.R.; Kozgar, M.I.; Tomlekova, N.; Khan, S.; Kazi, A.G.; Sheikh, S.A.; Ahmad, P. Mutation breeding: A novel technique for genetic improvement of pulse crops particularly Chickpea (*Cicer arietinum* L.). In *Improvement of Crops in the Era of Climatic Changes*; Springer: New York, NY, USA, 2014; pp. 217–248.

60. Mba, C.; Afza, R.; Shu, Q.Y.; Forster, B.P.; Nakagawa, H. Mutagenic radiations: X-rays, ionizing particles and ultraviolet. *Plant Mutat. Breed. Biotechnol.* **2012**, 83–90.

61. Acreche, M.M.; Briceño-Félix, G.; Sánchez, J.A.M.; Slafer, G.A. Physiological bases of genetic gains in Mediterranean bread wheat yield in Spain. *Eur. J. Agron.* **2008**, *28*, 162–170. [CrossRef]

62. Sadras, V.O.; Lawson, C. Genetic gain in yield and associated changes in phenotype, trait plasticity and competitive ability of South Australian wheat varieties released between 1958 and 2007. *Crop Pasture Sci.* **2011**, *62*, 533–549. [CrossRef]

63. Araus, J.L.; Cairns, J.E. Field high-throughput phenotyping: The new crop breeding frontier. *Trends Plant Sci.* **2014**, *19*, 52–61. [CrossRef] [PubMed]

64. Tardieu, F.; Cabrera-Bosquet, L.; Pridmore, T.; Bennett, M. Plant Phenomics, From Sensors to Knowledge. *Curr. Biol.* **2017**, *27*, R770–R783. [CrossRef] [PubMed]

65. Crossa, J.; Pérez, P.; de los Campos, G.; Mahuku, G.; Dreisigacker, S.; Magorokosho, C. Genomic selection and prediction in plant breeding. *J. Crop Improv.* **2011**, *25*, 239–261. [CrossRef]

66. Blum, A. Genomics for drought resistance-getting down to earth. In *Functional Plant Biology*; CSIRO Publishing: Melbourne, Australia, 2014; Volume 41, pp. 1191–1198.

67. Deery, D.; Jimenez-Berni, J.; Jones, H.; Sirault, X.; Furbank, R. Proximal remote sensing buggies and potential applications for field-based phenotyping. *Agronomy* **2014**, *4*, 349–379. [CrossRef]

68. White, J.W.; Conley, M.M. A flexible, low-cost cart for proximal sensing. *Crop Sci.* **2013**, *53*, 1646–1649. [CrossRef]

69. Saxena, K.; Saxena, R.K.; Varshney, R.K. Use of immature seed germination and single seed descent for rapid genetic gains in pigeonpea. *Plant Breed.* **2017**, *136*, 954–957. [CrossRef]

70. Shakoor, N.; Lee, S.; Mockler, T.C. High throughput phenotyping to accelerate crop breeding and monitoring of diseases in the field. *Curr. Opin. Plant Biol.* **2017**, *38*, 184–192. [CrossRef]

71. Janila, P.; Variath, M.T.; Pandey, M.K.; Desmae, H.; Motagi, B.N.; Okori, P.; Manohar, S.S.; Rathnakumar, A.L.; Radhakrishnan, T.; Liao, B.; et al. Genomic tools in groundnut breeding program: Status and perspectives. *Front. Plant Sci.* **2016**, *7*. [CrossRef]

72. Sarutayophat, T.; Nualsri, C. The efficiency of pedigree and single seed descent selections for yield improvement at generation 4 (F4) of two yardlong bean populations. *Kasetsart J. Nat. Sci.* **2010**, *44*, 343–352.

73. Holbrook, C.C.; Timper, P.; Culbreath, A.K.; Kvien, C.K. Registration of "Tifguard" Peanut. *J. Plant Regist.* **2008**, *2*, 92. [CrossRef]

74. Huang, X. From Genetic Mapping to Molecular Breeding: Genomics Have Paved the Highway. *Mol. Plant* **2016**. [CrossRef] [PubMed]

75. Jung, C. *Chapter 3 Molecular Tools for Plant Breeding*; Springer: New York, NY, USA, 2000; pp. 25–37.

76. Schaart, J.G.; van de Wiel, C.C.M.; Lotz, L.A.P.; Smulders, M.J.M. Opportunities for Products of New Plant Breeding Techniques. *Trends Plant Sci.* **2016**. [CrossRef] [PubMed]

77. Vilanova, S.; Cañizares, J.; Pascual, L.; Blanca, J.M.; Díez, M.J.; Prohens, J.; Picó, B. Application of Genomic Tools in Plant Breeding. *Curr. Genomics* **2012**, *13*, 179–195.

78. Wendler, N.; Mascher, M.; Nöh, C.; Himmelbach, A.; Scholz, U.; Ruge-Wehling, B.; Stein, N. Unlocking the secondary gene-pool of barley with next-generation sequencing. *Plant Biotechnol. J.* **2014**, *12*, 1122–1131. [CrossRef]

79. Metzker, M.L. Sequencing technologies—The next generation. *Nat. Rev. Genet.* **2010**, *11*, 31. [CrossRef]

80. Dwivedi, S.; Perotti, E.; Ortiz, R. Towards molecular breeding of reproductive traits in cereal crops. *Plant Biotechnol. J.* **2008**, *6*, 529–559. [CrossRef]

81. Wang, T.L.; Uauy, C.; Robson, F.; Till, B. TILLING in extremis. *Plant Biotechnol. J.* **2012**, *10*, 761–772. [CrossRef]

82. Zou, C.; Wang, P.; Xu, Y. Bulked sample analysis in genetics, genomics and crop improvement. *Plant Biotechnol. J.* **2016**, *14*, 1941–1955. [CrossRef]

83. Bolger, M.E.; Weisshaar, B.; Scholz, U.; Stein, N.; Usadel, B.; Mayer, K.F.X. Plant genome sequencing—Applications for crop improvement. *Curr. Opin. Biotechnol.* **2014**, *26*, 31–37. [CrossRef] [PubMed]

84. Edwards, D.; Batley, J. Plant genome sequencing: Applications for crop improvement. *Plant Biotechnol. J.* **2010**, *8*, 2–9. [CrossRef] [PubMed]

85. Dhingani, R.M.; Umrania, V.V.; Tomar, R.S.; Parakhia, M.V.; Golakiya, B. Introduction to QTL mapping in plants. *Ann. Plant Sci* **2015**, *4*, 1072–1079.

86. McCallum, C.M.; Comai, L.; Greene, E.A.; Henikoff, S. Targeted screening for induced mutations. *Nat. Biotechnol.* **2000**, *18*, 455. [CrossRef] [PubMed]

87. Nadeem, M.A.; Nawaz, M.A.; Shahid, M.Q.; Doğan, Y.; Comertpay, G.; Yıldız, M.; Hatipoğlu, R.; Ahmad, F.; Alsaleh, A.; Labhane, N.; et al. DNA molecular markers in plant breeding: Current status and recent advancements in genomic selection and genome editing. *Biotechnol. Biotechnol. Equip.* **2018**, *32*, 261–285. [CrossRef]

88. Lloyd, A.; Plaisier, C.L.; Carroll, D.; Drews, G.N. Targeted mutagenesis using zinc-finger nucleases in Arabidopsis. *Proc. Natl. Acad. Sci. USA* **2005**. [CrossRef]

89. Symington, L.S.; Gautier, J. Double-strand break end resection and repair pathway choice. *Annu. Rev. Genet.* **2011**, *45*, 247–271. [CrossRef]

90. Wood, A.J.; Lo, T.W.; Zeitler, B.; Pickle, C.S.; Ralston, E.J.; Lee, A.H.; Amora, R.; Miller, J.C.; Leung, E.; Meng, X.; et al. Targeted genome editing across species using ZFNs and TALENs. *Science* **2011**, *333*, 307. [CrossRef]

91. Sprink, T.; Metje, J.; Hartung, F. Plant genome editing by novel tools: TALEN and other sequence specific nucleases. *Curr. Opin. Biotechnol.* **2015**. [CrossRef]

92. Mao, Y.; Zhang, H.; Xu, N.; Zhang, B.; Gou, F.; Zhu, J.K. Application of the CRISPR-Cas system for efficient genome engineering in plants. *Mol. Plant* **2013**. [CrossRef]

93. Schneider, K.; Schiermeyer, A.; Dolls, A.; Koch, N.; Herwartz, D.; Kirchhoff, J.; Fischer, R.; Russell, S.M.; Cao, Z.; Corbin, D.R. Targeted gene exchange in plant cells mediated by a zinc finger nuclease double cut. *Plant Biotechnol. J.* **2016**, *14*, 1151–1160. [CrossRef] [PubMed]

94. De Pater, S.; Pinas, J.E.; Hooykaas, P.J.J.; van der Zaal, B.J. ZFN-mediated gene targeting of the Arabidopsis protoporphyrinogen oxidase gene through Agrobacterium-mediated floral dip transformation. *Plant Biotechnol. J.* **2013**, *11*, 510–515. [CrossRef] [PubMed]

95. Li, T.; Huang, S.; Jiang, W.Z.; Wright, D.; Spalding, M.H.; Weeks, D.P.; Yang, B. TAL nucleases (TALNs): Hybrid proteins composed of TAL effectors and FokI DNA-cleavage domain. *Nucleic Acids Res.* **2010**, *39*, 359–372. [CrossRef] [PubMed]

96. Char, S.N.; Unger-Wallace, E.; Frame, B.; Briggs, S.A.; Main, M.; Spalding, M.H.; Vollbrecht, E.; Wang, K.; Yang, B. Heritable site-specific mutagenesis using TALENs in maize. *Plant Biotechnol. J.* **2015**, *13*, 1002–1010. [CrossRef]

97. Mahfouz, M.M.; Piatek, A.; Stewart, C.N. Genome engineering via TALENs and CRISPR/Cas9 systems: Challenges and perspectives. *Plant Biotechnol. J.* **2014**, *12*, 1006–1014. [CrossRef]

98. Zhang, Y.; Xie, X.; Liu, Y.G.; Zhang, Y.; Xie, X.; Liu, Y.G.; Ma, X. *CRISPR/Cas9-Based Genome Editing in Plants*, 1st ed; Elsevier Inc.: Amsterdam, The Netherlands, 2017; Volume 149.

99. Zaidi, S.S.-e.-A.; Mahfouz, M.M.; Mansoor, S. CRISPR-Cpf1: A New Tool for Plant Genome Editing. *Trends Plant Sci.* **2017**, *22*, 550–553. [CrossRef]

100. Li, S.; Zhang, X.; Wang, W.; Guo, X.; Wu, Z.; Du, W.; Zhao, Y.; Xia, L. Expanding the scope of CRISPR/Cpf1-mediated genome editing in rice. *Mol. Plant* **2018**, *11*, 995–998. [CrossRef]

101. Qin, L.; Li, J.; Wang, Q.; Xu, Z.; Sun, L.; Alariqi, M.; Manghwar, H.; Wang, G.; Li, B.; Ding, X.; et al. High Efficient and Precise Base Editing of C•G to T•A in the Allotetraploid Cotton (*Gossypium hirsutum*) Genome Using a Modified CRISPR /Cas9 System. *Plant Biotechnol. J.* **2020**, *18*, 45–56. [CrossRef]

102. Bortesi, L.; Zhu, C.; Zischewski, J.; Perez, L.; Bassié, L.; Nadi, R.; Forni, G.; Lade, S.B.; Soto, E.; Jin, X.; et al. Patterns of CRISPR/Cas9 activity in plants, animals and microbes. *Plant Biotechnol. J.* **2016**, *14*, 2203–2216. [CrossRef]

103. Lee, K.; Zhang, Y.; Kleinstiver, B.P.; Guo, J.A.; Aryee, M.J.; Miller, J.; Malzahn, A.; Zarecor, S.; Lawrence-Dill, C.J.; Joung, J.K.; et al. Activities and specificities of CRISPR/Cas9 and Cas12a nucleases for targeted mutagenesis in maize. *Plant Biotechnol. J.* **2019**, *17*, 362–372. [CrossRef]

104. Zhang, Y.; Liang, Z.; Zong, Y.; Wang, Y.; Liu, J.; Chen, K.; Qiu, J.L.; Gao, C. Efficient and transgene-free genome editing in wheat through transient expression of CRISPR/Cas9 DNA or RNA. *Nat. Commun.* **2016**. [CrossRef] [PubMed]

105. Jinek, M.; Chylinski, K.; Fonfara, I.; Hauer, M.; Doudna, J.A.; Charpentier, E. A programmable dual-RNA–guided DNA endonuclease in adaptive bacterial immunity. *Science (80-.).* **2012**, *337*, 816–821. [CrossRef]

106. Wang, X.; Tu, M.; Wang, D.; Liu, J.; Li, Y.; Li, Z.; Wang, Y.; Wang, X. CRISPR/Cas9-mediated efficient targeted mutagenesis in grape in the first generation. *Plant Biotechnol. J.* **2018**, *16*, 844–855. [CrossRef] [PubMed]

107. Murugan, K.; Babu, K.; Sundaresan, R.; Rajan, R.; Sashital, D.G. The Revolution Continues: Newly Discovered Systems Expand the CRISPR-Cas Toolkit. *Mol. Cell* **2017**, *68*, 15–25. [CrossRef]

108. Ma, X.; Chen, X.; Jin, Y.; Ge, W.; Wang, W.; Kong, L.; Ji, J.; Guo, X.; Huang, J.; Feng, X.H.; et al. Small molecules promote CRISPR-Cpf1-mediated genome editing in human pluripotent stem cells. *Nat. Commun.* **2018**. [CrossRef]

109. Riesenberg, S.; Maricic, T. Targeting repair pathways with small molecules increases precise genome editing in pluripotent stem cells. *Nat. Commun.* **2018**. [CrossRef]

110. Yang, F.; Li, Y. The new generation tool for CRISPR genome editing: CRISPR/Cpf1. *Sheng wu gong cheng xue bao= Chinese J. Biotechnol.* **2017**, *33*, 361–371.

111. Maruyama, T.; Dougan, S.K.; Truttmann, M.C.; Bilate, A.M.; Ingram, J.R.; Ploegh, H.L. Corrigendum: Increasing the efficiency of precise genome editing with CRISPR-Cas9 by inhibition of nonhomologous end joining. *Nat. Biotechnol.* **2016**, *34*, 210. [CrossRef] [PubMed]

112. Ishii, T.; Araki, M. A future scenario of the global regulatory landscape regarding genome-edited crops. *GM Crop. Food* **2017**, *8*, 44–56. [CrossRef] [PubMed]

113. Wolt, J.D.; Wang, K.; Yang, B. The Regulatory Status of Genome-edited Crops. *Plant Biotechnol. J.* **2016**, *14*, 510–518. [CrossRef]

114. Callaway, E. CRISPR plants now subject to tough GM laws in European Union. *Nature* **2018**, *560*, 16. [CrossRef] [PubMed]

115. Sprink, T.; Eriksson, D.; Schiemann, J.; Hartung, F. Regulatory hurdles for genome editing: Process- vs. product-based approaches in different regulatory contexts. *Plant Cell Rep.* **2016**, *35*, 1493–1506. [CrossRef] [PubMed]

116. Slama-Ayed, O.; Bouhaouel, I.; Ayed, S.; De Buyser, J.; Picard, E.; Amara, H.S. Efficiency of three haplomethods in durum wheat (Triticum turgidum subsp. durum Desf.): Isolated microspore culture, gynogenesis and wheat× maize crosses. *Czech J. Genet. Plant Breed.* **2019**, *55*, 101–109. [CrossRef]

117. Chiurugwi, T.; Kemp, S.; Powell, W.; Hickey, L.T.; Powell, W. Speed breeding orphan crops. *Theor. Appl. Genet.* **2018**. [CrossRef] [PubMed]

118. Weller, J.L.; Beauchamp, N.; Kerckhoffs, L.H.J.; Platten, J.D.; Reid, J.B. Interaction of phytochromes A and B in the control of de-etiolation and flowering in pea. *Plant J.* **2001**, *26*, 283–294. [CrossRef] [PubMed]

119. Giliberto, L.; Perrotta, G.; Pallara, P.; Weller, J.L.; Fraser, P.D.; Bramley, P.M.; Fiore, A.; Tavazza, M.; Giuliano, G. Manipulation of the blue light photoreceptor cryptochrome 2 in tomato affects vegetative development, flowering time, and fruit antioxidant content. *Plant Physiol.* **2005**, *137*, 199–208. [CrossRef]

120. Croser, J.S.; Pazos-Navarro, M.; Bennett, R.G.; Tschirren, S.; Edwards, K.; Erskine, W.; Creasy, R.; Ribalta, F.M. Time to flowering of temperate pulses in vivo and generation turnover in vivo–in vitro of narrow-leaf lupin accelerated by low red to far-red ratio and high intensity in the far-red region. *Plant Cell. Tissue Organ Cult.* **2016**, *127*, 591–599. [CrossRef]

121. Ribalta, F.M.; Pazos-Navarro, M.; Nelson, K.; Edwards, K.; Ross, J.J.; Bennett, R.G.; Munday, C.; Erskine, W.; Ochatt, S.J.; Croser, J.S. Precocious floral initiation and identification of exact timing of embryo physiological maturity facilitate germination of immature seeds to truncate the lifecycle of pea. *Plant Growth Regul.* **2017**, *81*, 345–353. [CrossRef]

122. Moe, R.; Heins, R. Control of plant morphogenesis and flowering by light quality and temperature. *Acta Hortic.* **1990**, 81–90. [CrossRef]

123. Ausín, I.; Alonso-Blanco, C.; Martínez-Zapater, J.M. Environmental regulation of flowering. *Int. J. Dev. Biol.* **2005**, *49*, 689–705. [CrossRef]

124. Ochatt, S.J.; Sangwan, R.S.; Marget, P.; Assoumou Ndong, Y.; Rancillac, M.; Perney, P. New approaches towards the shortening of generation cycles for faster breeding of protein legumes. *Plant Breed.* **2002**, *121*, 436–440. [CrossRef]

125. Ochatt, S.J.; Sangwan, R.S. In vitro shortening of generation time in Arabidopsis thaliana. *Plant Cell. Tissue Organ Cult.* **2008**, *93*, 133–137. [CrossRef]

126. Heuschele, D.J.; Case, A.; Smith, K.P. Evaluation of Fast Generation Cycling in Oat (*Avena sativa*). *Cereal Res. Commun.* **2019**, *47*, 626–635. [CrossRef]

127. Yao, Y.; Zhang, P.; Liu, H.; Lu, Z.; Yan, G. A fully in vitro protocol towards large scale production of recombinant inbred lines in wheat (*Triticum aestivum* L.). *Plant Cell. Tissue Organ Cult.* **2017**, *128*, 655–661. [CrossRef]

128. Sysoeva, M.I.; Markovskaya, E.F.; Shibaeva, T.G. Plants under continuous light: A review. *Plant Stress* **2010**, *4*, 5–17.

129. Achigan-Dako, E.G.; Sogbohossou, O.E.; Maundu, P. Current knowledge on *Amaranthus* spp.: research avenues for improved nutritional value and yield in leafy amaranths in sub-Saharan Africa. *Euphytica* **2014**, *197*, 303–317. [CrossRef]

130. Collard, B.C.Y.; Beredo, J.C.; Lenaerts, B.; Mendoza, R.; Santelices, R.; Lopena, V.; Verdeprado, H.; Raghavan, C.; Gregorio, G.B.; Vial, L.; et al. Revisiting rice breeding methods–evaluating the use of rapid generation advance (RGA) for routine rice breeding. *Plant Prod. Sci.* **2017**, *20*, 337–352. [CrossRef]

131. Wolter, F.; Schindele, P.; Puchta, H. Plant breeding at the speed of light: The power of CRISPR/Cas to generate directed genetic diversity at multiple sites. *BMC Plant Biol.* **2019**, *19*, 1–8. [CrossRef]

132. Bartley, G. Wheat (*Triticum aestivum*) residue management before growing soybean (*Glycine max*) in Manitoba. Master's Thesis, Department of Plant Science, University of Manitoba, Winnipeg, Manitoba, 2019.

133. Wilton, M. A Broad-Scale Characterization of Corn (*Zea mays*)-Soybean (*Glycine max*) Intercropping as a Sustainable-Intensive Cropping Practice. Ph.D. Thesis, University of Waterloo, Ontario, Canada, 2019.

134. Borlaug, N.E. Sixty-two years of fighting hunger: Personal recollections. *Euphytica* **2007**, *157*, 287–297. [CrossRef]

135. Ferrie, A.M.R.; Möllers, C. Haploids and doubled haploids in Brassica spp. for genetic and genomic research. *Plant Cell Tissue Organ. Cult.* **2011**, *104*, 375–386. [CrossRef]

136. Lübberstedt, T.; Frei, U.K. Application of doubled haploids for target gene fixation in backcross programmes of maize. *Plant Breed.* **2012**, *131*, 449–452. [CrossRef]

137. Dirks, R.; Van Dun, K.; De Snoo, C.B.; Van Den Berg, M.; Lelivelt, C.L.C.; Voermans, W.; Woudenberg, L.; De Wit, J.P.C.; Reinink, K.; Schut, J.W.; et al. Reverse breeding: A novel breeding approach based on engineered meiosis. *Plant Biotechnol. J.* **2009**, *7*, 837–845. [CrossRef] [PubMed]

138. Kebede, A.Z.; Dhillon, B.S.; Schipprack, W.; Araus, J.L.; Bänziger, M.; Semagn, K.; Alvarado, G.; Melchinger, A.E. Effect of source germplasm and season on the in vivo haploid induction rate in tropical maize. *Euphytica* **2011**, *180*, 219–226. [CrossRef]

139. Castillo, A.M.; Cistué, L.; Vallés, M.P.; Soriano, M. Chromosome Doubling in Monocots. In *Advances in Haploid Production in Higher Plants*; Springer: New York, NY, USA, 2009; pp. 329–338.

140. Prigge, V.; Melchinger, A.E. Production of haploids and doubled haploids in maize. In *Plant Cell Culture Protocols*; Springer: New York, NY, USA, 2012; pp. 161–172.

141. Raju, S.K.K.; Shao, M.R.; Sanchez, R.; Xu, Y.Z.; Sandhu, A.; Graef, G.; Mackenzie, S. An epigenetic breeding system in soybean for increased yield and stability. *Plant Biotechnol. J.* **2018**, *16*, 1836–1847. [CrossRef]

142. Halpin, C. Gene stacking in transgenic plants—The challenge for 21st century plant biotechnology. *Plant Biotechnol. J.* **2005**, *3*, 141–155. [CrossRef]

143. Belhaj, K.; Chaparro-Garcia, A.; Kamoun, S.; Nekrasov, V. Plant genome editing made easy: Targeted mutagenesis in model and crop plants using the CRISPR/Cas system. *Plant Methods* **2013**. [CrossRef]

144. Low, J.W.; Mwanga, R.O.M.; Andrade, M.; Carey, E.; Ball, A.-M. Tackling vitamin A deficiency with biofortified sweetpotato in sub-Saharan Africa. *Glob. Food Sec.* **2017**, *14*, 23–30. [CrossRef]

145. Møller, I.S.; Gilliham, M.; Jha, D.; Mayo, G.M.; Roy, S.J.; Coates, J.C.; Haseloff, J.; Tester, M. Shoot Na+ exclusion and increased salinity tolerance engineered by cell type–specific alteration of Na+ transport in Arabidopsis. *Plant Cell* **2009**, *21*, 2163–2178. [CrossRef]

146. Singh, N.K.; Gupta, D.K.; Jayaswal, P.K.; Mahato, A.K.; Dutta, S.; Singh, S.; Bhutani, S.; Dogra, V.; Singh, B.P.; Kumawat, G.; et al. The first draft of the pigeonpea genome sequence. *J. Plant Biochem. Biotechnol.* **2012**, *21*, 98–112. [CrossRef]

147. Jackson, S.A. Rice: The First Crop Genome. *Rice* **2016**, 9. [CrossRef]

148. Egan, A.N.; Schlueter, J.; Spooner, D.M. Applications of next-generation sequencing in plant biology. *Am. J. Bot.* **2012**, *99*, 175–185. [CrossRef] [PubMed]

149. Bernier, G.; Périlleux, C. A physiological overview of the genetics of flowering time control. *Plant Biotechnol. J.* **2005**, *3*, 3–16. [CrossRef] [PubMed]

150. Kondić-špika, A.; Kobiljski, B. Biotechnology in Modern Breeding and Agriculture. In Proceedings of the International Conference on BioScience: Biotechnology and Biodiversity-Step in the Future. The Fourth Joint UNS-PSU Conference, Novi Sad, Serbia, 18–20 June 2012.

151. Liang, Z.; Chen, K.; Li, T.; Zhang, Y.; Wang, Y.; Zhao, Q.; Liu, J.; Zhang, H.; Liu, C.; Ran, Y.; et al. Efficient DNA-free genome editing of bread wheat using CRISPR/Cas9 ribonucleoprotein complexes. *Nat. Commun.* **2017**. [CrossRef] [PubMed]

152. Hedden, P. The genes of the Green Revolution. *TRENDS Genet.* **2003**, *19*, 5–9. [CrossRef]

153. Knoch, D.; Abbadi, A.; Grandke, F.; Meyer, R.C.; Samans, B.; Werner, C.R.; Snowdon, R.J.; Altmann, T. Strong temporal dynamics of QTL action on plant growth progression revealed through high-throughput phenotyping in canola. *Plant Biotechnol. J.* **2020**, *18*, 68–82. [CrossRef]

154. Yin, K.; Gao, C.; Qiu, J.L. Progress and prospects in plant genome editing. *Nat. Plants* **2017**, *3*, 1–6. [CrossRef]

155. Waltz, E. *CRISPR-Edited Crops Free to Enter Market, Skip Regulation*; Nature Publishing Group: Berlin, Germany, 2016.

156. Eş, I.; Gavahian, M.; Marti-Quijal, F.J.; Lorenzo, J.M.; Mousavi Khaneghah, A.; Tsatsanis, C.; Kampranis, S.C.; Barba, F.J. The application of the CRISPR-Cas9 genome editing machinery in food and agricultural science: Current status, future perspectives, and associated challenges. *Biotechnol. Adv.* **2019**. [CrossRef]

157. Xu, R.; Yang, Y.; Qin, R.; Li, H.; Qiu, C.; Li, L.; Wei, P.; Yang, J. Rapid improvement of grain weight via highly efficient CRISPR/Cas9-mediated multiplex genome editing in rice. *J. Genet. Genomics* **2016**, *43*, 529–532. [CrossRef]

158. Zhang, J.; Zhang, H.; Botella, J.R.; Zhu, J. Generation of new glutinous rice by CRISPR/Cas9-targeted mutagenesis of the Waxy gene in elite rice varieties. *J. Integr. Plant Biol.* **2018**, *60*, 369–375. [CrossRef]

159. Nekrasov, V.; Wang, C.; Win, J.; Lanz, C.; Weigel, D.; Kamoun, S. Rapid generation of a transgene-free powdery mildew resistant tomato by genome deletion. *Sci. Rep.* **2017**. [CrossRef]

160. Wang, F.; Wang, C.; Liu, P.; Lei, C.; Hao, W.; Gao, Y.; Liu, Y.G.; Zhao, K. Enhanced rice blast resistance by CRISPR/ Cas9-Targeted mutagenesis of the ERF transcription factor gene OsERF922. *PLoS ONE* **2016**. [CrossRef] [PubMed]

161. Kumar, M.; Aslam, M.; Manisha, Y.; Manoj, N. An Update on Genetic Modification of Chickpea for Increased Yield and Stress Tolerance. *Mol. Biotechnol.* **2018**, *60*, 651–663. [CrossRef] [PubMed]

162. Domoney, C.; Knox, M.; Moreau, C.; Ambrose, M.; Palmer, S.; Smith, P.; Christodoulou, V.; Isaac, P.G.; Hegarty, M.; Blackmore, T.; et al. Exploiting a fast neutron mutant genetic resource in Pisum sativum (pea) for functional genomics. *Funct. Plant Biol.* **2013**, *40*, 1261. [CrossRef]

163. Raman, H.; Raman, R.; Kilian, A.; Detering, F.; Carling, J.; Coombes, N.; Diffey, S.; Kadkol, G.; Edwards, D.; Mccully, M.; et al. Genome-Wide Delineation of Natural Variation for Pod Shatter Resistance in Brassica napus. *PLoS ONE* **2014**, *9*, e101673. [CrossRef] [PubMed]

164. Panigrahi, R.; Kariali, E.; Panda, B.; Lafarge, T.; Mohapatra, P.K. Controlling the trade-off between spikelet number and grain filling; the hierarchy of starch synthesis in spikelets of rice panicle in relation to hormone dynamics. *Funct. Plant Biol.* **2019**, *46*, 507–523. [CrossRef] [PubMed]

165. Witcombe, J.R.; Gyawali, S.; Subedi, M.; Virk, D.S.; Joshi, K.D. Plant breeding can be made more efficient by having fewer, better crosses. *BMC Plant Biol.* **2013**, *13*, 22. [CrossRef]

166. Basnet, B.R.; Crossa, J.; Dreisigacker, S.; Perez-Rodriguez, P.; Manes, Y.; Singh, R.P.; Rosyara, U.R.; Camarillo-Castillo, F.; Murua, M. Hybrid Wheat Prediction Using Genomic, Pedigree, and Environmental Covariables Interaction Models. *Plant Genome* **2019**, *12*. [CrossRef]

167. Xu, Y.; Babu, R.; Skinner, D.J.; Vivek, B.S.; Crouch, J.H. Maize Mutant opaque2 and the Improvement of Protein Quality through Conventional and Molecular Approaches. In Proceedings of the International Symposium on Induced Mutation in Plants, Vienna, Austria, 2–15 August 2008.

168. Shuwen, S.; Lianghong, L.; Jiangsheng, W.; Yongming, Z. In vitro screening stem rot resistant (tolerant) materials in Brassica napus L. *Chin. J. Oil Crop Sci.* **2003**, *25*, 5–8.

169. Pathirana, R. Plant mutation breeding in agriculture. *Plant Sci. Rev.* **2011**, 107–126. [CrossRef]

170. International Atomic Energy Agency. Proceedings of the International Symposium on Plant Mutation Breeding and Biotechnology, Vienna, Austria, 27–31 August 2018.

171. Ceballos, H.; Sanchez, T.; Denyer, K.; Tofino, A.P.; Rosero, E.A.; Dufour, D.; Smith, A.; Morante, N.; Perez, J.C.; Fahy, B. Induction and identification of a small-granule, high-amylose mutant in cassava (Manihot esculenta Crantz). *J. Agric. Food Chem.* **2008**, *56*, 7215–7222. [CrossRef]

172. Hamid, M.A.; Azad, M.A.K.; Howelider, M.A.R. Development of Three Groundnut Varieties with Improved Quantitative and Qualitative Traits through Induced Mutation. *Plant Mutat. reports* **2006**, *1*, 14–16.

173. Newell-McGloughlin, M. Nutritionally improved agricultural crops. *Plant Physiol.* **2008**, *147*, 939–953. [CrossRef] [PubMed]

174. Dunwell, J.M. Transgenic approaches to crop improvement. *J. Exp. Bot.* **2000**, *51*, 487–496. [CrossRef] [PubMed]

175. Toojinda, T.; Baird, E.; Booth, A.; Broers, L.; Hayes, P.; Powell, W.; Thomas, W.; Vivar, H.; Young, G. Introgression of quantitative trait loci (QTLs) determining stripe rust resistance in barley: An example of marker-assisted line development. *Theor. Appl. Genet.* **1998**, *96*, 123–131. [CrossRef]

176. Gimode, W.; Clevenger, J.; McGregor, C. Fine-mapping of a major quantitative trait locus Qdff3-1 controlling flowering time in watermelon. *Mol. Breed.* **2020**, *40*, 1–12. [CrossRef]

177. Joshi, S.G.; Schaart, J.G.; Groenwold, R.; Jacobsen, E.; Schouten, H.J.; Krens, F.A. Functional analysis and expression profiling of HcrVf1 and HcrVf2 for development of scab resistant cisgenic and intragenic apples. *Plant Mol. Biol.* **2011**, *75*, 579–591. [CrossRef]

178. Würdig, J.; Flachowsky, H.; Saß, A.; Peil, A.; Hanke, M.V. Improving resistance of different apple cultivars using the Rvi6 scab resistance gene in a cisgenic approach based on the Flp/FRT recombinase system. *Mol. Breed.* **2015**, *35*. [CrossRef]

179. Holme, I.B.; Wendt, T.; Holm, P.B. Intragenesis and cisgenesis as alternatives to transgenic crop development. *Plant Biotechnol. J.* **2013**, *11*, 395–407. [CrossRef]

180. Gadaleta, A.; Giancaspro, A.; Blechl, A.E.; Blanco, A. A transgenic durum wheat line that is free of marker genes and expresses 1Dy10. *J. Cereal Sci.* **2008**, *48*, 439–445. [CrossRef]

181. Cardi, T. Cisgenesis and genome editing: Combining concepts and efforts for a smarter use of genetic resources in crop breeding. *Plant Breed.* **2016**, *135*, 139–147. [CrossRef]

182. Jo, K.-R.; Kim, C.-J.; Kim, S.-J.; Kim, T.-Y.; Bergervoet, M.; Jongsma, M.A.; Visser, R.G.F.; Jacobsen, E.; Vossen, J.H. Development of late blight resistant potatoes by cisgene stacking. *BMC Biotechnol.* **2014**, *14*, 50. [CrossRef] [PubMed]

183. Schaart, J.G. *Towards Consumer-Friendly Cisgenic Strawberries which are Less Susceptible to Botrytis Cinerea*; Research Gate: Berlin, Germany, 2004; ISBN 908504104X.

184. Sawai, S.; Ohyama, K.; Yasumoto, S.; Seki, H.; Sakuma, T.; Yamamoto, T.; Takebayashi, Y.; Kojima, M.; Sakakibara, H.; Aoki, T.; et al. Sterol side chain reductase 2 is a key enzyme in the biosynthesis of cholesterol, the common precursor of toxic steroidal glycoalkaloids in potato. *Plant Cell* **2014**, *26*, 3763–3774. [CrossRef] [PubMed]

185. Luo, M.; Li, H.; Chakraborty, S.; Morbitzer, R.; Rinaldo, A.; Upadhyaya, N.; Bhatt, D.; Louis, S.; Richardson, T.; Lahaye, T.; et al. Efficient TALEN mediated gene editing in wheat. *Plant Biotechnol. J.* **2019**, *17*, 2026–2028. [CrossRef] [PubMed]

186. Zhou, J.; Peng, Z.; Long, J.; Sosso, D.; Liu, B.; Eom, J.-S.; Huang, S.; Liu, S.; Vera Cruz, C.; Frommer, W.B.; et al. Gene targeting by the TAL effector PthXo2 reveals cryptic resistance gene for bacterial blight of rice. *Plant J.* **2015**, *82*, 632–643. [CrossRef] [PubMed]

187. Svitashev, S.; Young, J.K.; Schwartz, C.; Gao, H.; Falco, S.C.; Cigan, A.M. Targeted Mutagenesis, Precise Gene Editing, and Site-Specific Gene Insertion in Maize Using Cas9 and Guide RNA. *Plant Physiol.* **2015**, *169*, 931–945. [CrossRef] [PubMed]

188. Zhou, X.; Jacobs, T.B.; Xue, L.-J.; Harding, S.A.; Tsai, C.-J. Exploiting SNPs for biallelic CRISPR mutations in the outcrossing woody perennial Populus reveals 4-coumarate:CoA ligase specificity and redundancy. *New Phytol.* **2015**, *208*, 298–301. [CrossRef] [PubMed]

189. Brooks, C.; Nekrasov, V.; Lippman, Z.B.; Van Eck, J. Efficient gene editing in tomato in the first generation using the clustered regularly interspaced short palindromic repeats/CRISPR-associated9 system. *Plant Physiol.* **2014**, *166*, 1292–1297. [CrossRef]

190. Li, Z.; Liu, Z.-B.; Xing, A.; Moon, B.P.; Koellhoffer, J.P.; Huang, L.; Ward, R.T.; Clifton, E.; Falco, S.C.; Cigan, A.M. Cas9-Guide RNA Directed Genome Editing in Soybean. *Plant Physiol.* **2015**, *169*, 960–970. [CrossRef]

191. Zhao, X.; Meng, Z.; Wang, Y.; Chen, W.; Sun, C.; Cui, B.; Cui, J.; Yu, M.; Zeng, Z.; Guo, S.; et al. Pollen magnetofection for genetic modification with magnetic nanoparticles as gene carriers. *Nat. Plants* **2017**, *3*, 956–964. [CrossRef]

192. Han, J.; Guo, B.; Guo, Y.; Zhang, B.; Wang, X.; Qiu, L.-J. Creation of Early Flowering Germplasm of Soybean by CRISPR/Cas9 Technology. *Front. Plant Sci.* **2019**, *10*, 1–10. [CrossRef]

193. Sánchez-León, S.; Gil-Humanes, J.; Ozuna, C.V.; Giménez, M.J.; Sousa, C.; Voytas, D.F.; Barro, F. Low-gluten, nontransgenic wheat engineered with CRISPR/Cas9. *Plant Biotechnol. J.* **2018**, *16*, 902–910. [CrossRef] [PubMed]

194. Farhat, S.; Jain, N.; Singh, N.; Sreevathsa, R.; Das, P.K.; Rai, R.; Yadav, S.; Kumar, P.; Sarkar, A.; Jain, A. CRISPR-cas 9 directed genome engineering for enhancing salt stress tolerance in rice. In *Proceedings of the Seminars in Cell & Developmental Biology*; Elsevier: Amsterdam, The Netherlands, 2019.

195. Zaman, Q.U.; Chu, W.; Hao, M.; Shi, Y.; Sun, M.; Sang, S.-F.; Mei, D.; Cheng, H.; Liu, J.; Li, C. CRISPR/Cas9-Mediated Multiplex Genome Editing of JAGGED Gene in Brassica napus L. *Biomolecules* **2019**, *9*, 725. [CrossRef] [PubMed]

196. Veillet, F.; Perrot, L.; Chauvin, L.; Kermarrec, M.-P.; Guyon-Debast, A.; Chauvin, J.-E.; Nogué, F.; Mazier, M. Transgene-free genome editing in tomato and potato plants using agrobacterium-mediated delivery of a CRISPR/Cas9 cytidine base editor. *Int. J. Mol. Sci.* **2019**, *20*, 402. [CrossRef] [PubMed]

197. Hsu, C.-T.; Cheng, Y.-J.; Yuan, Y.-H.; Hung, W.-F.; Cheng, Q.-W.; Wu, F.-H.; Lee, L.-Y.; Gelvin, S.B.; Lin, C.-S. Application of Cas12a and nCas9-activation-induced cytidine deaminase for genome editing and as a non-sexual strategy to generate homozygous/multiplex edited plants in the allotetraploid genome of tobacco. *Plant Mol. Biol.* **2019**. [CrossRef]

198. Yin, X.; Anand, A.; Quick, P.; Bandyopadhyay, A. Editing a Stomatal Developmental Gene in Rice with CRISPR/Cpf1. In *Plant Genome Editing with CRISPR Systems*; Springer: New York, NY, USA, 2019; pp. 257–268.

199. Malzahn, A.A.; Tang, X.; Lee, K.; Ren, Q.; Sretenovic, S.; Zhang, Y.; Chen, H.; Kang, M.; Bao, Y.; Zheng, X.; et al. Application of CRISPR-Cas12a temperature sensitivity for improved genome editing in rice, maize, and Arabidopsis. *BMC Biol.* **2019**, *17*, 1–14. [CrossRef]

200. Oconnor, D.; Wright, G.; George, D.; Hunter, M. Development and Application of Speed Breeding Technologies in a Commercial Peanut Breeding Program Development and Application of Speed Breeding Technologies in a Commercial Peanut Breeding Program. *Peanut Sci.* **2013**, *40*, 107–114. [CrossRef]

201. Riaz, A. Unlocking new sources of adult plant resistance to wheat leaf rust. Ph.D. Thesis, The University of Queensland, Queensland, Australia, 2018; pp. 1–241.

202. Wang, X.; Xuan, H.; Evers, B.; Shrestha, S.; Pless, R.; Poland, J. High-throughput phenotyping with deep learning gives insight into the genetic architecture of flowering time in wheat. *bioRxiv* **2019**, 527911.

203. Danzi, D.; Briglia, N.; Petrozza, A.; Summerer, S.; Povero, G.; Stivaletta, A.; Cellini, F.; Pignone, D.; de Paola, D.; Janni, M. Can high throughput phenotyping help food security in the mediterranean area? *Front. Plant. Sci.* **2019**, *10*, 15.

2

WB1, a Regulator of Endosperm Development in Rice, is Identified by a Modified MutMap Method

Hong Wang [1], Yingxin Zhang [1], Lianping Sun [1], Peng Xu [1], Ranran Tu [1], Shuai Meng [1], Weixun Wu [1], Galal Bakr Anis [1,2], Kashif Hussain [1], Aamiar Riaz [1], Daibo Chen [1], Liyong Cao [1,*], Shihua Cheng [1,*] and Xihong Shen [1,*]

[1] Key Laboratory for Zhejiang Super Rice Research, State Key Laboratory of Rice Biology, China National Rice Research Institute, Hangzhou 311400, Zhejiang, China; wjiyinh@126.com (H.W.); zyxrice@163.com (Y.Z.); slphongjun8868@126.com (L.S.); cnrri_pengxu@163.com (P.X.); 18883948050@163.com (R.T.); mengrice@163.com (S.M.); wuweixun@caas.cn (W.W.); galalanis5@gmail.com (G.B.A.); king3231251@gmail.com (K.S.); aamirriaz33@gmail.com (A.R.); cdb840925@163.com (D.C.)

[2] Rice Research and Training Center, Field Crops Research Institute, Agriculture Research Center, Kafr Elsheikh 33717, Egypt

* Correspondence: caoliyong@caas.cn (L.C.); shcheng@mail.hz.zj.cn (S.C.); xihongshen@126.com (X.S.)

Abstract: Abnormally developed endosperm strongly affects rice (*Oryza sativa*) appearance quality and grain weight. Endosperm formation is a complex process, and although many enzymes and related regulators have been identified, many other related factors remain largely unknown. Here, we report the isolation and characterization of a recessive mutation of *White Belly 1* (*WB1*), which regulates rice endosperm development, using a modified MutMap method in the rice mutant *wb1*. The *wb1* mutant develops a white-belly endosperm and abnormal starch granules in the inner portion of white grains. Representative of the white-belly phenotype, grains of *wb1* showed a higher grain chalkiness rate and degree and a lower 1000-grain weight (decreased by ~34%), in comparison with that of Wild Type (WT). The contents of amylose and amylopectin in *wb1* significantly decreased, and its physical properties were also altered. We adopted the modified MutMap method to identify 2.52 Mb candidate regions with a high specificity, where we detected 275 SNPs in chromosome 4. Finally, we identified 19 SNPs at 12 candidate genes. Transcript levels analysis of all candidate genes showed that *WB1* (*Os04t0413500*), encoding a cell-wall invertase, was the most probable cause of white-belly endosperm phenotype. Switching off *WB1* with the CRISPR/cas9 system in Japonica cv. Nipponbare demonstrates that *WB1* regulates endosperm development and that different mutations of *WB1* disrupt its biological function. All of these results taken together suggest that the *wb1* mutant is controlled by the mutation of *WB1*, and that the modified MutMap method is feasible to identify mutant genes, and could promote genetic improvement in rice.

Keywords: *Oryza sativa*; endosperm development; rice quality; *WB1*; the modified MutMap method

1. Introduction

Rice (*Oryza sativa*), one of the most important food crops in the world, provides more than 21% of human caloric needs [1]. With the improvement of living standards, there is increasing demand for high-quality rice, with greater quality of exterior, eating, and processing. Quality of rice appearance and yield are negatively affected by abnormally developed endosperm, which leads to grains with decreased weight and floury endosperm [2–8], shrunken endosperm [9–13], and great chalkiness [14,15].

Therefore, elucidating the mechanisms of endosperm development will be conducive to cultivating rice varieties with better appearance and higher yield.

Previous studies have shown that abnormality of rice endosperm can be caused by disorder of starch biosynthesis in the endosperm. Starch in the endosperm is composed of amylopectin (α-1,6-branched polyglucan) and amylose (α-1,4-polyglucan) [16]. In recent years, many key genes involved in starch biosynthesis have been identified in rice endosperm. The primary substrate of starch biosynthesis in rice endosperm comes from sucrose in the cell during photosynthesis [17], and it must be transported to the endosperm before being converted to glucose and fructose utilized for starch synthesis [18]. Several key genes involved in this process have been identified, including *OsSUT2*, which encodes a sucrose transporter and plays a vital role in transporting sucrose from source to sinks [19], *GIF1*, which encodes cell-wall invertase and is essential for the hydrolysis and uploading of sucrose [14], *OsSWEET4*, which encodes a hexose transporter and enhances sugar import into the endosperm from maternal phloem [20], and some genes (*SUS2, SUS3, SUS4*) from the sucrose synthase (SUS) genes family, which play an important role in the hydrolysis of sucrose [21]. However, glucose and fructose are not the direct substrate for starch synthesis: both need to be further converted to glucose 1-phosphate (G1P) under the catalysis of a series of enzymes [18]. The reaction of G1P with ATP (Adenosine 5'-triphosphate) produces the activated glucosyl donor ADP (Adenosine diphosphate)-glucose (ADPG), which is catalyzed by the enzyme ADP-glucose pyrophosphoryase (AGPase). In rice, the AGP gene family is made up of six subunit genes: two small subunit genes, *OsAGPS1* and *OsAGPS2* (*OsAGPS2a, OsAGPS2b*), and four large subunit genes, *OsAGPL1, OsAGPL2, OsAGPL3* and *OsAGPL4*. *OsAGPS1, OsAGPS2b, OsAGPL1* and *OsAGPL2* mainly function in rice endosperm [9,11,22]. In addition, pyruvate orthophosphate dikinase (PPDK) which is encoded by *OsPPDKB*, is involved in activating fructose [23]. In rice, loss-of-function mutants of these genes show abnormally developed endosperm, thus causing negative impacts on rice appearance quality and grain weight.

The activated substrates must cross the membrane of the amyloplasts before amylose and amylopectin are synthesized in the amyloplasts of the endosperm cells. During this transportation from cytoplasm to amyloplast, the major ADP-glucose transporter encoded by the *Brittle1* (*BT1*) imports the ADPG into amyloplasts; mutants with a defect in *BT1* develop shrunken endosperm [12,13]. When the activated substrates have been transported from the cytoplasm to the amyloplast, the gene *Waxy* encodes granule-bound starch synthase I (GBSS I), which primarily controls amylose synthesis [24]. Other genes control amylopectin synthesis in rice endosperm, including *SSI* [25], *SSIIa* [26], and *OsSSIIIa* [27], which encode starch synthase, *ISA1* [28,29] encoding isoamylase-type DBE isoamylase 1, and *OsBEIIb* [30] encoding starch branching enzymeIIb. Loss-of-function mutants of these genes severely disrupt the normal development of the endosperm.

Some regulators involved in starch synthesis during endosperm development have also been identified. *FLO2* mediates a protein-protein interaction, with a mutation of *FLO2* resulting in a floury endosperm [2]. *FLO6* directly interacts with *ISA1*, which affects the formation of starch granules during development of the rice endosperm [4]. *FLO7* encodes a regulator involved in starch synthesis and amylopectin development of the peripheral endosperm [8]. *Rice Starch Regulator1* (*RSR1*), a member of the AP2/EREBP family of transcription factors, negatively regulates starch synthesis [31]. *SUBSTANDARD STARCH GRAIN4* (*SSG4*) regulates the size of starch grains (SGs) in rice endosperm [6].

Deformity of the endosperm can also result from dysregulation of development of the protein bodies and storage proteins in the rice endosperm. *Chalk5*, which encodes a vacuolar H^+-translocating pyrophosphatase with inorganic pyrophosphate hydrolysis and H^+-translocation activity, is a major quantitative trait locus (QTL) which controls grain chalkiness [15]. The rice basic leucine Zipper factor (RISBZ1) and rice prolamin box binding factor (RPBF) are transcriptional activators, which coordinate to regulate the expression of SSP (seed storage protein) genes [32]. Decreased expression of *RISBZ1* (*OsbZIP58*) and *RPBF* in transgenic plants causes opaque endosperm.

Several methods are currently used for gene isolation. The most commonly used one is positional cloning (map-based cloning), by which many rice genes were isolated. However, map-based cloning is more time- and labor-intensive for isolating genes, especially QTLs. Therefore, many researchers have been exploring new methods for genes isolation. MutMap (Figure 1a) [33], based on next-generation sequencing (NGS) [34], is a recently developed method of rapid gene isolation. The MutMap method has been used to isolate some rice genes, including *OsRR22*, a gene responsible for the salinity-tolerant phenotype of *hst1* [35] and *Pii*, a gene enhancing rice blast resistance [36].

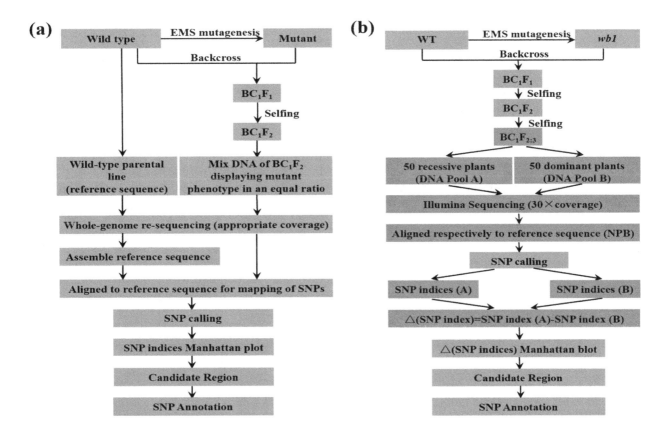

Figure 1. The steps of MutMap method applied to rice. (**a**) Common scheme of MutMap method applied to rice following the protocol described as previously reported [33]; (**b**) The scheme of gene mapping used in this study. The $BC_1F_{2:3}$ progeny formed mapping population. DNA of 50 recessive and 50 dominant plants from mapping population are mixed separately in an equal ratio to form the DNA Pool (A) and Pool (B) followed by the construction of DNA library and Illumina sequencing with 30×coverage, and then the treated sequencing data were aligned with the reference sequence followed by single nucleotide polymorphisms (SNP) calling. The reference sequence is the publicly available Nipponbare rice genome sequence [37]. For each identified SNP, SNP index (A) was obtained from Pool (A) and SNP index (B) corresponds with Pool (B). SNP index (A) minus SNP index (B) is Δ (SNP index) which is used for Manhattan plot, and we can obtain candidate region followed by SNP annotation. The pink color represents the different steps compared to the common scheme of MutMap. EMS: Ethane methyl sulfonate.

Although the understanding of the molecular mechanisms of the formation of rice endosperm has made great progress, rice endosperm is a very complex agronomic trait. Hence, it is still necessary to identify more functional genes and to describe their molecular mechanisms in order to enable systematic and comprehensive understanding of the inheritance of rice endosperm formation. In this study, we isolated *WB1*, which controlled rice endosperm development, via the modified MutMap method, and found that *WB1* played an important role in the regulation of starch synthesis during rice endosperm development. We also verified the target gene by CRISPR/Cas9 system. Our study also played a crucial role in explaining the molecular mechanisms of the formation of rice endosperm and the exploration of new methods for gene mapping.

2. Results

2.1. Phenotypic Characterization and Genetic Analysis of the wb1 Mutant

To identify new regulators of endosperm development, we recovered an endosperm development defective mutant named *wb1* from a mutant pool (in the *Japonica* variety ChangLiGeng background). The *wb1* mutant showed no apparent differences from WT throughout the vegetative stage. Plant height and the number of panicles per plant of *wb1* plants were similar to those of the WT at the mature stage. The number of spikelets per panicle, number of grains per panicle, seed-setting rate and 1000-grain weight of *wb1* were all showed a marked decreased compared with those of the WT (Table S1). Unlike WT, the glume of *wb1* grains showed brown color (Figure 2a–c), and *wb1* displayed markedly more grain chalkiness in the grain belly (Figure 2d,e).

Grain chalkiness rate and grain chalkiness degree were 94.8% and 47.6% in *wb1* grains, while those of WT grains were 2.8% and 0.6%(Figure 2i,j). Scanning electron microscope images clearly indicated that the endosperm from grains of *wb1* developed abnormally as a result of loosely packed, spherical starch granules, in contrast to the densely packed, irregularly polyhedral starch granules of the normal endosperm from the grains of WT (Figure 2f,g). Grain size measurements showed that the seed length, width, and thickness were all significantly reduced in *wb1* grains (Figure 2h), resulting in a smaller grain size than that of WT, even occasionally in a shriveled phenotype. We also measured amylose and amylopectin content of the mature grains of *wb1* and WT. Amylose and amylopectin content were remarkably decreased in *wb1* grains (Figure 2k,l), suggesting that the starch accumulation in *wb1* grains was severely disrupted.

All these results collectively reveal that mutation of *WB1* caused a defect in the endosperm development, which led to the higher grain chalkiness degree and a significant reduction of 1000-grain weight in *wb1* grains. We also analyzed physical properties of *wb1* and WT grains, including gel consistency (Figure 2n), brown rice rate (Figure 2o), milled rice rate (Figure 2p) and head rice rate (Figure 2q). Each of these was significantly reduced in the mutant, suggesting that dramatic physical changes have occurred in the *wb1* grains, which will further affect rice processing and eating quality.

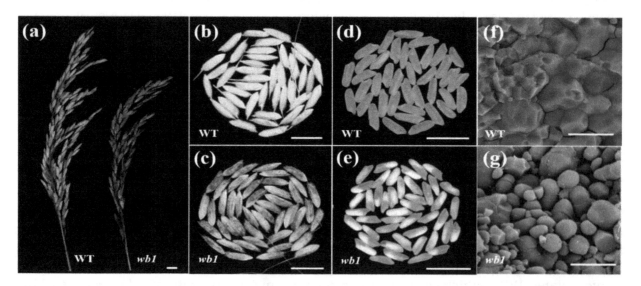

Figure 2. *Cont.*

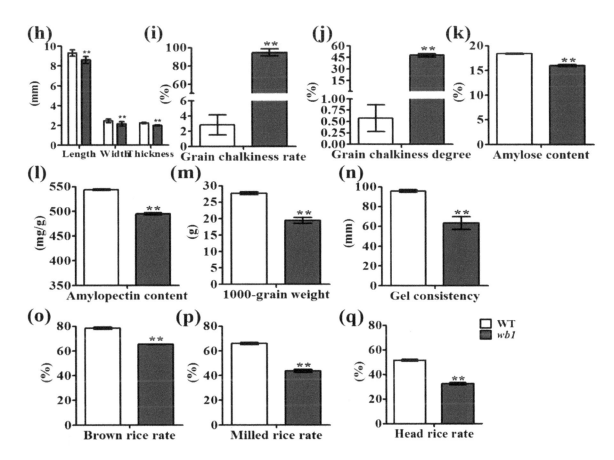

Figure 2. Phenotypic analyses of *wb1*. (**a**) Comparison of representative WT and *wb1* plant panicles; (**b,c**) Appearance of WT (**b**) and *wb1* (**c**) grains; (**d,e**) Appearance of WT (**d**) and *wb1* (**e**) white grains; (**f,g**) Scanning electron microscope images of WT (**f**) and *wb1* (**g**) seed endosperm. Magnification, ×2000; (**h**) Measurements of seed length, width, and thickness of WT and *wb1* grains (*n* = 20); (**i**) Grain chalkiness rate comparison of WT and *wb1* grains (*n* = 6), and grain chalkiness rate is the rate of chalky grains in total grains; (**j**) Grain chalkiness degree comparison of WT and *wb1* grains (*n* = 6), and grain chalkiness degree is the grain chalkiness rate multiplied by grain chalkiness area (the percent area of chalk in a grain); (**k**) Amylose content comparison of WT and *wb1* grains (0.05 g grain powder each, *n* = 3); (**l**) Amylopectin content comparison of WT and *wb1* grains (0.01 g grain powder each, *n* = 3); (**m**) Comparison of 1000-grain weight of WT and *wb1* (*n* = 10); (**n**) Gel consistency comparison of WT and *wb1* grains (*n* = 4); (**o,p,q**) Comparisons of brown rice, milled rice and head rice rates of WT and *wb1* (25 g paddy each, *n* = 3). Data are given as means ± SD (standard deviation). The asterisks represent statistical significance between WT and *wb1*, determined by a student's *t*-test (** $p \leq 0.01$). Scale bars: (**a–e**) 10 mm; (**f,g**) 30 μm.

To verify that this locus associated with the *wb1* phenotype was controlled by a single recessive gene, genetic analysis was conducted to examine the phenotype of all plants from BC_1F_1 and F_1 progeny, and of 1087 and 1000 plants from BC_1F_2 and F_2 progeny, respectively. The results showed that BC_1F_1 and F_1 seeds exhibited the wild-type phenotype, while the segregation model of normal to chalky grains fitted well to the expected ratio of a single inheritance, 3:1 (820:267, 745:255), in the BC_1F_2 and F_2 progeny (Table S2).

2.2. Candidate Region of the WB1 Gene Obtained through the Modified MutMap Method

A modified MutMap method (Figure 1b) was applied to isolate the WB1 gene. After re-sequencing for Pool A and Pool B, we obtained 125,252,285 (SRA accession SRP135580) and 120,484,878 (SRA accession SRP135578) cleaned reads for Pool A and Pool B, respectively, corresponding to >20 Gb of total read length with $30\times$ coverage of the rice genome (370 Mb; Table S3). After these cleaned reads were aligned separately to the Nipponbare reference sequence by the BWA software, we obtained 110,119,455 and 105,488,179 unique mapped reads for Pool A and Pool B, respectively, corresponding to 87.55% and 87.92% coverage of the rice genome (Table S4). Then we calculated Δ (SNP indices) or *Fst* value based on the sliding window of the whole genome scan following by plotting the Δ (SNP indices) for all 12 chromosomes of rice (Figure 3b). As we expected, Δ (SNP indices) were distributed randomly around 0 for most parts of the genome (Figure 3b). Finally, we obtained the candidate region of 2.52 Mb (Figure 3b).

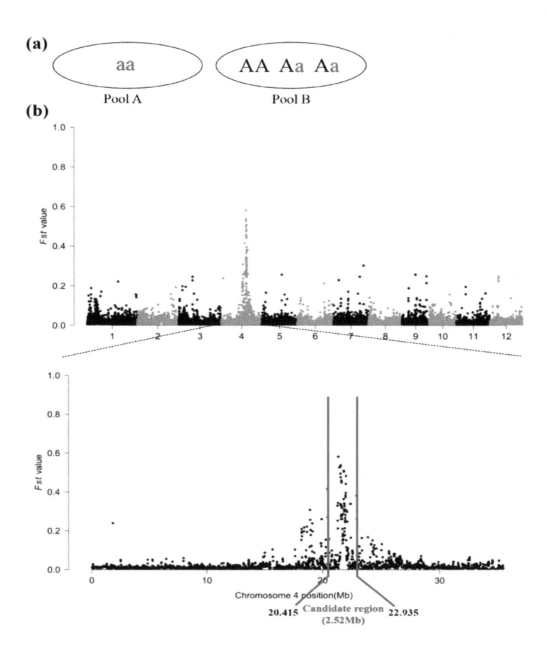

Figure 3. Candidate region of wb1 obtained by the modified MutMap method. (**a**) An example for explanation of Δ (SNP index) for the casual SNP. Theoretically, SNP index (A) would be 1, SNP index (B) 0.333 (1/3), and thus Δ (SNP index) would be 0.667 (1 minus 1/3); (**b**) Δ (SNP indices) Manhattan plot. *Fst* value, defined as the proportion of genetic diversity due to allele frequency differences among populations described by the previous report [38]. Δ (SNP indices) and *Fst* values have the same meaning in this study.

2.3. Screening the SNPs Detected in the Candidate Region

From the Δ (SNP indices) plot (Figure 3b), we obtained the candidate region of 2.52 Mb where we detected 275 SNPs in chromosome 4 followed by gene annotation (Table S5). To identify the true causal SNP, we screened these SNPs with three steps: (i) retaining the SNPs in which Δ (SNP indices) ranged from 0.6 to 0.8; (ii) removing SNPs located in the intergenic region and SNPs which resulted in synonymous substitutions; and (iii) detecting the SNPs between WT and wb1 by sequencing. As a result, we obtained nineteen SNPs, which were located in twelve candidate genes (Table 1).

Table 1. Nineteen SNPs in twelve candidate genes.

Δ (SNP Index)	Accession	Location (bp)	Reference Base (WT)	Altered Base in *wb1*	Type of Mutation	Gene Annotation
0.758		21550665	T	G	Missense (T to P)	
0.754		21550664	G	T	Missense (T to K)	Helicase conserved
0.663	ORF1	21550888	C	T	Intron mutation	C-terminal domain
0.655		21550286	T	A	Missense (T to S)	containing protein
0.649		21551279	G	A	Intron mutation	
0.754		21539737	A	G	3′-UTR mutation	Protein of unknown
0.612	ORF2	21539457	G	T	Splice region mutation	function DUF668 family protein
0.743	ORF3	21331260	G	A	Missense (D to N)	40S ribosomal protein S10
0.734		21514382	C	T	Intron mutation	Similar to
0.708	ORF4	21513793	A	G	Intron mutation	H0315E07.10 protein
0.734		21612944	C	A	Missense (K to N)	CENP-E-like kinetochore
0.663	ORF5	21610862	C	A	Missense (S to I)	protein
0.733	ORF6	20423829	G	A	Missense (A to T)	Glycosyl hydrolases
0.733	ORF7	21795109	G	A	Missense (L to F)	Expressed protein
0.672	ORF8	21493980	G	A	Intron mutation	Similar to H0315E07.7 protein
0.639	ORF9	21897538	C	T	3′-UTR mutation	Nonsense-mediated decay UPF3
0.634	ORF10	21970357	C	T	5′-UTR mutation	Peptide transporter PTR2
0.631	ORF11	21710470	C	G	Intron mutation	Conserved hypothetical protein
0.61	ORF12	21734385	C	T	Nonsense (R to *)	No apical meristem protein

The asterisk indicates the stop codon.

2.4. Identification of the Casual SNP

We detected the expression levels of twelve candidate genes in endosperm tissues at various development stages (5, 10, 15 and 20 DAF) (Figure 4). We successfully detected all genes transcript levels except for that of *ORF8*. *ORF6* maintained relatively high expression level in comparison with other genes and its transcript level changed significantly during the four stages of endosperm development between WT and *wb1* (Figure 4). Although some genes demonstrated higher transcript levels at the DAF15 and DAF20 stages (Figure 4c,d) compared with the DAF5 and DAF10 stages (Figure 4a,b), and the transcript levels of several genes were also significantly altered between WT and *wb1* (Figure 4), all other genes showed low expression levels on the whole in contrast to the transcript levels of *ORF6*. Therefore, we may conclude that the mutation of *ORF6* (*Os04t0413500* or *Os04g33740*) played a major role in the defect of *wb1*.

SNP-20423829 were G to A transitions, presumably caused by EMS mutagenesis [39], and it was located at the site 1659 bp of the third exon of *ORF6* encoding a glycosyl hydrolase. This SNP led to an A159T mutation (codon GCG to ACG; Figure 5a,b). Moreover, results of digestion of restriction endonuclease *Hae* II confirmed this mutant site (Figure 5c). Accordingly, we hypothesized that *wb1* was caused by a missense substitution in *ORF6*. We also found that *ORF6* was a novel allele of *GIF1* (*Os04g33740*) which controlled rice grain filling and yield [14].

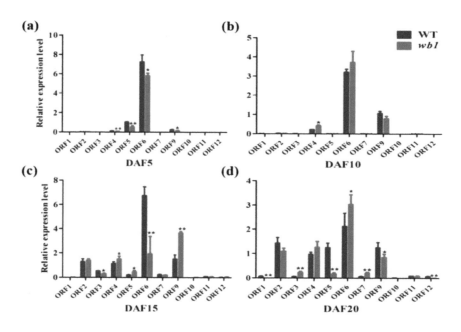

Figure 4. Relative expression analysis of 11 candidate genes based on real-time quantitative PCR (qPCR) at four stages of endosperm development between WT and *wb1*. (**a**) Relative expression analysis of 11 candidate genes at DAF5 stage; (**b**) Relative expression analysis of 11 candidate genes at DAF10 stage; (**c**) Relative expression analysis of 11 candidate genes at DAF15 stage; (**d**) Relative expression analysis of 11 candidate genes at DAF20 stage. All data were compared with transcript levels of WT by Student's *t*-test (* $p \leq 0.05$, ** $p \leq 0.01$). Values were means $\pm$ SD ($n = 3$).

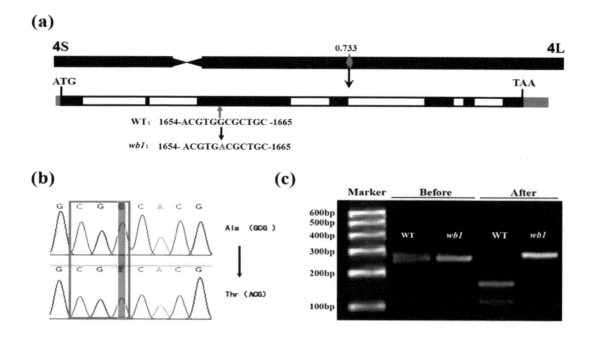

Figure 5. Further verification of causal SNP in *wb1*. (**a,b**) Sequencing validation of the causal SNP and the type of mutation; the red arrow indicates the mutant site, and the black arrows indicate the alternations; (**c**) Digestion of restriction endonuclease *Hae*II. "Before" represents the non-treated PCR product and "After" represents the *Hae*II-treated PCR product.

2.5. Function Verification of the WB1 Gene (ORF6) through the CRISPR/Cas9 System in Reverse

To verify our hypothesis, we created six novel alleles of *WB1* through the CRISPR/Cas9 system in *Japonica* cultivar Nipponbare (NPB). We found two target sequences in the third exon of *WB1* corresponding to CRISPR/Cas9 system and obtained six different mutants in T1 lines (Figure 6a). Grains of six mutants displayed brown glumes and grain chalkiness in the grain belly compared with the common grains of NPB (Figure 6b). SEM images distinctly showed that endosperm of six mutants developed abnormally compared to the normal endosperm of NPB (Figure 6b). The phenotypes of six mutants were similar to that of *wb1* (Figure 2b–g). Those results further proved that *WB1* was the target gene responsible for the *wb1* phenotype.

In NPB and six mutant lines, we measured 1000-grain weight (Figure 6h) and the main factors affecting 1000-grain weight, including grain length (Figure 6c), width (Figure 6d), and thickness (Figure 6e), grain chalkiness rate (Figure 6f) and degree (Figure 6g). Duncan's test indicated that the grain chalkiness rate degree were the major factors causing the significant reduction of 1000-grain weight of six mutant lines. The differences in grain length, width, and thickness between six mutant lines and NPB were not similar to the differences between WT and *wb1* (Figure 2h). This discrepancy was probably caused by the longer grain length of WT (~9.3 mm, Figure 2h) compared with that of NPB (~6.6 mm, Figure 6c) and the different mutations of WB1 between wb1 (Figure 5a) and the six mutant lines (Figure 6a).

Interestingly, some differences were also found among the six mutant lines (Figure 6c–h). Those results suggested that the grain chalkiness rate and grain chalkiness degree collectively determined the 1000-grain weight (Figure 6f–h), especially in the mutant line *nc-3*, where grain chalkiness rate and grain chalkiness degree showed significant decreases as compared to the other mutant lines, corresponding to its higher 1000-grain weight. To test whether the differences in the grain chalkiness rate and grain chalkiness degree among the six mutant lines were caused by different mutations of WB1, we performed multiple comparison of WB1 sequences by MEGA 5.0 software (Figure 7). The findings indicated that the six mutant lines showed different mutations which disrupted the substrate binding site and the active site of WB1, except in the mutant line *nc-5* (Figure 7).

2.6. Expression Analysis of Starch Metabolism-Related Genes in Endosperm

We performed qPCR analysis of total RNA extracted from the seed endosperm of WT and *wb1* at various stages (DAF5, DAF10, DAF15, and DAF20) and detected the transcript levels of some genes involved in starch synthesis. As shown in Figure 8, transcript levels of those genes were all altered during development of the rice endosperm. During the critical stages (DAF10 and DAF15) of grain filling, transcript levels of all genes showed a striking contrast between WT and *wb1*. This suggests that altered transcript levels of starch synthesis-related genes are probably involved in the abnormal development of rice endosperm. The higher transcript levels of *WB1*, *OsAPS1*, *OsAPL1*, *OsPPDKB*, and *FLO6* at the mature stage (DAF20) in the *wb1* mutant were probably caused by different maturity of seeds between WT and *wb1*.

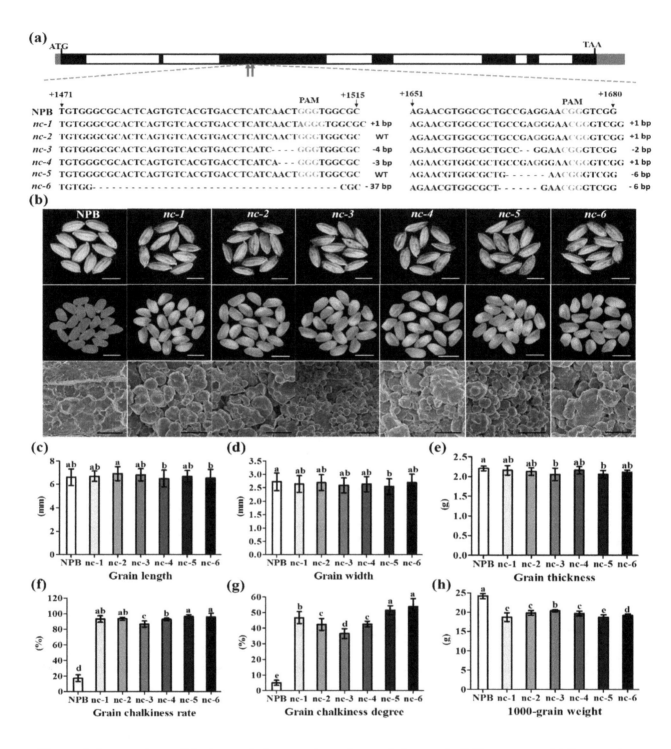

Figure 6. Sequencing validation and phenotypic analyses of six novel allelic mutants. (**a**) Sequencing validation of six novel allelic mutants. Blue color represents the PAM sequence of CRISPR/Cas9 system; red color represents insert bases; "-" represents deletion bases; the red arrows indicate the mutant sites mediated by CRISPR/Cas9 system. (**b**) Appearance and SEM of NPB and mutants grains. Magnification, ×1000; (**c–h**) Measurements of grain length (*n* = 30), width (*n* = 30), thickness (*n* = 10), grain chalkiness rate (*n* = 10), grain chalkiness degree (*n* = 10) and 1000-grain weight (*n* = 10) of NPB and mutant lines. Different letters indicate the statistical difference at $p \leq 0.05$ by Duncan's test. Values were means ± SD. Scale bars: bars of grains figures 5 mm; bars of SEM figures 10 μm.

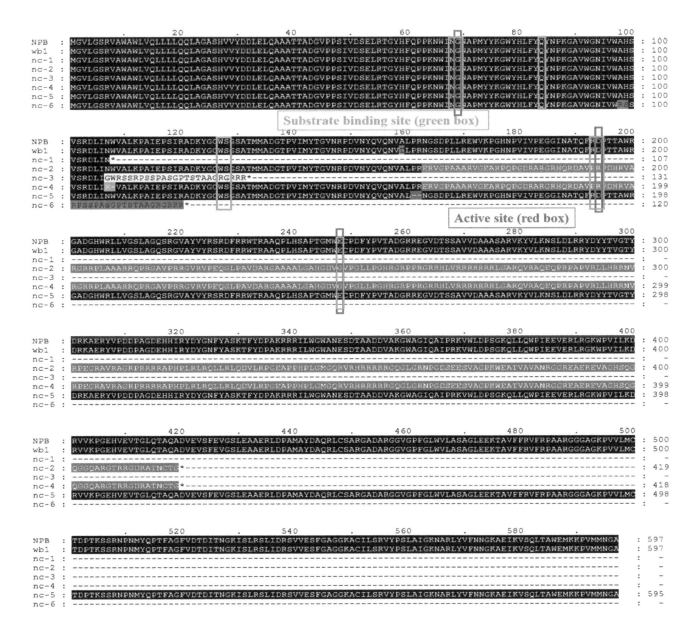

Figure 7. WB1 alignments of NPB (WT) with six mutant lines and *wb1*. Analysis performed with MEGA 5.0 software. The substrate binding site (eight residues, green boxes) and the active site (three residues, red boxes) are indicated by the Blast search program. Black color indicates a sequence that is consistent with that of WT. Except for the black color, the same color represents the same sequence, and different colors represent the different mutations among the mutant lines.

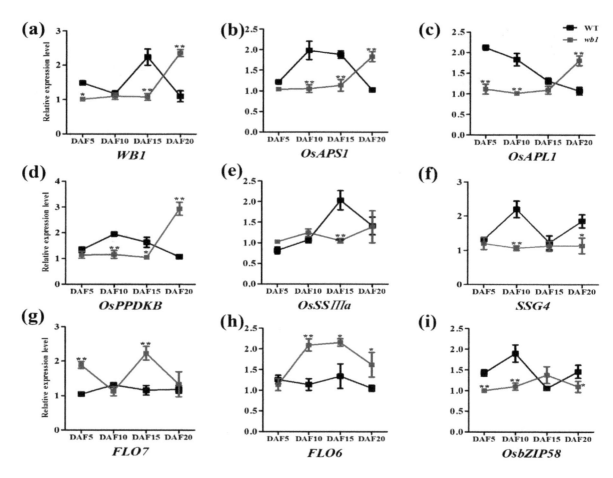

Figure 8. Relative expression analysis of genes associated with starch synthesis by qPCR at four stages of WT and *wb1* endosperm development. (**a–i**) Relative expression analysis of *WB1* (**a**), *OsAPS1* (**b**), *OsAPL1* (**c**), *OsPPDKB* (**d**), *OsSSIIIa* (**e**), *SSG4* (**f**), *FLO7* (**g**), *FLO6* (**h**), *OsbZIP58* (**i**) in WT and the wb1 mutant. All data were compared with the relative expression levels of WT by Student's *t*-test (* $p \leq 0.05$, ** $p \leq 0.01$). Values were means ± SD ($n = 3$).

3. Discussion

3.1. WB1 Controls Rice Endosperm Development

Grain chalkiness is one of the most important factors leading to low grain weight [15] and affecting rice appearance and milling, cooking, and eating quality [40,41]. Grain chalkiness is controlled by complex quantitative trait loci and by climatic conditions during rice grain filling, especially high temperature [42]. *Chalk5*, a major quantitative trait locus controlling grain chalkiness, and affecting head rice rate, is the only one that has been identified and characterized up to now. *Chalk5* is involved in the biogenesis of protein bodies in the endosperm cells [15]. Grain chalkiness can be an indicator of abnormally developed endosperm [2–8]. The major component of rice endosperm is a starch that is mainly composed of amylose and amylopectin. Many genes directly involved in the biosynthesis of amylose and amylopectin in rice endosperm cells have been identified and characterized, such as *Waxy*, *SSI*, *SSIIa*, *OsSSIIIa*, *ISA1*, *OsBEIIb* and *OsPPDKB* [23–28,30]. Loss of function of these genes can result in an abnormally developed endosperm, displaying more grain chalkiness and low grain weight.

Sucrose is produced by the source organ or photosynthetic organ and used as a carbon source for starch biosynthesis in endosperm cells. Sucrose must be transported from source organs into sink organs, which occurs via apoplast and/or sympast. Accordingly, in addition to the genes directly involved in starch biosynthesis of endosperm cells, other genes involved in this process can also affect endosperm development. In the apoplastic pathway, sucrose can be converted by cell-wall invertases into glucose and fructose, which are transported into cells by hexose transporters. Sucrose can also be directly taken by sucrose transporters into sink cells where it is hydrolyzed into glucose and fructose by sucrose hydrolases, including SUS2, SUS3, and SUS4 [21]. *GIF1* encodes a cell-wall invertase that mainly functions in the hydrolysis and uploading of sucrose during early grain-filling. The *gif1* mutant shows slower grain filling, ~24% lower final grain weight, and lower contents of amylose and amylopectin, and markedly more grain chalkiness as a result of abnormally developed and loosely packed starch granules [14]. *OsSWEET4* encodes a hexose transporter that is responsible for transferring hexoses across the BETL (basal endosperm transfer layer) to sustain of rice endosperm, downstream of a cell-wall invertase. The *ossweet4-1* mutant shows incomplete grain filling and significantly decreased grain weight [20]. *OsSUT2* encodes a sucrose transporter that functions in sucrose uptake from the vacuole. The *ossut2* mutant has significantly decreased sugar export ability and 1000-grain weight [19]. In our study, *WB1* encoded the *cell-wall invertase 2* (*OsCIN2*) and was a novel allele of the *GIF1* (*Os04t0413500* or *Os04g33740*) gene which controlled rice grain-filling and thus affected rice endosperm development [14]. Due to the *WB1* gene mutation (Figure 5) which led to grain incomplete filling [14], the physical and chemical properties of grain endosperm of the *wb1* mutant have been altered, like higher grain chalkiness rate (Figure 2i), higher grain chalkiness degree (Figure 2j), markedly more grain chalkiness as a result of loosely packed, spherical granules (Figure 2e,g), lower contents of amylose and amylopectin (Figure 2k,l), and ~30.0% lower 1000-grain weight (Figure 2m). In addition, transcript levels of the starch synthesis-related genes in our study varied during rice endosperm development (Figure 8). All of these observations suggest that the *wb1* mutant exhibits a defect in endosperm development, thus leading to the white-belly endosperm with altered phy-chemical property.

3.2. Different Mutations of WB1 Can Disrupt Its Biological Function

Many genes make up a large regulatory framework that regulates life activity in higher plants. These genes encode active proteins that are responsible for the major functions in this global regulatory framework. The function of each of active protein is determined by its own primary, secondary, and tertiary structure. In rice, mutations of a gene can result in its encoding protein with structural alterations which can affect its biological function followed by phenotypic changes. The *sd1* gene, well known as the genetic basis for the first "green revolution" in rice, encodes a GA 20-oxidase involved in the GA biosynthesis pathway; the *sd1* gene controls the plant height of rice, and mutations (*sd1-d*, *sd1-r*, *sd1-c*, *sd1-j*) in this locus cause the dwarfism of rice to different degrees [43–45].

In our study, six mutants of novel alleles of *WB1* displayed the same phenotype as the *wb1* mutant (Figures 2b–e and 6b), primarily showing higher chalkiness rates (Figure 6f), higher chalkiness degrees (Figure 6g), and lower 1000-grain weight (Figure 6h). The sequence analysis showed that six different mutations have occurred in the WB1 locus (Figure 6a), leading to alterations of the amino acid sequence of the WB1 protein in different types (Figure 7). In the previous research of the *gif1* mutant, the *GIF1* gene revealed a 1-nt deletion in the coding region, causing the premature GIF1 protein which disrupt its biological function (incomplete grain-filling) [14]. The *WB1* gene of *nc-1*, *nc-3* and *nc-6* also caused different premature WB1 proteins with altered substrate binding site and active site (Figure 7). The frame shift mutation of the *WB1* gene of *nc-2* and *nc-4* also disrupted its biological function (Figures 6b and 7). Interestingly, the single amino acid substitution (A159T) of WB1 of *wb1*, and Proline-161 and Arginine-162 deletion of WB1 of *nc-5* led to the dysfunction of WB1 without altered substrate binding site and active site (Figures 2, 6 and 7), suggesting that Alanine-159, Proline-161 and Arginine-162 are required for activity of WB1. Moreover, grain chalkiness degrees and 1000-grain weight showed significant differences among some mutants. However, several mutants showed no differences in grain chalkiness degree and 1000-grain weight (Figure 6g,h); these results suggest that different mutations probably affect the formation of grain chalkiness in different degrees, and further research still needs to be conducted to explain the molecular mechanism. In summary, seven novel alleles (including the *wb1* mutant) had different mutations which disrupted their biological functions.

3.3. The Modified MutMap Method Applied to Isolate WB1 Is Feasible for Gene Mapping

MutMap is a new method used for gene identification [33]. Using MutMap method, researchers can isolate mutant genes and QTLs rapidly, accurately, and conveniently compared to conventional map-based cloning [33,35,36]. Through the MutMap method, researchers only sequence the DNA pool from recessive individuals of F_2 population based on second-generation sequencing, followed by aligning to the assembled whole-genome sequence of wild-type. The population used for the MutMap method is BC_1F_2 population, which can show unequivocal segregation between the mutant and wild-type phenotype. Notably, the MutMap method requires assembling the whole-genome sequence of wild type accurately used as the reference sequence.

Previously, many genes have been identified by the MutMap method in rice. *OsRR22*, responsible for the salinity-tolerance phenotype for the *hst1* mutant, has been identified by a MutMap method: the sequence depth and the average coverage of wild-type are $28.7\times$ and 59.9%; the number of BC_1F_2 individuals is 20 and the sequence depth is $18.4\times$ [35]. Two mutant genes regarding pale green leaf have been identified: the sequence depth and the average coverage of wild-type are $>12\times$ and $\sim95.5\%$; the number of BC_1F_2 individuals is 20 and the sequence depths are $12.5\times$ and $24.1\times$ [33]. Four mutant genes regarding semi-dwarf phenotype also have been identified: the sequence depth and the average coverage of wild-type are $>12\times$ and $82.4\%\sim84.2\%$; the number of BC_1F_2 individuals is 20 and the sequence depths are $14.2\times\sim16.6\times$ [33].

The MutMap method is subject to high error rates because of multiple factors, including difficulty in determining the number of F_2 progeny showing the mutant phenotype, the average coverage (depth) of genome sequencing, and classification of phenotypes between wild and mutant phenotype [33,46]. Therefore, the greater the number of F_2 progeny showing the mutant phenotype to be bulked, the deeper the average depth of genome to be sequenced, and more accurate classification of phenotypes between wild and mutant type, the lower the rate of false positives [33].

In our study, a modified MutMap method (Figure 1b) was applied to successfully isolate the *WB1* gene related to endosperm development in rice. Compared with the MutMap method, our modified MutMap method has some advantages. Firstly, the individuals of bulked DNA Pools used for sequencing were from $BC_1F_{2:3}$, which has a more stable genetic background than the BC_1F_2 population; because of the reduced effect from other gene mutations on the target phenotype, it was easier to distinguish plants between mutant type and wild type; Secondly, the appropriately elevated number of $BC_1F_{2:3}$ individuals (50, Figure 1b) and sequence depth ($30\times$, Figure 1b) ensured relatively high coverage (87.92% and 87.55%; Table S4); Lastly, we sequenced the DNA pools not only from recessive individuals but also from dominant individuals followed by aligning to the reference sequence Nipponbare, respectively; therefore, it was not necessary to sequence and assemble the whole-genome sequence of wild type used as reference sequence. We directly used the Nipponbare genomic sequence as our reference sequence, so that we greatly reduced the costs required for sequencing and assembling reference sequence. Therefore, the *WB1* gene mapping result showed higher specificity with the single peak in the chromosome 4 (Figure 3b) compared to that by the MutMap method [33]. Besides, the modified MutMap method also has a deficiency that is the significant difference in the whole genome sequence between Nipponbare and Wild-type (reference sequence). Delightedly, several rice genome sequences have been published, like *indica* cultivar 9311 [47], Zhenshan97, Minghui63 [48] and Shuhui498 [49] which can be used as reference sequence directly, expanding the application scope of the modified MutMap method. Overall, the modified MutMap method showed a low error rate, a relatively low cost and a high specificity, and could promote the development of rice genetics.

4. Materials and Methods

4.1. Plant Materials and Growth Condition

The *wb1* (mutant) was initially identified from the ethyl methanesulfonate (EMS)-treated *Japonica* rice variety ChangLiGeng (CLG, Wild Type) M_2 population. The *wb1*, as the pollen acceptor, was crossed with WT and ZhongHui8015 (ZH8015, *Indica*), respectively. The resulting first filial generation (BC$_1$F$_1$, F$_1$) plant was self-pollinated, and the second generation (BC$_1$F$_2$, F$_2$) was used as the genetic analysis population. We collected seeds of 100 individuals from BC$_1$F$_2$ population, and then cropped the seeds to obtain 100 pedigrees (BC$_1$F$_{2:3}$) which were used as the mapping population for the modified MutMap method (Figure 1b). All plants were grown in an experimental paddy field at China National Rice Research Institute (Hangzhou, Zhejiang province and Lingshui, Hainan province, in China) under natural open-air condition.

4.2. Grain Quality Analysis

Scanning electron microscopy was performed as described previously [50]. Measurements of amylose content of mature grains (0.05 g powder) were conducted by HPSEC-MALLS-RI following the method of Fujita et al. (2003) [51]. Quantitative amylopectin content was determined by processing 0.01 g powder of the mature grain, according to a method from a previous report [52]. Each measurement was repeated three times ($n = 3$).

The paddy rice of WT and wb1 were dried to moisture content of 12–14% and were maintained at room temperature at least three months before measuring the brown rice rate, milled rice rate, and head rice rate by grain polisher AH001151 (KETT, Tokyo, Japan), performed as Zhou et al. (2015) [53]. Each measurement of 25 g paddy rice was performed with three replicates ($n = 3$, total 75 g paddy rice). Grain chalkiness rate and grain chalkiness degree were determined using SC-E Rice Quality Inspection and Analysis System (WanShen, Hangzhou, China). The white grains from 12 plants (6 plants from WT and wb1, respectively, $n = 6$) were used for grain chalkiness rate and degree measurements. Gel consistency was measured ($n = 4$) following the protocol described in Li et al. (2014) [15].

4.3. PCR, RNA Isolation and Real-time Quantitative PCR (qPCR)

PCR amplifications of candidate genes were performed using KOD FX DNA Polymerase (TOYOBO). The PCR product of the reaction of restriction enzyme *HaeII* digestion was amplified by KOD-Plus-Neo (TOYOBO). The primer pairs designed for this study are listed in Table S6.

Total RNA was prepared from grains of WT and *wb1* at 5, 10, 15, and 20 DAF (days after flowering) using the TIANGEN RNAprep Pure Plant Kit (Tiangen Biotech, Beijing, China). The first cDNA strand was synthesized from DNase I-treated RNA using Oligo-dT (18) primers in a 20 μL reaction system based on a SuperScriptIII Reverse Transcriptase Kit (TOYOBO). qPCR was performed on a Roche Light Cycle 480 device using THUNDERBIRD SYBR qPCR Mix (TOYOBO). Each reaction was performed with three replicates ($n = 3$). The primers used in this analysis are listed in Table S6.

4.4. DNA Template Preparation, DNA Library Construction, and Re-Sequencing

Genomic DNA was extracted (large scale) from young leaf tissues following the modified hexadecyl trimethylammonium bromide (CTAB) method [54]. Young leaves (total 5 g, 0.1 g per plant) were obtained from 50 plants displaying the mutant or wild phenotype in the BC$_1$F$_{2:3}$ population and were used to prepare the pooled genomic DNA (Pool A and Pool B, respectively) which was used for illumina sequencing. The DNA concentration was measured by Nanodrop 2000 spectrophotometer. ~1 μg, each for both Pool A and Pool B, of total high-quality pooled DNA samples (1.8 ≤ OD260:OD280 ≤ 2.0) was used for re-sequencing library construction. Two libraries with the target insert size of 300 bp were generated by the Illumina Gnomic DNA sample kit according to the manufacturer's instruction. The quality of two libraries was controlled by qPCR. Two libraries were re-sequenced through the Illumina HiSequation 2500 at the BeiJing Berry Genomics Biotechnology Co., Ltd. (Beijing, China) to generate 125 nt paired-end short sequence reads (raw reads) for each pools.

4.5. Re-Sequencing Analysis

The FastQC program was used to evaluate the quality of raw reads (http://www.bioinformatics. babraham.ac.uk/projects/fastqc/).The Illumina paired-end adapters' sequence of raw reads was removed using the FASTX toolkit program (http://hannonlab.cshl.edu/fastx_toolkit/index.html). Removal of low-quality bases (Illumina phred quality score Q < 20) [55] and ≤40 bp of reads was completed using SolexaQA software [56]. The cleaned reads from Pool (A) and Pool (B) have been submitted to the SRA database of NCBI (SRA accessions are SRP135580 and SRP135578, respectively).

The cleaned reads were aligned separately with BWA software (Burrows-Wheeler aligner) [57] to the Nipponbare reference sequence. Alignments were filtered based on the Illumina phred quality score of ≥30, corresponding to 0.1% of the error rate, to obtain the unique mapped reads. Alignment files were converted to SAM files through SAMtools [58], and applied to GATK Pipeline [59] to identify reliable SNPs based on the reference genomic sequence.

4.6. Calculation of Δ (SNP Indices)

Average SNP indices of Pool (A) and Pool (B) were estimated via the sliding window method (sliding window 50 Kb; walking 10 Kb) and the Δ (SNP indices) Manhattan plot was obtained using a custom script written in R version 3.1.1 (https://www.r-project.org/). According to the MutMap method, SNP index (A) would be 1 for the causal SNP or for closely linked SNPs and 0.5 for unlinked loci for each identified SNP in the whole genome sequence, while the SNP index (B) would be 0.333 (1/3) for the causal SNP or closely linked SNPs and 0.5 for unlinked loci. Therefore, Δ (SNP index) would be 0.667 (2/3) for the causal or closely-linked SNPs and 0 for unlinked SNPs.

4.7. Restriction Endonuclease Digestion Analysis

The restriction endonuclease HaeII site of the target gene was identified using the primer premier 5.0 software (Premier, Ottawa, ON, Canada). Two pairs of primers were designed (W-H for WT and M-H for wb1, see Table S6) to generate 337 bp of DNA fragments by polymerase chain reaction (PCR). These PCR products were used for restriction endonuclease HaeII digestion. A total of 150 μL of this reaction system contained 3 μL HaeII (20 units per 1 μL), 15 μL NEBuffer (1×), 50 μL DNA template (2.5 μg), and 82 μL ddH$_2$O. Two reaction systems were incubated at 37 °C for 15 min followed by a 2.0% agarose gel electrophoresis.

4.8. Vector Construction for CRISPR/Cas9-Mediated Mutation

Six novel allelic mutants were created in *Japonica* cv Nipponbare by a CRISPR/Cas-targeted genome editing tool. The pBWA(V)H_cas9i2-CRISPR/Cas9 plasmid (Figure S1) was constructed according the method described in Shan et al. (2013) [60]. To generate pBWA(V)H_cas9i2-CRISPR/Cas9 targeting vector, we used the pBWA(V)H_cas9i2 vector containing codon-optimized Cas9 driven by the 35S promoter, the OsU3 promoter and sgRNA scaffolds, as well as the Cas9 expression backbone vector. The targeting sequence primer pair was ACGTGACCTCATCAACTGGGTGG and AACGTGGCGCTGCCGAGGAACGG. The OsU3 promoter was used to drive the sgRNA expression, and the 35S promoter was used to drive the Cas9 expression. Both the OsU3::gRNA and 35S::Cas9 fragments were cloned into pBWA(V)H_cas9i2 binary vector which was introduced into *Agrobacterium* strain EHA105. Transformed calli were induced from Nipponbare seeds for *Agrobacterium*-mediated transformation as previously described [61]. The T$_0$ transgenic mutant plants regenerated from hygromycin-resistant calli were examined for the presence of transgene using primer pair Cas-seq (Table S6).

5. Conclusions

Breeding of rice with high quality of appearance and high yield is important for rice cultivation. In this study, we isolate and characterize a candidate recessive gene *WB1* that regulates rice endosperm development using a modified MutMap method. The candidate gene *WB1* is further verified by CRISPR/Cas9 system. The *wb1* mutant, as well as six mutants mediated by CRISPR-Cas9 system, all cause a defect in the endosperm development, which lead to the higher grain chalkiness rate and degree and a significant reduction of 1000-grain weight in comparison with that of wild-type plants.

Relative expression analysis of genes associated with starch synthesis by qPCR also suggests that loss of function of *WB1* leads to disorder of starch metabolism-related genes expression, resulting in the abnormal endosperm development. In particular, the modified MutMap method used in this study shows a low error rate, a relatively low cost and a high specificity, and could promote the development of rice genetics. Overall, the gene *WB1* involved in rice endosperm development affects rice quality of appearance and yield, and therefore, it can be used by rice breeders through molecular breeding to improve rice quality of appearance and yield in Green Super Rice.

Author Contributions: H.W. and Y.Z. contributed equally to this work. Conceptualization, H.W.; Data curation, H.W. and Y.Z.; Formal analysis, H.W. and Y.Z.; Funding acquisition, L.C., S.C. and X.S.; Investigation, H.W., P.X., R.T., S.M. and D.C.; Methodology, H.W., Y.Z., L.S., L.C., S.C. and X.S.; Project administration, L.C., S.C. and X.S.; Resources, L.C., S.C. and X.S.; Supervision, L.C., S.C. and X.S.; Validation, Y.Z., L.C., S.C. and X.S.; Visualization, H.W. and R.T.; Writing—original draft, H.W.; Writing—review & editing, Y.Z., L.S., P.X., R.T., W.W., G.B.A., K.H., A.R., L.C., S.C. and X.S.

Acknowledgments: We would like to thank associate professor Lu Lu (Graduate School of Chinese Academy of Agricultural Sciences, Beijing, China) for critical proofreading, feedback, and editing of the manuscript.

References

1. Fitzgerald, M.A.; McCouch, S.R.; Hall, R.D. Not just a grain of rice: The quest for quality. *Trends Plant Sci.* **2009**, *14*, 133–139. [CrossRef] [PubMed]

2. She, K.C.; Kusano, H.; Koizumi, K.; Yamakawa, H.; Hakata, M.; Imamura, T.; Fukuda, M.; Naito, N.; Tsurumaki, T.; Yaeshima, M.; et al. A novel factor *FLOURY ENDOSPERM2* is involved in regulation of rice grain size and starch quality. *Plant Cell* **2010**, *22*, 3280–3294. [CrossRef] [PubMed]

3. Wang, Y.; Ren, Y.; Liu, X.; Jiang, L.; Chen, L.; Han, X.; Jin, M.; Liu, S.; Liu, F.; Lv, J.; et al. *OsRab5a* regulates endomembrane organization and storage protein trafficking in rice endosperm cells. *Plant J.* **2010**, *64*, 812–824. [CrossRef] [PubMed]

4. Peng, C.; Wang, Y.H.; Liu, F.; Ren, Y.; Zhou, K.; Lv, J.; Zheng, M.; Zhao, S.; Zhang, L.; Wang, C.; et al. *FLOURYE DOSPERM6* encodes a CBM48 domain-containing protein involved in compound granule formation and starch synthesis in rice endosperm. *Plant J.* **2014**, *77*, 917–930. [CrossRef] [PubMed]

5. Ren, Y.; Wang, Y.; Liu, F.; Zhou, K.; Ding, Y.; Zhou, F.; Wang, Y.; Liu, K.; Gan, L.; Ma, W.; et al. *GLUTELIN PRECURSOR ACCUMULATION3* encodes a regulator of post-Golgi vesicular traffic essential for vacuolar protein sorting in rice endosperm. *Plant Cell* **2014**, *26*, 410–425. [CrossRef] [PubMed]

6. Matsushima, R.; Maekawa, M.; Kusano, M.; Kondo, H.; Fujita, N.; Kawagoe, Y.; Sakamoto, W. Amyloplast-localized SUBSTANDARD STARCH GRAIN4 protein influences the size of starch grains in rice endosperm. *Plant Physiol.* **2014**, *164*, 623–636. [CrossRef] [PubMed]

7. Wen, L.; Fukuda, M.; Sunada, M.; Ishino, S.; Ishino, Y.; Okita, T.W.; Ogawa, M.; Ueda, T.; Kumamaru, T. Guanine nucleotide exchange factor 2 for Rab5 proteins coordinated with GLUP6/GEF regulates the intracellular transport of the proglutelin from the Golgi apparatus to the protein storage vacuole in rice endosperm. *J. Exp. Bot.* **2015**, *66*, 6137–6147. [CrossRef] [PubMed]

8. Zhang, L.; Ren, Y.; Lu, B.; Yang, C.; Feng, Z.; Liu, Z.; Chen, J.; Ma, W.; Wang, Y.; Yu, X.; et al. *FLOURY ENDOSPERM 7* encodes a regulator of starch synthesis and amyloplast development essential for peripheral endosperm development in rice. *J. Exp. Bot.* **2016**, *67*, 633–647. [CrossRef] [PubMed]

9. Lee, S.K.; Hwang, S.K.; Han, M.; Eom, J.S.; Kang, H.G.; Han, Y.; Choi, S.B.; Cho, M.H.; Bhoo, S.H.; An, G.; et al. Identification of the ADP-glucose pyrophosphorylase isoforms essential for starch synthesis in the leaf and seed endosperm of rice (*Oryza sativa* L.). *Plant Mol. Biol.* **2007**, *65*, 531–546. [CrossRef] [PubMed]

10. Satoh, H.; Shibahara, K.; Tokunaga, T.; Nishi, A.; Tasaki, M.; Hwang, S.; Okita, T.W.; Kaneko, N.; Fujita, N.; Yoshida, M.; et al. Mutation of the plastidial alpha-glucan phosphorylase gene in rice affects the synthesis and structure of starch in the endosperm. *Plant Cell* **2008**, *20*, 1833–1849. [CrossRef] [PubMed]

11. Tuncel, A.; Kawaguchi, J.; Ihara, Y.; Matsusaka, H.; Nishi, A.; Nakamura, T.; Kuhara, S.; Hirakawa, H.; Nakamura, Y.; Cakir, B.; et al. The rice endosperm ADP-glucose pyrophosphorylase large subunit is essential

for optimal catalysis and allosteric regulation of the heterotetrameric enzyme. *Plant Cell Physiol.* **2014**, *55*, 1169–1183. [CrossRef] [PubMed]

12. Cakir, B.; Shiraishi, S.; Tuncel, A.; Matsusaka, H.; Satoh, R.; Singh, S.; Crofts, N.; Hosaka, Y.; Fujita, N.; Hwang, S.K.; et al. Analysis of the rice ADP-glucose transporter (*OsBT1*) indicates the presence of regulatory processes in the amyloplast stroma that control ADP-glucose flux into starch. *Plant Physiol.* **2016**, *170*, 1271–1283. [CrossRef] [PubMed]

13. Li, S.; Wei, X.; Ren, Y.; Qiu, J.; Jiao, G.; Guo, X.; Tang, S.; Wan, J.; Hu, P. *OsBT1* encodes an ADP-glucose transporter involved in starch synthesis and compound granule formation in rice endosperm. *Sci. Rep.* **2017**, *7*, 40124. [CrossRef] [PubMed]

14. Wang, E.; Wang, J.; Zhu, X.; Hao, W.; Wang, L.; Li, Q.; Zhang, L.; He, W.; Lu, B.; Lin, H.; et al. Control of rice grain-filling and yield by a gene with a potential signature of domestication. *Nat. Genet.* **2008**, *40*, 1370–1374. [CrossRef] [PubMed]

15. Li, Y.; Fan, C.; Xing, Y.; Yun, P.; Luo, L.; Yan, B.; Peng, B.; Xie, W.; Wang, G.; Li, X.; et al. Chalk5 encodes a vacuolar H+- translocating pyrophosphatase influencing grain chalkiness in rice. *Nat. Genet.* **2014**, *46*, 398–404. [CrossRef] [PubMed]

16. Hannah, L.C.; James, M. The complexities of starch biosynthesis in cereal endosperms. *Curr. Opin. Biotechnol.* **2008**, *19*, 160–165. [CrossRef] [PubMed]

17. Sauer, N. Molecular physiology of higher plant sucrose transporters. *FEBS Lett.* **2007**, *581*, 2309–2317. [CrossRef] [PubMed]

18. Toyota, K.; Tamura, M.; Ohdan, T.; Nakamura, Y. Expression profiling of starch metabolism-related plastidic translocator genes in rice. *Planta* **2006**, *223*, 248–257. [CrossRef] [PubMed]

19. Eom, J.S.; Cho, J.I.; Reinders, A.; Lee, S.W.; Yoo, Y.; Tuan, P.Q.; Choi, S.B.; Bang, G.; Park, Y.I.; Cho, M.H.; et al. Impaired function of the tonoplast-localized sucrose transporter in rice, *OsSUT2*, limits the transport of vacuolar reserve sucrose and affects plant growth. *Plant Physiol.* **2011**, *157*, 109–119. [CrossRef] [PubMed]

20. Sosso, D.; Luo, D.; Li, Q.B.; Sasse, J.; Yang, J.; Gendrot, G.; Suzuki, M.; Koch, K.E.; McCarty, D.R.; Chourey, P.S.; et al. Seed filling in domesticated maize and rice depends on SWEET-mediated hexose transport. *Nat. Genet.* **2015**, *47*, 1489–1493. [CrossRef] [PubMed]

21. Tatsuro, H.; Grahamn, S.; Tomio, T. An expression analysis profile for the entire sucrose synthase gene family in rice. *Plant Sci.* **2008**, *174*, 534–543. [CrossRef]

22. Wei, X.J.; Jiao, G.A.; Lin, H.Y.; Sheng, Z.H.; Shao, G.N.; Xie, L.H.; Tang, S.Q.; Xu, Q.; Hu, P.S. *GRAIN INCOMPLETE FILLING* 2 regulates grain filling and starch synthesis during rice caryopsis development. *J. Integr. Plant Biol.* **2017**, *59*, 134–153. [CrossRef] [PubMed]

23. Kang, H.; Park, S.; Matsuoka, M.; An, G. White-core endosperm floury endosperm-4 in rice is generated by knockout mutations in the C4-type pyruvate orthophosphate dikinase gene (*OsPPDKB*). *Plant J.* **2005**, *42*, 901–911. [CrossRef] [PubMed]

24. Tian, Z.; Qian, Q.; Liu, Q.; Yan, M.; Liu, X.; Yan, C.; Liu, G.; Gao, Z.; Tang, S.; Zeng, D.; et al. Allelic diversities in rice starch biosynthesis lead to a diverse array of rice eating and cooking qualities. *Proc. Natl. Acad. Sci. USA* **2009**, *106*, 21760–21765. [CrossRef] [PubMed]

25. Fujita, N.; Yoshida, M.; Asakura, N.; Ohdan, T.; Miyao, A.; Hirochika, H.; Nakamura, Y. Function and characterization of *starch synthase I* using mutants in rice. *Plant Physiol.* **2006**, *140*, 1070–1084. [CrossRef] [PubMed]

26. Zhang, G.; Cheng, Z.; Zhang, X.; Gao, X.; Su, N.; Jiang, L.; Mao, L.; Wan, J. Double repression of soluble starch synthase genes *SSIIa* and *SSIIIa* in rice (*Oryza sativa* L.) uncovers interactive effects on the physicochemical properties of starch. *Genome* **2011**, *54*, 448–459. [CrossRef] [PubMed]

27. Fujita, N.; Yoshida, M.; Kondo, T.; Saito, K.; Utsumi, Y.; Tokunaga, T.; Nishi, A.; Satoh, H.; Park, J.H.; Jane, J.L.; et al. Characterization of *SSIIIa*-deficient mutants of rice: The function of *SSIIIa* and pleiotropic effects by *SSIIIa* deficiency in the rice endosperm. *Plant Physiol.* **2007**, *144*, 2009–2023. [CrossRef] [PubMed]

28. Kawagoe, Y.; Kubo, A.; Satoh, H.; Takaiwa, F.; Nakamura, Y. Roles of isoamylase and ADP-glucose pyrophosphorylase in starch granule synthesis in rice endosperm. *Plant J.* **2005**, *42*, 164–174. [CrossRef] [PubMed]

29. Kubo, A.; Fujita, N.; Harada, K.; Matsuda, T.; Satoh, H.; Nakamura, Y. The starch-debranching enzymes isoamylase and pullulanase are both involved in amylopectin biosynthesis in rice endosperm. *Plant Physiol.* **1999** *121*, 399–410. [CrossRef] [PubMed]

30. Tanaka, N.; Fujita, N.; Nish, A.; Satoh, H.; Hosaka, Y.; Ugaki, M.; Kawasaki, S.; Nakamura, Y. The structure of starch can be manipulated by changing the expression levels of *starch branching enzyme IIb* in rice endosperm. *Plant Biotechnol. J.* **2004**, *2*, 507–516. [CrossRef] [PubMed]

31. Fu, F.F.; Xue, H.W. Coexpression analysis identifies *Rice Starch Regulator1*, a rice AP2/EREBP family transcription factor, as a novel rice starch biosynthesis regulator. *Plant Physiol.* **2010**, *154*, 927–938. [CrossRef] [PubMed]

32. Kawakatsu, T.; Yamamoto, M.P.; Touno, S.M.; Yasuda, H.; Takaiwa, F. Compensation and interaction between RISBZ1 and RPBF during grain filling in rice. *Plant J.* **2009**, *59*, 908–920. [CrossRef] [PubMed]

33. Abe, A.; Kosugi, S.; Yoshida, K.; Natsume, S.; Takagi, H.; Kanzaki, H.; Matsumura, H.; Yoshida, K.; Mitsuoka, C.; Tamiru, M.; et al. Genome sequencing reveals agronomically important loci in rice using MutMap. *Nat. Biotechnol.* **2012**, *30*, 174–178. [CrossRef] [PubMed]

34. Mardis, E.R. Next-Generation DNA Sequencing Method. *Annu. Rev. Genom. Hum. Genet.* **2008**, *9*, 387–402. [CrossRef] [PubMed]

35. Takagi, H.; Tamiru, M.; Abe, A.; Yoshida, K.; Uemura, A.; Yaeqashi, H.; Obara, T.; Oikawa, K.; Utsushi, H.; Kanzaki, E.; et al. MutMap accelerates breeding of a salt-tolerant rice cultivar. *Nat. Biotechnol.* **2015**, *33*, 445–449. [CrossRef] [PubMed]

36. Takagi, H.; Uemura, A.; Yaegashi, H.; Tamiru, M.; Abe, A.; Mitsuoka, C.; Utsushi, H.; Natsume, S.; Kanzaki, H.; Matsumura, H.; et al. MutMap-Gap: Whole-genome resequencing of mutant F_2 progeny bulk combined with de novo assembly of gap regions identifies the rice blast resistance gene *Pii*. *New Phytol.* **2013**, *200*, 276–283. [CrossRef] [PubMed]

37. Matsumoto, T.; Wu, J.; Kanamori, H.; Katayose, Y.; Fujisawa, M.; Namiki, N.; Mizuno, H.; Yamamoto, K.; Antonio, B.A.; Baba, T.; et al. The map-based sequence of the rice genome. *Nature* **2005**, *2436*, 793–800. [CrossRef]

38. Holsinger, K.E.; Weir, B.S. Genetics in geographically structured populations: Defining, estimating and interpreting F(ST). *Nat. Rev. Genet.* **2009**, *10*, 639–650. [CrossRef] [PubMed]

39. Bökel, C. EMS screens: From mutagenesis to screening and mapping. *Methods Mol. Biol.* **2008**, *420*, 119–138. [CrossRef] [PubMed]

40. Singh, N.; Sodhi, N.S.; Kaur, M.; Saxena, S.K. Physico-chemical, morphological, thermal, cooking and textural properties of chalky and translucent rice kernels. *Food Chem.* **2003**, *82*, 433–439. [CrossRef]

41. Cheng, F.M.; Zhong, L.J.; Wang, F.; Zhang, G.P. Differences in cooking and eating properties between chalky and translucent parts in rice grains. *Food Chem.* **2005**, *90*, 39–46. [CrossRef]

42. Lisle, A.J.; Martin, M.; Fitzgerald, M.A. Chalky and translucent rice grains differ in starch composition and structure and cooking properties. *Cereal Chem.* **2000**, *77*, 627–632. [CrossRef]

43. Sasaki, A.; Ashikari, M.; Ueguchi-Tanaka, M.; Itoh, H.; Nishimura, A.; Swapan, D.; Ishiyama, K.; Saito, T.; Kobayashi, M.; Khush, G.S.; et al. Green revolution: A mutant gibberellin-synthesis gene in rice. *Nature* **2002**, *416*, 701–702. [CrossRef] [PubMed]

44. Monna, L.; Kitazawa, N.; Yoshino, R.; Suzuki, J.; Masuda, H.; Maehara, Y.; Tanji, M.; Sato, M.; Nasu, S.; Minobe, Y. Positional cloning of rice semidwarfing gene, *sd-1*: Rice "green revolution gene" encodes a mutant enzyme involved in gibberellin synthesis. *DNA Res.* **2002**, *9*, 11–17. [CrossRef] [PubMed]

45. Spielmeyer, W.; Ellis, M.H.; Chandler, P.M. Semidwarf (*sd-1*), "green revolution" rice, contains a defective gibberellin 20-oxidase gene. *Proc. Natl. Acad. Sci. USA* **2002**, *99*, 9043–9048. [CrossRef] [PubMed]

46. Sims, D.; Sudbery, I.; Ilott, N.E.; Heger, A.; Ponting, C.P. Sequencing depth and coverage: Key considerations in genomic analyses. *Nat. Rev. Genet.* **2014**, *15*, 121–132. [CrossRef] [PubMed]

47. Yu, J.; Hu, S.; Wang, J.; Wong, G.K.; Li, S.; Liu, B.; Deng, Y.; Dai, L.; Zhou, Y.; Zhang, X.; et al. A draft sequence of the rice genome (*Oryza sativa* L. ssp. *indica*). *Science* **2002**, *296*, 79–92. [CrossRef] [PubMed]

48. Zhang, J.W.; Chen, L.L.; Xing, F.; Kudrna, D.A.; Yao, W.; Copetti, D.; Mu, T.; Li, W.; Song, J.M.; Xie, W.; et al. Extensive sequence divergence between the reference genomes of two elite indica rice varieties Zhenshan 97 and Minghui 63. *Proc. Natl. Acad. Sci. USA* **2016**, *113*, E5163–E5171. [CrossRef] [PubMed]

49. Du, H.; Yu, Y.; Ma, Y.; Gao, Q.; Cao, Y.; Chen, Z.; Ma, B.; Qi, M.; Li, Y.; Zhao, X.; et al. Sequencing and de novo assembly of a near complete *indica* rice genome. *Nat. Commun.* **2017**, *8*, 15324. [CrossRef] [PubMed]

50. Zhang, P.; Zhang, Y.; Sun, L.; Sittipun, S.; Yang, Z.; Sun, B.; Xuan, D.; Li, Z.; Yu, P.; Wu, W.; et al. The Rice AAA-ATPase *OsFIGNL1* Is Essential for Male Meiosis. *Fron. Plant Sci.* **2017**, *8*, 1639. [CrossRef] [PubMed]

51. Fujita, N.; Kubo, A.; Suh, S.D.; Wong, K.S.; Jane, J.L.; Ozawa, K.; Takaiwa, F.; Inaba, Y.; Nakamura, Y.

Antisense inhibition of isoamylase alters the structure of amylopectin and the physicochemical properties of starch in rice endosperm. *Plant Cell Physiol.* **2003**, *44*, 607–618. [CrossRef] [PubMed]

52. Hovenkamphermelink, J.H.; Devries, J.N.; Adamse, P.; Jacobsen, E.; Witholt, B.; Feenstra, W.J. Rapid estimation of the amylose/amylopectin ratio in small amounts of tuber and leaf tissue of the potato. *Potato Res.* **1988**, *31*, 241–246. [CrossRef]

53. Zhou, L.; Liang, S.; Ponce, K.; Marundon, S.; Ye, G.; Zhao, X. Factors affecting head rice yield and chalkiness in *indica* rice. *Field Crop. Res.* **2015**, *172*, 1–10. [CrossRef]

54. Chen, D.H.; Ronald, P.C. A rapid DNA minipreparation method suitable for AFLP and other PCR applications. *Plant Mol. Biol. Rep.* **1999**, *17*, 53–57. [CrossRef]

55. Ewing, B.; Hillier, L.; Wendl, M.C.; Green, P. Base-calling of automated sequencer traces using phred. I. Accuracy assessment. *Genome Res.* **1988**, *8*, 175–185. [CrossRef]

56. Cox, M.P.; Peterson, D.A.; Biggs, P.J. SolexaQA: At-a-glance quality assessment of Illumina second-generation sequencing data. *BMC Bioinform.* **2010**, *11*, 485. [CrossRef] [PubMed]

57. Li, H.; Durbin, R. Fast and accurate short read alignment with Burrows-Wheeler Transform. *Bioinformatics* **2009**, *25*, 1754–1760. [CrossRef] [PubMed]

58. Li, H.; Handsaker, B.; Wysoker, A.; Fennell, T.; Ruan, J.; Homer, N. The Sequence Alignment/Map (SAM) format and SAM tools. *Bioinformatics* **2009**, *25*, 2078–2079. [CrossRef] [PubMed]

59. McKenna, A.L.; Hanna, M.; Banks, E.; Sivachenko, A.; Cibulskis, K.; Kernytsky, A.; Garimella, K.; Altshuler, D.; Gabriel, S.; Daly, M.; et al. The Genome Analysis Toolkit: A MapReduce framework for analyzing next-generation DNA sequencing data. *Genome Res.* **2010**, *20*, 1297–1303. [CrossRef] [PubMed]

60. Shan, Q.; Wang, Y.; Li, J.; Zhang, Y.; Chen, K.; Liang, Z.; Zhang, K.; Liu, J.; Xi, J.J.; Qiu, J.L.; et al. Targeted genome modification of crop plants using a CRISPR-Cas system. *Nat. Biotechnol.* **2013**, *31*, 686–688. [CrossRef] [PubMed]

61. Hiei, Y.; Ohta, S.; Komari, T.; Kumashiro, T. Efficient transformation of rice (*Oryza sativa* L.) mediated by Agrobacterium and sequence analysis of the boundaries of the T-DNA. *Plant J.* **1994**, *6*, 271–282. [CrossRef] [PubMed]

3

Particle Bombardment of the *cry2A* Gene Cassette Induces Stem Borer Resistance in Sugarcane

Shiwu Gao, Yingying Yang, Liping Xu *, Jinlong Guo, Yachun Su, Qibin Wu, Chunfeng Wang and Youxiong Que *

Key Laboratory of Sugarcane Biology and Genetic Breeding, Ministry of Agriculture and Key Laboratory of Crop Genetics and Breeding and Comprehensive Utilization, College of Crop Science, Fujian Agriculture and Forestry University, Ministry of Education, Fuzhou 350002, China; gaoshiwu2008@126.com (S.G.); yingyingyang13@163.com (Y.Y.); jl.guo@163.com (J.G.); syc2009mail@163.com (Y.S.); wqbaidqq@163.com (Q.W.); 18305999305@163.com (C.W.)
* Correspondence: xlpmail@fafu.edu.cn (L.X.); queyouxiong@fafu.edu.cn (Y.Q.)

Abstract: Sugarcane borer is the most common and harmful pest in Chinese sugarcane fields, and can cause damage to the whole plant during the entire growing season. To improve borer resistance in sugarcane, we constructed a plant expression vector pGcry2A0229 with the *bar* gene as the marker and the *cry2A* gene as the target, and introduced it into embryogenic calli of most widely cultivated sugarcane cultivar ROC22 by particle bombardment. After screening with phosphinothricin in vitro and Basta spray, 21 resistance-regenerated plants were obtained, and 10 positive transgenic lines harboring the *cry2A* gene were further confirmed by conventional PCR detection. Real-time quantitative PCR (RT-qPCR) analysis showed that the copy number of the *cry2A* gene varied among different transgenic lines but did not exceed four copies. Quantitative ELISA analysis showed that there was no linear relationship with copy number but negatively correlated with the percentage of borer-infested plants. The analysis of industrial and agronomic traits showed that the theoretical sugar yields of transgenic lines TR-4 and TR-10 were slightly lower than that of the control in both plant cane and ratoon cane; nevertheless, TR-4 and TR-10 lines exhibited markedly lower in frequency of borer-infested plants in plant cane and in the ratoon cane compared to the control. Our results indicate that the introduction of the *cry2A* gene via bombardment produces transgenic lines with obviously increased stem borer resistance and comparable sugar yield, providing a practical value in direct commercial cultivation and crossbreeding for ROC22 has been used as the most popular elite genitor in various breeding programs in China.

Keywords: sugarcane; *cry2A* gene; particle bombardment; stem borer; resistance

1. Introduction

Sugarcane is the most important sugar crop, with sucrose accounting for 80% of the total sugar production in the world and accounting for more than 92% of total sugar production in China. As a C$_4$ crop, sugarcane makes it one of the most important energy crops due to its high biomass, high fiber, and years of ratooning. Nearly 90% of biofuel ethanol is produced by sugarcane in the United States and Brazil [1]. The risk level of transgenic safety for sugarcane is low mainly due to the following three reasons. Firstly, sugarcane is an asexually propagated crop and generally does not bloom during field cultivation in China, indicating little chance of exogenous gene drift by flowering. Secondly, as an industrial raw material for sucrose and fuel ethanol, sugarcane is not a directly circulated food. The processing of sucrose requires as high as 107 °C for crystallization and crystal sucrose belongs to a purified carbohydrate, without protein ingredient. Besides, the fuel ethanol is not edible.

A similar opinion of high food safety level of transgenic sugarcane can be ascribed to the decomposition of protein expressed by the exogenous gene during process of sucrose crystallization [2]. Thirdly, transgenic sugarcane does not affect the microbial community diversity and has no significant effect on enzyme activities in rhizosphere soil, which means better ecological security [3]. To date, two cases involving genetically modified sugarcanes were approved for commercial planting, namely drought resistant transgenic sugarcane in Indonesia and insect-resistant transgenic sugarcane in Brazil.

There are currently five major species of stem borer thriving in Chinese sugarcane fields: *Chilo sacchariphagus* Bjojer, *Scirpophaga nivella* Fabricius, *C. infuscatellus* Snellen, *Argyroploce schistaceana* Snellen, and *Sesamia inferens* Walker. Because several generations occur in one planting season, together with several different species and overlap among generations, stem borer is the most common and harmful pests in the Chinese sugarcane industry. The percentage of dead heart seedlings is normally within the range of 10–20%, but can reach 60% in severely infected sugarcane fields, and the damage incurred during the mid-late stage leads to a significant reduction in sucrose content and the increasing of wind broken stalks [4,5]. Sugarcane, which is heterogeneous polyploid and aneuploid, has a complex genetic background [6,7]. As many as 120 chromosomes are available in modern sugarcane hybrids and there is a lack of stem borer resistance genes in the gene pool [8,9], it is extremely difficult to breed a cultivar with resistance to stem borers in traditional crossbreeding program. Chemical pesticides have long been the main method for preventing and controlling stem borers in Chinese sugarcane, which not only increases production cost but also pollutes the environment.

In 1987, Vaeck et al. successfully introduced the *cry1A(b)* gene into tobacco through *Agrobacterium*-mediated transformation and obtained stem borer-resistant transgenic tobacco [10]. Subsequently, the *Bt* gene was introduced into crops such as cotton [11–14], maize [15], rice [16–19], and tomato [20], resulting in effective improvement in stem borer resistance. In sugarcane, Arencibia et al. first described the transformation of *cry1A* gene to improve stem borer resistance in 1997 [8], followed by numerous reports on the improvement of insect resistance in transgenic sugarcanes such as *cry1A(b)* [21–23], *GNA* [24,25], *cry1Aa3* [26], *cry1Ac* [27–29] and proteinase inhibitor [30,31]. Besides, researchers also attempted to use RNAi technology to control pest damage in sugarcane [32], with success in other crops [33–35]. Our previous study showed that the application of insect-resistant transgenic sugarcane can economically and effectively solve the problem of stem borers in the sugarcane industry [36]. *cry2A*, which has low homology (<45%) with *cry1A*, is another Bt protein. Previous researchers demonstrated that cry2A protein is toxic to several lepidopteran pests, indicating its feasibility to be used as a bio-insecticide [37,38]. However, there is no report about the application of *cry2A* in sugarcane.

Compared to other screening marker genes such as *npt II*, *bar* gene screening can be performed at the early stage of genetic transformation of sugarcane, thereby reducing the workload involving tissue culture and increasing efficiency [39]. Thus, the *bar* gene as a screening marker gene has an obvious advantage in eliminating of pseudotyped transformants during selection of putative transformants after bombardment, as sugarcane is phosphinothricin (PPT)-sensitive. The antibiotics and PPT resistance screening tests of two sugarcane genotypes, FN81-745 (*Saccharum* spp. hybrid) and Badila (*Saccharum officinarum*), showed that the effective concentration of both G418 and Hyg was 30.0 mg/L, while only 0.75 mg/L and 1.0 mg/L for PPT, respectively [39]. In plant genetic transformation, the *bar* gene has been widely used as a screening marker gene [40,41], and its application to sugarcane has also been described in several reports [22,29,42,43].

In the present study, to obtain *cry2A* transgenic sugarcane, a plant expression vector pGcry2A0229 with the *bar* gene as a screening marker and *cry2A* as a target gene was constructed and genetically transformed into sugarcane by particle bombardment. PPT and Basta resistance screening and PCR validation were conducted to confirm the positive *cry2A* gene transgenic sugarcane plants. Then, the copy number of the *cry2A* gene and its protein expression in transgenic lines were determined by Real-time quantitative PCR (RT-qPCR) and quantitative ELISA detection of protein, respectively. Finally, several transgenic lines with better comprehensive traits based on a field survey of industrial and agronomic traits were identified, which provides a transgenic line of potential commercial cultivation and the foundation for crossbreeding of stem borer-resistant traits in sugarcane.

2. Results

2.1. Construction and Verification of the Plant Vector pGcry2A0229

The construction scheme of plant expression vector pGcry2A0229 is depicted in Figure 1. Single-enzyme digestion of pGcry2A0229 using restriction endonuclease *Hind* III generated the expected single band with a size of 8213 bp. Electrophoresis analysis of double enzyme digestion with *Hind* III and *EcoR* I showed the expected two bands with sizes of about 4436 and 3777 bp, respectively (Figure 2). The sequencing results confirmed that the pGcry2A0229 was the expected positive recombinant plasmid.

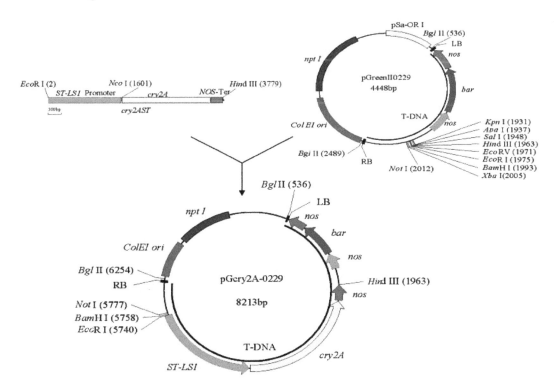

Figure 1. Construction roadmap of the plant expression vector pGcry2A0229.

2.2. Particle Bombardment and Resistance Screening

Micro-bombs were prepared using the tungsten particles and pGcry2A0229 DNA, and bombardment transformation was conducted with the embryogenic calli of sugarcane cultivar ROC22 as the receptor material (Figure 3a). Based on our preliminary experimental results, the embryogenic calli after bombardment was subcultured and differentiated with PPT of 0.8 mg/L. Some tissues gradually differentiated into regenerated plantlets with PPT resistance, whereas most wild-type calli gradually became brown and died (Figure 3b). The negative control died. When the resistant regenerated plantlets grew up to a height of 4–5 cm, they were transferred into the rooting medium without PPT for rooting (Figure 3c). Finally, 95 resistant regenerated plants were obtained.

The resistant regenerated plantlets were transplanted into nutrient pots to ensure their survival. After spray screening with 3.0‰ (*v*/*v*) Basta, most plants gradually turned yellow, wilted, and died after 15–20 days, and finally only 21 plants survived (Figure 3d).

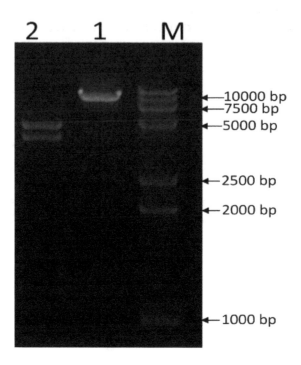

Figure 2. The products of recombinant plasmid pGcry2A0229 digested by restriction enzymes: M, DL15,000 + 2, 000 DNA ladder; 1, The products of pGcry2A0229 digested by *Hin*d III; 2, The products of pGcry2A0229 digested by *Hin*d III and *Eco*R I.

Figure 3. Putative recombinant screening: (**a**) wild-type calli on medium without PPT; (**b**) PPT-resistant plantlets at the differentiation stage on selection medium; (**c**) regenerated plantlets at the stage of rooting culture; and (**d**) spraying screening of resistant plantlets with 3.0‰ Basta.

2.3. PCR Identification of the cry2A Gene in Resistant Regenerated Plants

A total of 21 resistant regenerated plantlets obtained by PPT and Basta screening were verified by PCR amplification of the *cry2A* gene. The results showed that a single band was amplified from 10 samples; the position of the band was consistent with that of the positive plasmid and showed an approximate size of 600 bp. The sequencing results were also consistent with the partial sequence of the *cry2A* gene, whereas no band was amplified from the non-transgenic negative control and ddH$_2$O blank control (Figure 4). Therefore, PCR analysis verified that 10 positive *cry2A* transgenic plants were successfully obtained.

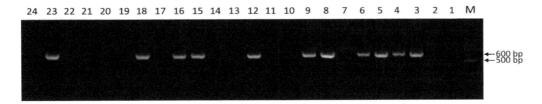

Figure 4. Electrophoretic analysis of PCR amplification products of a putative *cry2A* gene for transgenic sugarcane plants: M, DNA Marker; 1, Blank control of ddH$_2$O; 2, Negative control (non-transgenic sugarcane without bombardment); 3, Positive control (plasmid pGcry2A0229); 4–24, Herbicide Basta-resistant plants.

2.4. RT-qPCR Detection of the cry2A Gene and the Copy Number Estimation in Transgenic Lines

Ten PCR-positive transgenic sugarcane lines were tested by RT-qPCR technique and the copy number of the *cry2A* gene was estimated. The RT-qPCR quantitative standard curve of *cry2A* gene was constructed using the following equation: $y = -3.593x + 43.082$, $R^2 = 0.994$, where the y-axis represents the C_t value, the x-axis represents the logarithm of the initial template copy number. A good correlation between C_t values (18–40) and initial template copy number (10^1–10^8) was observed. According to the linear equation, x, the total copy number of the *cry2A* gene in the sample, was determined, and the number of exogenous *cry2A* copies of a single cell (Table 1) was calculated using the followed formula: Copies/genome = $10^x / [25 \text{ ng} \times 10^{-9} \times 6.02 \times 10^{23} / (10,000 \times 10^6 \times 660)]$. Table 1 shows that the *cry2A* gene copy number of different transgenic sugarcane lines varied, wherein three lines had two copies, five lines had three copies, and two lines had four copies.

Table 1. Estimated *cry2A* gene copy number of different transgenic lines.

Line	C_t I	C_t II	C_t III	C_t Mean	Copy Number
TR-1	29.96	29.84	30.19	30.00 ± 0.10	1.92
TR-2	29.10	29.39	29.22	29.24 ± 0.08	3.13
TR-3	29.78	30.06	30.20	30.01 ± 0.12	1.90
TR-4	29.95	30.11	30.09	30.05 ± 0.05	1.86
TR-5	29.08	28.97	29.10	29.05 ± 0.04	3.53
TR-6	29.75	29.67	29.63	29.68 ± 0.04	2.35
TR-7	29.70	29.39	29.70	29.60 ± 0.10	2.49
TR-8	29.48	29.51	29.62	29.54 ± 0.04	2.58
TR-9	29.68	29.68	29.78	29.72 ± 0.03	2.30
TR-10	30.00	29.10	29.48	29.53 ± 0.26	2.60
Non-transgenic	37.97	38.22	40.03	38.74 ± 0.65	0.01

2.5. cry2A Protein Expression in the Transgenic Lines

cry2A protein expression in the mature leaves of 10 transgenic sugarcane lines was quantitated by ELISA, and a standard curve was constructed using the Bt protein standard in the kit as the following equation: $y = -1.156x + 3.865$, $R^2 = 0.999$, where the y-axis represents the OD$_{450}$ absorbance, the x-axis represents the concentration of the Bt protein standard. The absorbance value correlated well with the protein standard concentration ($R^2 = 0.999$). The amount of protein expression in the 10 samples was calculated according to the linear equation (Figure 5), which showed that the *cry2A* protein expression was observed in all 10 transgenic lines at levels within the range of 76.45–90.75 μg/FWg, of which three lines, namely, TR-4, TR-8, and TR-10 had higher protein expression levels (85.86, 82.49 and 90.75 μg/FWg, respectively) and the difference among the three lines was statistically significant.

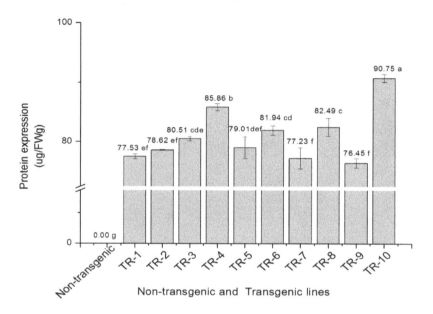

Figure 5. The *cry2A* protein expression in the leaves of non- transgenic and 10 different transgenic sugarcane lines detected by quantitative ELISA. The value is the average of three replicate experiments ± standard deviation (*n* = 3), and the different letters indicate significant difference at 0.05 level.

2.6. Survey of Industrial and Agronomic Traits of the cry2A Transgenic Sugarcane Lines

According to protein expression and field performance, three *cry2A* transgenic sugarcane lines, TR-4, TR-8, and TR-10, were selected for the further field experiment using non-transgenic recipient ROC22 as a control, and plant height, stem diameter, brix, effective stalk number, and other indicators of industrial and agronomic traits at maturity were determined. The results were then subjected to univariate statistical analysis, and the results are shown in Table 2.

In the plant cane, when refer to the plant height of three transgenic lines, both TR-4 and TR-10 were slightly lower than the control, while the significant lower was observed in line TR-8. Although the stem diameters of the three lines were lower than the control, but not statistically significant. The brix (of the three lines) was higher than that of the control, although not statistically significant. The lines TR-4 and TR-8 had slightly lower number of effective stalk per hectare than the control, while line TR-10 is slightly higher than the control, but both had no significant difference. The theoretical sugar yield of TR-4 and TR-10 was comparable to that of the control, whereas TR-8 was significantly lower than the control.

In the ratoon cane, the height and stem diameter of the three transgenic lines were lower than the control, however, only the plant height of TR-8 and the stem diameter of TR-10 were significantly lower than the control. The brix of the three transgenic lines was higher than that of the control, and TR-8 and TR-10 were significantly higher than the control. Although the number of effective stalk per hectare of TR-8 and TR-10 were higher than that of the control, this difference was not statistically significant. Similar to that in the plant cane, the theoretical sugar yield of TR-4 and TR-10 were comparable to that of the control, while significantly lower was observed in TR-8. Compared to the plant cane, the brix of the transgenic lines in the ratoon cane increased, of which TR-8 and TR-10 increased by more than 1.0, while the control ROC22 decreased by 0.43.

Table 2. Industrial and agronomic traits of different *cry2A* transgenic sugarcane lines during the plant and ratoon cane.

Crop Season	Line	Plant Height, H	Stem Diameter, D	Brix, Bx	Effective Stems (stem/ha)	Theoretical Sugar Yield (t/ha)
Plant cane	TR-4	270.23 ± 3.23 [a]	2.33 ± 0.16 [a]	20.74 ± 0.26 [a]	64,989.74 ± 1982.53 [a]	11.02 ± 0.34 [a]
	TR-8	240.27 ± 1.88 [b]	2.30 ± 0.07 [a]	20.39 ± 0.36 [a]	64,989.74 ± 799.89 [a]	9.33 ± 0.13 [b]
	TR-10	258.67 ± 5.06 [a]	2.34 ± 0.12 [a]	20.33 ± 0.32 [a]	69,692.95 ± 709.90 [a]	11.07 ± 0.18 [a]
	Non-transgenic	272.70 ± 4.71 [a]	2.38 ± 0.06 [a]	19.98 ± 0.30 [a]	66,700.00 ± 1385.47 [a]	11.26 ± 0.19 [a]
Ratoon cane	TR-4	271.13 ± 8.05 [a,b]	2.50 ± 0.04 [a,b]	20.91 ± 0.47 [a,b]	54,646.15 ± 1397.39 [a]	10.87 ± 0.28 [a]
	TR-8	248.57 ± 3.48 [b]	2.46 ± 0.02 [a,b]	21.69 ± 0.11 [a]	52,084.61 ± 1597.40 [a]	9.70 ± 0.21 [b]
	TR-10	264.33 ± 5.02 [a,b]	2.36 ± 0.07 [b]	21.85 ± 0.15 [a]	58,061.53 ± 1597.39 [a]	10.73 ± 0.17 [a]
	Non-transgenic	279.43 ± 2.96 [a]	2.66 ± 0.10 [a]	19.85 ± 0.53 [b]	51,230.77 ± 2091.48 [a]	10.96 ± 0.20 [a]

Data followed with the different letters ([a] and [b]) indicate significant difference at 0.05 level, but the same letter ([a] or [b]) indicates that the difference is not significant.

2.7. Insect Resistance Identification of the cry2A Transgenic Sugarcane Lines

Under natural field conditions, the percentage of borer-infested plants of the three lines and the control ROC22 in the plant cane and ratoon cane were investigated. The leaves and stems of the transgenic sugarcane lines showed more pronounced insect resistance compared to that in the non-transgenic control ROC22 (Figure 6). The survey results (Table 3) showed that the percentage of borer-infested plants in the three transgenic lines was lower than the control. In the plant cane, the percentage of borer-infested plants in the TR-10 line was only 26.67%, but as high as 80.0% in the control, and the difference was statistically significant. Although the TR-4 and TR-8 lines compared to the control did not reach significant level, these were only 36.67% and 53.33%, respectively. After one year of ratooning, the percentage of borer-infested plants decreased in the three transgenic lines, but slightly increased in the control, and the percentage of borer-infested plants was 30.0% and 16.67% in TR-4 and TR-10 lines, respectively, which was significantly lower than the 83.33% of the control. Especially, the transgenic lines affected by stem borers only incurred damages in the cane stem cortex (Figure 6b), whereas the control line exhibited more serious damage with injuries in the entire stem (Figure 6c).

Table 3. The percentage of borer-infested plants of different *cry2A* transgenic sugarcane lines in plant and ratoon cane.

Line	Percentage of Borer-Infested Plants (%)	
	Plant Cane	Ratoon Cane
TR-4	36.67 ± 12.02 [a,b]	30.00 ± 11.55 [b]
TR-8	53.33 ± 17.64 [a,b]	50.00 ± 10.00 [a,b]
TR-10	26.67 ± 5.77 [b]	16.67 ± 3.33 [b]
Non-transgenic	80.00 ± 6.67 [a]	83.33 ± 6.67 [a]

Data followed with the different letters ([a] and [b]) indicate significant difference at 0.05 level, but the same letter ([a] or [b]) indicates that the difference is not significant.

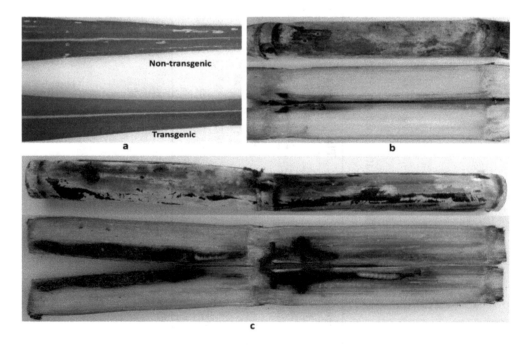

Figure 6. Stem borer damage in sugarcane under natural field conditions: **(a)** symptoms of the transgenic and non-transgenic sugarcane leaf; **(b)** symptoms of the transgenic sugarcane stem (only in cortex); and **(c)** symptoms of non-transgenic sugarcane stem.

3. Discussion and Conclusions

Sugarcane is a perennial crop and, to save costs, ratooning is usually conducted for over three years or even up to five years or more, during which sucrose content reduction and wind broken stalk increases can be caused by stem borer [4,5]. Arencibia et al. (1997) transformed the *cry1A(b)* gene into sugarcane by the cell electroporation, and improved the stem borer resistance [8]. Arvinth et al. (2010) introduced the *cry1Ab* gene into sugarcane, which significantly reduced the percentage of dead heart sugarcane seedlings [22]. The *GNA* gene was integrated into sugarcane genome via *Agrobacterium*-mediated transformation by Zhangsun et al. (2007), and the results showed that transgenic sugarcane plants had a significant resistance to the woolly aphid [25]. Falco et al. (1997) introduced the soybean bowman-birk inhibitor into sugarcane callus using particle bombardment, and it demonstrated that, compared to larvae fed on leaf tissue from untransformed ones, the growth of larvae feeding on leaf tissue from transgenic plants was significantly retarded, however the retardation was not sufficient to prevent the "dead heart" symptom [30]. Weng et al. (2011) [28] and Gao et al. (2016) [29] introduced the *cry1Ac* gene into different sugarcane varieties, and the transgenic sugarcane plants showed much better resistance to stem borer than the non-transgenic ones. ROC22, the most widely cultivated cultivar accounting for more than 60% of Chinese sugarcane acreage in the past 15 years, was used as the receptor in this study. A field test comparing the three transgenic lines with non-transgenic control found that the percentage of borer-infested plants of the transgenic lines in the ratoon cane decreased compared to that in the plant cane, whereas contrarily, slightly increased in the control. In addition, compared to the control, line TR-4, exhibiting markedly lower in frequency of borer-infested plants in the ratoon cane (30.0% vs. 83.3%) and much lower in plant cane (36.67% vs. 80.0%) indicating that the stress of stem borers gradually shifted to non-transgenic control.

Assessment of copy number of transgenic lines is essential to phenotypic studies and investigations on genetic stability. The traditional method for copy number identification is the Southern blot, which is highly cumbersome and strongly operation dependent [44], and various external factors may influence visualization of hybridization bands and thus are often underestimated. Previous research has shown that Southern blotting was not able to accurately determine the number copy numbers of exogenous genes in sugarcane b, whereas RT-qPCR is characterized by high specificity and high sensitivity, and thus more accurate [29,45]. RT-qPCR has been widely used to identify exogenous gene copy number [44–49], even for transgene copy number from 3 to >50 [45]. Sugarcane has a complex genetic background and is a highly heterogeneous polyploid or aneuploid crop, with genome sizes of up to 10 Gb [50]. In the present study, an RT-qPCR assay standard curve for the *cry2A* gene with a slope of −3.593 and a correlation coefficient of 0.994 was established, which indicated that PCR amplification efficiency and C_t values correlate well with the initial template copy number. Based on the standard curve, the *cry2A* gene copy number in the 10 transgenic sugarcane lines was determined, which revealed that the copy number of each transgenic line did not exceed four copies, which is discrepant to the findings of our previous study on *cry1Ac* transgenic sugarcane [29]. These may be related to different exogenous genes introduced and different genotypes of receptor materials.

In the present study, ELISA was used to quantitatively determine the cry2A protein expression levels in the leaves of 10 obtained transgenic lines, which ranged from 76.45 to 90.75 μg/FWg, with significant differences in some lines. However, no clear linear relationship between protein expression and copy number was observed, which was similar to that observed in previous studies [51–53]. The expression of exogenous Bt protein can effectively improve the insect resistance of transgenic plants [18,54]. Weng et al. (2011) introduced a modified *cry1Ac* gene into sugarcane cultivars ROC16 and YT79-177 by particle bombardment, and 17 transgenic plants were positive for Western blot. It also demonstrated that the expression of water-soluble proteins in leaves ranged from 2.2 ng/mg to 50 ng/mg, and when the expression exceeded 9 ng/mg (9 μg/FWg), insect resistance was observed, with the content of cry1Ac protein in transgenic sugarcane positively correlated with its insect resistance [28]. Arvinth et al. (2010) found that the total soluble cry1Ab protein expression in the obtained transgenic sugarcane leaves ranged from 0.007% to 1.73%, and protein expression was negatively correlated to the percentage of dead heart seedlings [22]. In our previous research, the cry1Ac protein expression in *cry1Ac* transgenic sugarcane leaves ranged from 0.85 to 70.9 µg/FWg, and the higher the protein expression, the lower the percentage of borer-infested plants, which exhibited a significant negative correlation [29]. Here, again, we observed that the higher the protein expression, the better the insect-resistant effect, which is consistent with the results of our previous investigation [29] and with other reports [22,28].

Weng et al. (2011) generated a ubiquitin (ubi) initiated *cry1Ac* transgenic sugarcane, and the assessment of industrial and agronomic traits showed that the agronomic traits such as plant height and stem diameter were greatly affected. However, the industrial indicators such as sucrose content and brix exhibited no significant difference compared to the control [28]. Our group previously conducted a field survey on double 35 s initiated *cry1Ac* transgenic sugarcane cultivar FN15, and it showed that only 2 of 14 transgenic lines had slightly greater plant heights than the control, although not statistically significant, whereas the other lines were lower than the control, and all the stem diameters (of the transgenic lines) were lower than that of the control. However, both higher and lower brix than that of the control was observed, and the calculated theoretical sugar yields (of the transgenic lines) were all lower than that of the control though three lines are unobvious [29]. Wang et al. (2017) introduced *cry1Ab* gene into sugarcane cultivar ROC22 by *Agrobacterium*, and investigated the industrial and agronomic traits of five single-copy transgenic lines. The result showed that plant height, stem diameter, brix, effective stalk number of several transgenic lines was only slightly lower than that of the control, while calculated theoretical sugar yield was significantly lower than that of the control [23]. Besides, a three-year field performance trial of transgenic sugarcane with *npt II* gene showed a reduction in growth and cane yield, but, when individual events were analyzed separately, the yields of several transgenic events were comparable to that of no transformants [53]. The present study conducted on plant cane and ratoon cane in the field using the obtained transgenic lines TR-4, TR-8, and TR-10, and their theoretical sugar yields were all lower (9.33–11.07 t/ha for plant cane, and 9.70–10.87 t/ha for ratoon cane) than that of the control (11.26 and 10.96 t/ha). It indicates that the introduction of exogenous *cry2A* gene into sugarcane increased stem borer resistance and reduced the percentage of infested plants, while the expression of the Bt protein consumes energy, thereby resulting in a decrease in sugar yield in generally, though comparable sugar yield of transgenic lines can be obtained, such as TR-4 and TR-10 in this study, which is in line with the results of two previous researches [29,53].

In conclusion, the introduction of the *cry2A* gene via particle bombardment produces the transgenic lines with obviously increased stem borer resistance and comparable sugar yield, providing a practical value in direct commercial cultivation, and crossbreeding for ROC22 has been used as the most popular elite parent in various breeding programs in China.

4. Materials and Methods

4.1. Materials

The plant expression vector pGreenII0229 was obtained from John Innes Center in Norwich, Norfolk, UK, and the clone 2AST1305.1 containing the *cry2A* gene was a gift from Professor Illimar Altosaar of the University of Ottawa in Canada. The pGreen plasmid can help plant genetic transformation because it was a versatile and flexible binary vector [55], and the pGreenII0229 vector contains the *bar* gene as the screening marker gene. The receptor material used for genetic transformation was ROC22, the most widely cultivated sugarcane cultivar in mainland China, which was provided by Key Laboratory of Sugarcane Biology and Genetics and Breeding, Ministry of Agriculture, China.

4.2. Plant Vector Construction of the cry2A Gene

The *cry2A* gene plant expression vector was constructed using the directional cloning strategy. First, plasmid DNA of the 2AST1305.1 cloning vector that harbored the exogenous *cry2A* gene was digested with restriction endonucleases *Eco*R I and *Hin*d III. The exogenous gene expression cassette containing the ST-LS1 promoter, the *cry2A* gene, and the *nos* terminator was recovered. Meanwhile, the plasmid DNA of the plant expression vector pGreenII0229 was digested with restriction enzymes *Eco*R I and *Hin*d III, and the target fragment containing the *bar* gene as a screening marker gene was recovered. Finally, the two recovered fragments were ligated with T4-DNA ligase to obtain a new *cry2A* gene plant expression vector pGcry2A0229.

4.3. Transformation and Screening

Shoots of ROC22 sugarcane plants that showed robust growth in the field were selected. The leaves were collected from the shoots and disinfected with 75% alcohol, and the outer leaf sheaths were stripped under aseptic conditions. Then, the heart lobe above the growth point was removed and sliced into about 2-mm thick discs, cultured in the dark at 26–28 °C for 2–4 weeks, and then subjected to particle bombardment transformation after callus generation [29]. Before bombardment, the tungsten particles (Bio-Rad, Foster City, CA, USA, 0.7) were coated by the plasmid of pGcry2A0229 DNA as the micro-bombs, with 1.0 μg of DNA each bombardment. The operation was performed according to the protocol of the PDS-1000/He gene gun (Bio-Rad, Hercules, CA, USA). The bombarded and transformed material was restored culture in subculture medium, then subjected to a screening culture using 0.8 mg/L PPT according to our preliminary experiment, until the plantlets had differentiated, which refers to literature for details [29]. Once developing roots, the plants were transplanted into a nutrient pot. Upon reaching a height of about 10 cm and on a sunny day, the plants were sprayed with 3.0‰ Basta solution (v/v) [29]. Calli were inducted on medium consisted of MS, 3.0 mg/L 2,4-D, 30 g/L sucrose, and 6.0 g/L agar powder, at a pH of 5.8. The subculture medium comprised MS, 2 mg/L 2,4-D, 30 g/L sucrose, and 6 g/L agar powder, at a pH of 5.8. The differentiation medium included MS, 1.5 mg/L 6-BA, 1.0 mg/L KT, 0.2 mg/L NAA, 30 g/L sucrose, and 6 g/L agarose, at a pH of 5.8. The rooting medium consisted of $\frac{1}{2}$ MS, 0.2 mg/L 6-BA, 3 mg/L NAA, 60 g/L sucrose, and 5.5 g/L agarose, at a pH of 5.8.

4.4. DNA Extraction and Primer Design

Genomic DNA was extracted from young leaves of resistant plants that survived the PPT and Basta screening and non-transgenic ROC22 negative control plants using a modified CTAB method [56]. Based on the *cry2A* gene sequence, Primer Premier 5 software was used to design PCR and RT-qPCR primers. The PCR primers were as follows: 2ast1178s: 5′-AACAGGCAACAACCCATAGAGG-3′ and 2ast1798r: 5′-AGGGAGCCCACCTTCTTGAG-3′, and the resulting amplified fragment was 620 bp in size. The RT-qPCR primers were as follows: forward primer: 5′-CAACCAGCAGGTGGACAACTT-3′, reverse primer: 5′-AAGAGCTGCTGCATGGTGTTC-3′, and probe: 5′-CTCAACCCGACCCAGAACCCGG-3′.

4.5. PCR Amplification of Putative Transgenic Sugarcane Lines

Using non-transformed ROC22 as the negative control, pGcry2A0229 plasmid DNA containing the *cry2A* gene as the positive control, and ddH$_2$O as a blank control, amplification and identification were performed using an Eppendorf 5331 PCR instrument (Eppendorf, Hamburg, Germany). Each PCR amplification system consisted of the following reagents: 2.5 μL of 10 × PCR buffer (Mg^{2+} Plus), 2.0 μL of a dNTP mixture (2.5 mmol/L each), 1.0 μL of the DNA template (50.0 ng/μL), 1 μL each of the upstream and downstream primers, 0.25 μL of Taq DNA polymerase (5 U/μL), and topped up to 25.0 μL with ddH$_2$O. The reaction conditions were as follows: pre-denaturation at 95 °C for 5 min; followed by 30 cycles of denaturation at 95 °C for 30 s, annealing at 57 °C for 30 s, and extension at 72 °C for 40 s; and a final extension at 72 °C for 10 min. After amplification, the PCR products were electrophoresed on a 1.5% agarose gel and photographed using a gel imaging system.

4.6. Copy Number Calculation in Transgenic Sugarcane Lines by RT-qPCR

The *cry2A* gene was quantitatively detected in the PCR-positive transgenic sugarcane lines using the designed and synthesized RT-qPCR primers. The fluorescence quantitative PCR instrument was an ABI PRISM 7500 Sequence Detection System (Foster City, CA, USA). The total volume of the detection system was 25.0 μL, which contained 12.5 μL of a FastStart Universal Probe Master Mix, 1.0 μL of gDNA (25.0 ng/μL), 1.0 μL (10.0 μmol/L) of the forward primer, 1.0 μL (10.0 μmol/L) of the reverse primer, 0.2 μL (10.0 μmol/L) of probe, and then topped up to a final volume of 25.0 μL with ddH$_2$O. The amplification conditions were as follows: 50 °C for 2 min; 95 °C for 10 min; 40 cycles of 95 °C for 15 s and 60 °C for 1 min; and a final cycle of 95 °C for 15 s, 60 °C for 15 s, and 95 °C for 15 s. Three replicates were used for each sample. At the same time, gradient dilutions of 10^8, 10^7, 10^6, 10^5, 10^4, 10^3, 10^2, and 10^1 copies/μL were prepared using pGcry2A0229 plasmid DNA. Plasmid copy number was calculated using the following equation: Plasmid copy number (copies/μL) = 6.02×10^{23} copies/mol × plasmid concentration (g/μL)/plasmid molecular weight (g/mol)/660 [57]. After the reaction, using log(plasmid copy number) as x-axis and the C_t value as y-axis, a standard curve was generated using the formula $y = kx + b$. Further, based on the C_t value (y) and linear equation, the total copy number (10^x) of the *cry2A* transgenic lines was determined, and then the single cell copy number of each sample was calculated using the following formula: Copies/genome = 10^x/[25 ng × 10^{-9} × 6.02×10^{23}/(10,000 × 10^6 × 660)] [58].

4.7. Quantitative ELISA of the cry2A Protein in Transgenic Sugarcane Lines

The *cry2A* protein in the leaves of PCR-positive transgenic sugarcane plants was detected using double-antibody sandwich enzyme linked immunosorbent assay (ELISA). Non-transformed ROC22 plants were used as the negative control and ddH$_2$O as a blank control. Gradient dilutions of cry2A protein reference standards in a Qualiplat kit for *cry2A* purchased from Envirologix (Portland, OR, USA) were prepared, with the y-axis representing the OD$_{450}$ absorbance and the x-axis representing the Bt standard protein concentration to construct a standard curve. Quantitative ELISA was conducted according to the protocol provided in the *cry2A* protein assay kit. Three replicates of each sample were prepared.

4.8. Field Trial Design and Assessment of Phenotype Traits of the Transgenic Sugarcane Lines

Three *cry2A* transgenic sugarcane lines with good performance in the field were selected used for further investigation, and non-transgenic ROC22 was used as the control in the field experiment. The experiment followed a randomized block design that consisted of triplicates. The length of the plot was 8.0 m, three rows with a row spacing of 1.3 m were used, the plot area was 31.2 m^2, and 13 buds per meter length. The present study applied common fertilizers at amounts routinely used in the sugarcane field: 345.0 kg/ha of nitrogen fertilizer (N), 240.0 kg/ha of phosphate fertilizer (P$_2$O$_5$), and 360.0 kg/ha of potassium fertilizer (K$_2$O), coupled with normal field management. The industrial and agronomic traits including plant height, stem diameter, brix, effective stalk number, and percentage of borer-infested plants at maturity were investigated, and 20 plants in each plot as the biological repeats in the plot were measured. At the same time, 5 m long and more evenly distributed sections in each plot were selected, and the effective stalk number was counted. The theoretical cane yield per mu and sugar yield per mu were calculated according to the formulae [59]:

Theoretical cane yield = Plant height × Stem diameter^2 × 0.785/1000 × Effective stalk number
Sucrose content (%) = Brix × 1.0825 − 7.703
Theoretical sugar yield = Theoretical cane yield × Sucrose content (%)
DPS analysis software and Tukey method were used for statistical analysis of the collected data.

Author Contributions: Conceived and designed the experiments: S.G., L.X. and Y.Q. Performed the experiments: S.G., Y.Y., J.G., Y.S., Q.W. and C.W. Analyzed the data: S.G., Y.Y., J.G., Y.S., Q.W. and C.W. Wrote the paper: S.G., L.X. and Y.Q. Revised and approved the final version of the paper: L.X., Y.Q. and S.G.

Acknowledgments: The authors especially thank Illimar Altosaar in University of Ottawa, Canada, for providing the cloning vector 2AST1305.1.

Abbreviations

ELISA	Enzyme-linked immuno sorbent assay
CTAB	Cetyltrimethyl ammonium bromide
G418	Geneticin
Hyg	Hygromycin
KT	Kinetin
MS	Murashige and shoog medium
NAA	Naphthalene acetic acid
PPT	Phosphinothricin
2,4-D	2,4-Dichlorophenoxy acetic acid
6-BA	6-Benzyladenine

References

1. Fisher, G.; Teixeira, E.; Hizsnyik, E.T.; Velthuizen, H.V. Land use dynamics and sugarcane production. In *Sugarcane Ethanol: Contributions to Climate Change Mitigation and the Environment*; Zuurbier, P., Vooren, J.V.D., Eds.; Wageningen Academic Publishers: Wageningen, The Netherlands, 2008; pp. 29–62.
2. Lakshmanan, P.; Geijskes, R.J.; Aitken, K.S.; Grof, C.L.P.; Bonnett, G.D.; Smith, G.R. Sugarcane biotechnology: The challenges and opportunities. *In Vitro Cell. Dev. Biol. Plant* **2005**, *41*, 345–363. [CrossRef]
3. Zhou, D.G.; Xu, L.P.; Gao, S.W.; Guo, J.L.; Luo, J.; You, Q.; Que, Y.X. Cry1Ac transgenic sugarcane does not affect the diversity of microbial communities and has no significant effect on enzyme activities in rhizosphere soil within one crop season. *Front. Plant Sci.* **2016**, *7*, 265. [CrossRef] [PubMed]
4. Huang, Y.K.; Li, W.F. The effective test of 5% carbosulfan G against *Sesamia inferens* walker and *Chilo infuscatellus* snellen. *Sugar Crops China* **2006**, *4*, 34–35.
5. Wang, Z.P.; Liu, L.; Jiang, H.T.; Zhang, G.M.; Huang, W.H.; Liang, Q.; Duan, W.X.; Li, Y.J.; Wei, J.J.; Qin, Z.Q. Field efficacy test of 22% *Fipronil FS* against sugarcane stem borers and thrips. *Plant Dis. Pests* **2016**, *7*, 23–26.
6. Aitken, K.S.; Jackson, P.A.; McIntyre, C.L. A combination of AFLP and SSR markers provides extensive map coverage and identification of homo(eo)logous linkage groups in a sugarcane cultivar. *Theor. Appl. Genet.* **2005**, *110*, 789–801. [CrossRef] [PubMed]
7. Piperidis, G.; Piperidis, N.; D'Hont, A. Molecular cytogenetic investigation of chromosome composition and transmission in sugarcane. *Mol. Genet. Genom.* **2010**, *284*, 65–73. [CrossRef] [PubMed]
8. Arencibia, A.; Vázquez, R.I.; Prieto, D.; Téllez, P.; Carmona, E.R.; Coego, A.; Hernandez, L.; de la Riva, G.A.; Selman-Housein, G. Transgenic sugarcane plants resistant to stem borer attack. *Mol. Breed.* **1997**, *3*, 247–255. [CrossRef]
9. Zhou, D.G.; Guo, J.L.; Xu, L.P.; Gao, S.W.; Lin, Q.L.; Wu, Q.B.; Wu, L.G.; Que, Y.X. Establishment and application of a loop-mediated isothermal amplification (LAMP) system for detection of Cry1Ac transgenic sugarcane. *Sci. Rep.* **2014**, *4*, 4912. [CrossRef] [PubMed]
10. Vaeck, M.; Reynaerts, A.; Höfte, H.; Jansens, S.; De Beuckeleer, M.; Dean, C.; Zabeau, M.; Van Montagu, M.; Leemans, J. Transgenic plants protected from insect attack. *Nature* **1987**, *328*, 33–37. [CrossRef]
11. Perlak, F.J.; Deaton, R.W.; Armstron, T.A.; Fuchs, R.L.; Sims, S.R.; Greenplate, J.T.; Fischhoff, D.A. Insect resistant cotton plants. *BioTechnology* **1990**, *8*, 939–943. [CrossRef] [PubMed]
12. Ribeiro, T.P.; Arraes, F.B.M.; Lourenço-Tessutti, I.T.; Silva, M.S.; Lisei-de-Sá, M.E.; Lucena, W.A.; Macedo, L.L.P.; Lima, J.N.; Amorim, R.M.S.; Artico, S.; et al. Transgenic cotton expressing Cry10Aa toxin confers high resistance to the cotton boll weevil. *Plant Biotechnol. J.* **2017**, *15*, 997–1009. [CrossRef] [PubMed]

13. Guo, X.; Huang, C.; Jin, S.X.; Liang, S.; Nie, Y.; Zhang, X.L. *Agrobacterium*-mediated transformation of *Cry1C*, *Cry2A* and *Cry9C* genes into *Gossypium hirsutum* and plant regeneration. *Biol. Plant.* **2007**, *51*, 242–248. [CrossRef]

14. Bakhsh, A.; Rao, A.Q.; Khan, G.A.; Rashid, B.; Shahid, A.A.; Husnain, T. Insect resistance studies of transgenic cotton cultivar harboring *Cry1Ac* and *Cry2A*. *Tabad Tarım Bilim. Arastırma Derg.* **2012**, *5*, 167–171.

15. Koziel, M.G.; Beland, G.L.; Bowman, C.; Carozzi, N.B.; Crenshaw, R.; Crossland, L.; Dawson, J.; Desai, N.; Hill, M.; Kadwell, S.; et al. Field performance of elite transgenic maize plants expressing an insecticidal protein derived from *Bacillus thuringiensis*. *BioTechnology* **1993**, *11*, 194–200. [CrossRef]

16. Fujimoto, H.; Itoh, K.; Yamamoto, M.; Kyozuka, J.; Shimamoto, K. Insect resistant rice generated by introduction of a modified δ-endotoxin gene of *Bacillus thuringiensis*. *BioTechnology* **1993**, *11*, 1151–1155. [CrossRef] [PubMed]

17. Bashir, K.; Husnain, T.; Fatira, T.; Latif, Z.; Mehdi, S.A.; Riazuddin, S. Field evaluation and risk assessment of transgenic indica basmati rice. *Mol. Breed.* **2004**, *13*, 301–312. [CrossRef]

18. Chen, H.; Tang, W.; Xu, C.G.; Li, X.H.; Lin, Y.J.; Zhang, Q.F. Transgenic indica rice plants harboring a synthetic *Cry2A* gene of *Bacillus thuringiensis* exhibit enhanced resistance against lepidopteran rice pests. *Theor. Appl. Genet.* **2005**, *111*, 1330–1337. [CrossRef] [PubMed]

19. Gunasekara, J.M.A.; Jayasekera, G.A.U.; Perera, K.L.N.S.; Wickramasuriya, A.M. Development of a Sri Lankan rice variety Bg 94-1 harbouring *Cry2A* gene of *Bacillus thuringiensis* resistant to rice leaffolder [*Cnaphalocrocis medinalis* (Guenée)]. *J. Natl. Sci. Found. Sri Lanka* **2017**, *45*, 143–157. [CrossRef]

20. Delannay, X.; LaVallee, B.J.; Proksch, R.K.; Fuchs, R.L.; Sims, S.R.; Greenplate, J.T.; Marrone, P.G.; Dodson, R.B.; Augustine, J.J.; Layton, J.G.; et al. Field performance of transgenic tomato plants expressing the *Bacillus thuringinesis* var. kurstaki insect control plant. *BioTechnology* **1989**, *7*, 1265–1269.

21. Arencibia, A.D.; Carmona, E.R.; Cornide, M.T.; Castiglione, S.; O'Relly, J.; Chinea, A.; Oramas, P.; Sala, F. Somaclonal variation in insect-resistant transgenic sugarcane (*Saccharum hybrid*) plants produced by cell electroporation. *Transgenic Res.* **1999**, *8*, 349–360. [CrossRef]

22. Arvinth, S.; Arun, S.; Selvakesavan, R.K.; Srikanth, J.; Mukunthan, N.; Ananda Kumar, P.; Premachandran, M.N.; Subramonian, N. Genetic transformation and pyramiding of aprotinin-expressing sugarcane with *cry1Ab* for shoot borer (*Chilo infuscatellus*) resistance. *Plant Cell Rep.* **2010**, *29*, 383–395. [CrossRef] [PubMed]

23. Wang, W.Z.; Yang, B.P.; Feng, X.Y.; Cao, Z.Y.; Feng, C.L.; Wang, J.G.; Xiong, G.R.; Shen, L.B.; Zeng, J.; Zhao, T.T.; et al. Development and characterization of transgenic sugarcane with insect resistance and herbicide tolerance. *Front. Plant Sci.* **2017**, *8*, 1535. [CrossRef] [PubMed]

24. Sétamou, M.; Bernal, J.S.; Legaspi, J.C.; Mirkov, T.E.; Legaspi, B.C. Evaluation of lectin-expressing transgenic sugarcane against stalkborers (Lepidoptera: Pyralidae): Effects on life history parameters. *J. Econ. Entomol.* **2002**, *95*, 469–477. [CrossRef] [PubMed]

25. Zhangsun, D.T.; Luo, S.L.; Chen, R.K.; Tang, K.X. Improved *agrobacterium*-mediated genetic transformation of GNA transgenic sugarcane. *Biologia* **2007**, *62*, 386–393. [CrossRef]

26. Kalunke, R.M.; Kolge, A.M.; Babu, K.H.; Prasad, D.T. *Agrobacterium* mediated transformation of sugarcane for borer resistance using *Cry1Aa3* gene and one-step regeneration of transgenic plants. *Sugar Technol.* **2009**, *11*, 355–359. [CrossRef]

27. Weng, L.X.; Deng, H.H.; Xu, J.L.; Li, Q.; Wang, L.H.; Jiang, Z.D.; Zhang, H.B.; Li, Q.; Zhang, L.H. Regeneration of sugarcane elite breeding lines and engineering of strong stem borer resistance. *Pest Manag. Sci.* **2006**, *62*, 178–187. [CrossRef] [PubMed]

28. Weng, L.X.; Deng, H.H.; Xu, J.L.; Li, Q.; Zhang, Y.Q.; Jiang, Z.D.; Li, Q.W.; Chen, J.W.; Zhang, L.H. Transgenic sugarcane plants expressing high levels of modified *Cry1Ac* provide effective control against stem borers in field trials. *Transgenic Res.* **2011**, *20*, 759–772. [CrossRef] [PubMed]

29. Gao, S.W.; Yang, Y.Y.; Wang, C.F.; Guo, J.L.; Zhou, D.G.; Wu, Q.B.; Su, Y.C.; Xu, L.P.; Que, Y.X. Transgenic sugarcane with a *Cry1ac* gene exhibited better phenotypic traits and enhanced resistance against sugarcane borer. *PLoS ONE* **2016**, *11*, e0153929. [CrossRef] [PubMed]

30. Falco, M.C.; Silva Filho, M.C. Expression of soybean proteinase inhibitors in transgenic sugarcane plants: Effects on natural defense against Diatraea saccharalis. *Plant Physiol. Biochem.* **2003**, *41*, 761–766. [CrossRef]

31. Nutt, K.A.; Allsopp, P.G.; Geijskes, R.J.; McKeon, M.G.; Smith, G.R.; Hogarth, D.M. Canegrub resistant sugarcane. *Proc. Int. Soc. Suger Cane Technol. Congr.* **2001**, *24*, 582–583.

32. Zhang, Y.L.; Huang, Q.X.; Zhang, S.Z.; Wang, J.H.; Wu, S.R.; Wang, Z.G.; Liu, Z.X. Use of bacterially mediated RNAi technology to study the molt-regulating transcription factor gene *CiHR3* of the sugarcane stem borer, *Chilo infuscatellus*. *Chin. J. Appl. Entomol.* **2013**, *50*, 1301–1310.

33. Mao, Y.B.; Cai, W.J.; Wang, J.W.; Hong, G.J.; Tao, X.Y.; Wang, L.J.; Huang, Y.P.; Chen, X.Y. Silencing a cotton bollworm P450 monooxygenase gene by plant-mediated RNAi impairs larval tolerance of gossypol. *Nat. Biotechnol.* **2007**, *25*, 1307–1313. [CrossRef] [PubMed]

34. Baum, J.A.; Bogaert, T.; Clinton, W.; Heck, G.R.; Feldmann, P.; Ilagan, O.; Johnson, S.; Plaetinck, G.; Munyikwa, T.; Pleau, M.; et al. Control of coleopteran insect pests through RNA interference. *Nat. Biotechnol.* **2007**, *25*, 1322–1326. [CrossRef] [PubMed]

35. Zha, W.J.; Peng, X.X.; Chen, R.Z.; Du, B.; Zhu, L.L.; He, G.C. Knockdown of midgut genes by ds RNA-transgenic plant-mediated RNA interference in the hemipteran insect *Nilaparvata lugens*. *PLoS ONE* **2011**, *6*, e20504. [CrossRef] [PubMed]

36. Ye, J.; Yang, Y.Y.; Xu, L.P.; Li, Y.R.; Que, Y.X. Economic Impact of Stem Borer-Resistant Genetically Modified Sugarcane in Guangxi and Yunnan Provinces of China. *Sugar Tech* **2016**, *18*, 1–9. [CrossRef]

37. Karim, S.; Dean, D.H. Toxicity and receptor binding properties of *Bacillus thuringiensis* δ-endotoxins to the midgut brush border membrane vesicles of the rice leaf folders, *Cnaphalocrocis medinalis* and *Marasmia patnalis*. *Curr. Microbiol.* **2000**, *41*, 276–283. [CrossRef] [PubMed]

38. Alcantara, E.P.; Aguda, R.M.; Curtiss, A.; Dean, D.H.; Cohen, M.B. *Bacillus thuringiensis* δ-endotoxin binding to brush border membrane vesicles of rice stem borers. *Arch. Insect Biochem. Physiol.* **2004**, *55*, 169–177. [CrossRef] [PubMed]

39. Luo, S.L.; Lin, J.Y.; Zhangsun, D.T. Selective test of antibiotics and PPT in different stages of sugarcane tissue culture. *J. Hainan Univ. Nat. Sci.* **2003**, *21*, 259–265.

40. Gordon-Kamm, W.J.; Spencer, T.M.; Mangano, M.L.; Adams, T.R.; Daines, R.J.; Start, W.G.; O'Brien, J.V.; Chambers, S.A.; Adams, W.R., Jr.; Willetts, N.G.; et al. Transformation of maize cells and regeneration of fertile transgenic plants. *Plant Cell* **1990**, *2*, 603–618. [CrossRef] [PubMed]

41. Vasil, V.; Castillo, A.M.; Fromm, M.E.; Vasil, I.K. Herbicide resistant fertile transgenic wheat plants obtained by microprojectile bombardment of regenerable embryogenic callus. *BioTechnology* **1992**, *10*, 667–674. [CrossRef]

42. Gallo-Meagher, M.; Irvine, J.E. Herbicide resistant transgenic sugarcane plants containing the *Bar* gene. *Crop Sci.* **1996**, *36*, 1367–1374. [CrossRef]

43. Manickavasagam, M.; Ganapathi, A.; Anbazhagan, V.R.; Sudhakar, B.; Selvaraj, N.; Vasudevan, A. *Agrobacterium*-mediated genetic transformation and development of herbicide-resistant sugarcane (*Saccharum* species hybrids) using axillary buds. *Plant Cell Rep.* **2004**, *23*, 134–143. [CrossRef] [PubMed]

44. Mason, G.; Provero, P.; Vaira, A.M.; Accotto, G.P. Estimating the number of integrations in transformed plants by quantitative real-time PCR. *BMC Biotechnol.* **2002**, *2*, 20. [CrossRef]

45. Casu, R.E.; Selivanova, A.; Perroux, J.M. High-throughput assessment of transgene copy number in sugarcane using realtime quantitative PCR. *Plant Cell Rep.* **2012**, *31*, 167–177. [CrossRef] [PubMed]

46. Song, P.; Cai, C.Q.; Skokut, M.; Kosegi, B.D.; Petolino, J.F. Quantitative real-time PCR as a screening tool for estimating transgene copy number in WHISKERS™-derived transgenic maize. *Plant Cell Rep.* **2002**, *20*, 948–954.

47. Yang, L.T.; Ding, J.Y.; Zhang, C.M.; Jia, J.W.; Weng, H.B.; Liu, W.X.; Zhang, D.B. Estimating the copy number of transgenes in transformed rice by real-time quantitative PCR. *Plant Cell Rep.* **2005**, *23*, 759–763. [CrossRef] [PubMed]

48. Yi, C.X.; Zhang, J.; Chan, K.M.; Liu, X.K.; Hong, Y. Quantitative realtime PCR assay to detect transgene copy number in cotton (*Gossypium hirsutum*). *Anal. Biochem.* **2008**, *375*, 150–152. [CrossRef] [PubMed]

49. Ji, Z.G.; Gao, X.J.; Ao, J.X.; Zhang, M.H.; Huo, N. Establishment of SYBR Green-base quantitative real-time PCR assay for determining transgene copy number in transgenic soybean. *J. Northeast Agric. Univ.* **2011**, *42*, 11–15.

50. Sun, Y.; Joyce, P.A. Application of droplet digital PCR to determine copy number of endogenous genes and transgenes in sugarcane. *Plant Cell Rep.* **2017**, *36*, 1775–1783. [CrossRef] [PubMed]

51. Dominguez, A.; Guerri, J.; Cambra, M.; Navarro, L.; Morenod, P.; Pena, L. Efficient production of transgenic citrus plants expressing the coat protein gene of citrus tristeza virus. *Plant Cell Rep.* **2000**, *19*, 427–433. [CrossRef]

52. Yang, X.J.; Liu, C.L.; Zhang, X.Y.; Li, F.G. Molecule characterization and expression of T-DNA integration in transformed plants. *Genom. Appl. Biol.* **2010**, *29*, 125–130.

53. Joyce, P.; Hermann, S.; O'Connell, A.; Dinh, Q.; Shumbe, L.; Lakshmanan, P. Field performance of transgenic sugarcane produced using *Agrobacterium* and biolistics methods. *Plant Biotechnol. J.* **2014**, *12*, 411–424. [CrossRef] [PubMed]

54. Fu, J.P.; Wang, B.; Liu, L.J.; Yang, J.Y.; Wang, X.X.; Xing, X.L.; Peng, D.X. Transgenic ramie with *Bt* gene mediated by *Agrobacterium tumefacien* and evaluation of its pest-resistance. *Acta Agron. Sin.* **2009**, *35*, 1771–1777. [CrossRef]

55. Hellens, R.P.; Edwards, E.A.; Leyland, N.R.; Bean, S.; Mullineaux, P.M. pGreen: A versatile and flexible binary Ti vector for Agrobacterium-mediated plant transformation. *Plant Mol. Biol.* **2000**, *42*, 819–832. [CrossRef] [PubMed]

56. Paterson, A.H.; Brubaker, C.L.; Wendel, J.F. A rapid method for extraction of cotton (*Gossypium* spp.) genomic DNA suitable for RFLP or PCR analysis. *Plant. Mol. Biol. Rep.* **1993**, *11*, 122–127. [CrossRef]

57. Li, H.F.; Li, L.; Zhang, L.J.; Xu, Y.L.; Li, W.F. Standard curve generation of *PepT1* gene for absolute quantification using real-time PCR. *J. Shanxi Agric. Univ.* **2010**, *30*, 332–334.

58. Xue, B.T.; Guo, J.L.; Que, Y.X.; Fu, Z.W.; Wu, L.G.; Xu, L.P. Selection of suitable endogenous reference genes for relative copy number detection in sugarcane. *Int. J. Mol. Sci.* **2014**, *15*, 8846–8862. [CrossRef] [PubMed]

59. Luo, J.; Pan, Y.B.; Xu, L.P.; Zhang, Y.Y.; Zhang, H.; Chen, R.K.; Que, Y.X. Photosynthetic and canopy characteristics of different varieties at the early elongation stage and their relationships with the cane yield in sugarcane. *Sci. World J.* **2014**, *2014*, 707095. [CrossRef] [PubMed]

4

Differentially Expressed Genes Associated with the Cabbage Yellow-Green-Leaf Mutant in the *ygl-1* Mapping Interval with Recombination Suppression

Xiaoping Liu [†], Hailong Yu [†], Fengqing Han, Zhiyuan Li, Zhiyuan Fang, Limei Yang, Mu Zhuang, Honghao Lv, Yumei Liu, Zhansheng Li, Xing Li and Yangyong Zhang *

Institute of Vegetables and Flowers, Chinese Academy of Agricultural Sciences, Key Laboratory of Biology and Genetic Improvement of Horticultural Crops, Ministry of Agriculture, Beijing 100081, China; 82101181120@caas.cn (X.L.); yuhailong@caas.cn (H.Y.); feng857142@163.com (F.H.); 82101181071@caas.cn (Z.L.); fangzhiyuan@caas.cn (Z.F.); yanglimei@caas.cn (L.Y.); zhuangmu@caas.cn (M.Z.); lvhonghao@caas.cn (H.L.); liuyumei@caas.cn (Y.L.); lizhansheng@caas.cn (Z.L.); xzdlixing@126.com (X.L.)
* Correspondence: zhangyangyong@caas.cn
† These authors contributed equally to this work.

Abstract: Although the genetics and preliminary mapping of the cabbage yellow-green-leaf mutant YL-1 has been extensively studied, transcriptome profiling associated with the yellow-green-leaf mutant of YL-1 has not been discovered. Positional mapping with two populations showed that the yellow-green-leaf gene *ygl-1* is located in a recombination-suppressed genomic region. Then, a bulk segregant RNA-seq (BSR) was applied to identify differentially expressed genes (DEGs) using an F_3 population (YL-1 × 11-192) and a BC_2 population (YL-1 × 01-20). Among the 37,286 unique genes, 5730 and 4118 DEGs were detected between the yellow-leaf and normal-leaf pools from the F_3 and BC_2 populations. BSR analysis with four pools greatly reduced the number of common DEGs from 4924 to 1112. In the *ygl-1* gene mapping region with suppressed recombination, 43 common DEGs were identified. Five of the DEGs were related to chloroplasts, including the down-regulated *Bo1g087310*, *Bo1g094360*, and *Bo1g098630* and the up-regulated *Bo1g059170* and *Bo1g098440*. The *Bo1g098440* and *Bo1g098630* genes were excluded by qRT-PCR. Hence, we inferred that these three DEGs (*Bo1g094360*, *Bo1g087310*, and *Bo1g059170*) in the mapping interval may be tightly associated with the development of the yellow-green-leaf mutant phenotype.

Keywords: cabbage; yellow-green-leaf mutant; recombination-suppressed region; bulk segregant RNA-seq; differentially expressed genes

1. Introduction

Yellow-green-leaf mutants have been extensively studied in many species, including *Arabidopsis thaliana* [1], barley [2], *Brassica napus* [3], rice [4–6], cabbage [7], and muskmelon [8]. Leaf color mutants are an ideal model for studying mechanisms of photosynthesis and light morphogenesis, since yellow-green-leaf mutants are commonly related to chlorophyll synthesis or degradation [9,10].

Chlorophyll is the most important pigment related to photosynthesis. In *Arabidopsis*, 27 genes involved in 15 steps in the pathway from glutamyl-tRNA to chlorophylls a and b have been identified. Leaf color mutants commonly result from blocking a portion of the chlorophyll synthesis pathway, such as the synthesis of 5-aminolevulinic acid (ALA) [11]. Runge et al. [12] isolated and classified some chlorophyll-deficient xantha mutants of *Arabidopsis thaliana* and found that some of the mutants were blocked at various steps of the chlorophyll pathway between ALA and protochlorophyllide (Pchlide), and the latter did not accumulate in the dark.

Bulked segregant analysis (BSA) is a powerful strategy that is commonly used in gene mapping [13]. Futschik and Schlötterer showed that sequencing of pools of samples from individuals are often more effective for Single Nucleotide Polymorphisms (SNP) discovery and provide more accurate allele frequency estimates [14]. Typically, two populations are used for BSA: a backcross (BC) population [15,16] and an F_2 population [17,18]. Mackay and Caligari [19] found that quantitative trait loci (QTLs) are more easily detected in BC populations than in F_2 populations.

In recent years, transcriptome analysis based on deep RNA sequencing (RNA-seq) has been used for the estimation of genome-wide gene expression levels [20,21]. Transcriptome sequencing encompasses mRNA transcript expression analysis. Combined RNA-seq analysis can be used for purposes such as novel transcript prediction, gene structure refinement, alternative splicing analysis, and SNP/InDel analysis [22]. Bulk segregant RNA-seq (BSR) has been applied to identify differentially expressed genes (DEGs) and trait-associated SNPs [23,24].

A yellow-green-leaf mutant (YL-1) was discovered in cabbage [10], and measurements of photosynthetic pigment contents, chloroplast ultrastructure, and chlorophyll fluorescence parameters indicated that YL-1 was deficient in its total chlorophyll content [10]. In a previous study, we mapped *ygl-1*, which controls the yellow-green-leaf phenotype, to chromosome C01 [7]. The linkage distance of the mapping interval was only 0.75 cM, but the physical distance in the reference genome TO1000 was ~10 Mb, indicating that recombination suppression existed in this interval. In this study, the recombination-suppressed region was identified by gene mapping. Two runs of BSR were performed using BC and F_3 populations, with the aim of obtaining DEGs associated with the yellow-green-leaf mutant.

2. Materials and Methods

2.1. Plant Materials

Group I: The F_2, BC_1P_1, and F_3 populations were constructed using as parents the yellow-green-leaf cabbage mutant YL-1 (P_1) and the normal green leaf cabbage inbred line 01-20 (P_2). The F_2, BC_1P_1 population was employed for *ygl-1* mapping.

Group II: The BC_1P_1 and BC_2P_1 populations were constructed using as parents the mutant YL-1 (P_1) and the normal green leaf Chinese kale inbred line 11–192 (P_3) (Supplementary Figure S1). The BC_2P_1 population was employed for *ygl-1* mapping.

The F_3 population in group I and the BC_2 population in group II were used for RNA-seq analysis. All plant materials came from the Cabbage and Broccoli Research Group, the Institute of Vegetables and Flowers (IVF), and the Chinese Academy of Agricultural Sciences (CAAS).

2.2. Identification of Recombination Suppression in the ygl-1 Gene-Mapping Interval

The sequences of 24 markers from the 02-12 reference genome (Supplementary Table S1) were aligned to chromosome C01 and the scaffold of the TO1000 reference genome [25] (Figure 1). Based on this alignment, we propose that possible assembly errors might exist in the 02-12 reference genome. Hence, InDel primers designed based on the TO1000 reference genome were applied for further mapping. The rates of recombination in the two populations were compared with the normal level in the cabbage genome (~600 kb/cM) to analyze the recombination-suppressed region.

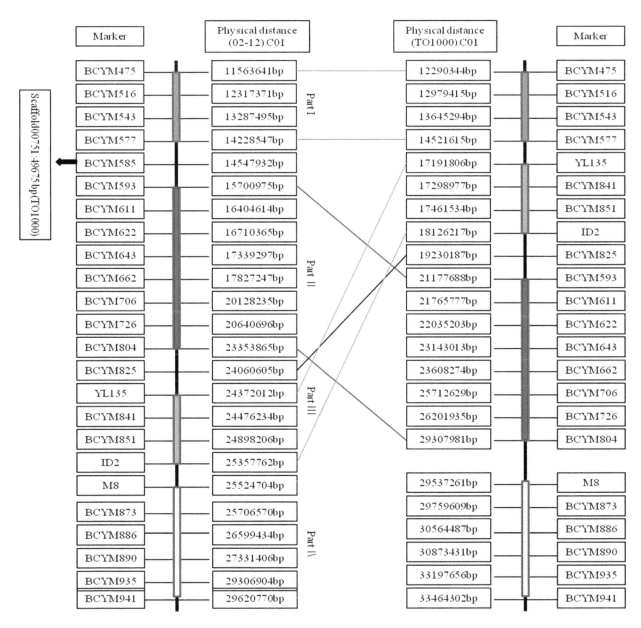

Figure 1. The physical distances of 24 InDel markers in the two reference genomes (02-12 and TO1000).

2.3. BSA, RNA Isolation, and Library Construction

Before RNA isolation, leaf samples from the two populations (the F_3 population in group I and the BC_2 population in group II) were harvested to prepare four bulk groups: Bulk F_yellow (consisting of equal amounts of leaf tissues from 20 yellow-green-leaf F_3 individuals), Bulk F_normal (20 normal-green-leaf F_3 individuals), BC_yellow (20 yellow-green-leaf BC_2 individuals), and BC_normal (20 normal-green-leaf BC_2 individuals).

Total RNA extraction was performed according to instructions of the manufacturer of the TIANGEN kit employed for extraction (Invitrogen, Carlsbad, CA, USA). RNA purity was determined using a NanoDrop spectrophotometer (Thermo Fisher Scientific Inc., Wilmington, DE, USA), 1% formaldehyde gel electrophoresis, and a 2100 Bioanalyzer (Agilent Technologies, Santa Clara, CA, USA).

A total amount of 1 μg of RNA per sample was employed for RNA sample preparation. Sequencing libraries were generated using the NEBNext® Ultra™ RNA Library Prep Kit for Illumina®

(Illumina, CA, USA) following the manufacturer's recommendations. The cDNA library products were sequenced in a paired-end flow cell using an Illumina HiSeq™ 2000 system.

3. Data Analysis

Reads containing adaptor sequences, low-quality reads (bases with more than 50% of quality scores ≤ 5), and unknown bases (>5% N bases) were removed from each dataset to obtain more reliable results, because such data negatively affect bioinformatics analyses. The sequencing reads were then aligned to the reference database for the *B. oleracea* genome (TO1000) (http://plants.ensembl. org/Brassica_oleracea/Info/Index) (accessed on 5 May 2017) [25] using HISAT [26]. Differential expression analysis to identify DEGs was performed using DESeq [27], with a threshold q value (or false discovery rate [FDR]) < 0.01 & $|\log_2(\text{fold change})| > 1$ for significant differential expression. DEGs were displayed using Circos v0.66 [28]. GO (http://www.geneontology.org/) (accessed on 7 May 2017) [29] enrichment analysis of the DEGs was implemented using GOseq, in which gene length bias was corrected. GO functional analysis provides GO functional classifications and annotations for DEGs. Various genes usually cooperate with each other to exercise their biological functions. A pathway-related database was therefore obtained based on Kyoto Encyclopedia of Genes and Genome (KEGG) results (http://www.genome.jp/kegg/) (accessed on 11 May 2017) [30].

Gene Expression Validation

DEGs associated with the yellow-green-leaf mutant were subjected to quantitative real-time RT-PCR (qRT-PCR) analysis. The primers designed according to the gene CDS sequences using DNAMAN are listed in Supplementary Table S6. Three technical replicates were performed for each gene. First-strand cDNA was synthesized using the PrimeScript™ RT reagent Kit (TAKARA BIO, Inc., Shiga, Japan). qRT-PCR was performed with the SYBR Premix Ex Taq™ Kit (Takara, Dalian, China) with the following cycling parameters: 95 °C for five min, followed by 40 cycles of 95 °C for 10 s and 55 °C for 30 s, with a final cycle of 95 °C for 15 s, 55 °C for 60 s, and 95 °C for 15 s. Relative transcription levels were analyzed using the $2^{-\Delta\Delta Ct}$ method [31]. qRT-PCR was performed in a BIO-RAD CFX96 system (Bio-Rad, Hercules, CA, USA), and the actin gene was employed as the internal control [32].

4. Results

4.1. Identification of the Recombination-Suppressed Region

In a previous study [7], we mapped *ygl-1*, which controls the yellow-green-leaf phenotype, to chromosome C01 using a population derived from YL-1 and 01-20. The *ygl-1* gene is flanked by the InDel markers ID2 and M8, and the interval between these two markers is 167 kb (C01: 25,357,762–25,524,704 bp) in the 02-12 reference genome.

However, these two markers are anchored to the TO1000 reference genome, in which the interval between ID2 (C01: 18,126,217 bp) and M8 (C01: 29,537,261 bp) is 11.41 Mb, which is approximately 680 times greater than the distance (167 kb) in the 02-12 reference genome. Then, 24 markers from the 02-12 reference genome (Supplementary Table S1) were aligned to chromosome C01 and the scaffold of the TO1000 reference genome (Figure 1). In the 02-12 reference genome, the physical interval between BCYM475 (11,563,641 bp) and BCYM941 (29,620,770 bp) could be divided into four parts [Part I: BCYM475 (11,563,641 bp) to BCYM577 (14,228,547 bp); Part II: BCYM593 (15,700,975 bp) to BCYM804 (23,353,865 bp); Part III: YL135 (24,372,012 bp) to ID2 (25,357,762 bp); and Part IV: BCYM873 (25,706,570 bp) to BCYM941 (29,620,770 bp)]. The physical locations of Part I and Part IV in the two reference genomes were parallel. However, the physical locations of Part II and Part III were opposite. The makers' order of linkage map was consistent with the physical map order of TO1000 reference genome but not 02-12 reference genome. Therefore, we proposed that an assembly error might exist in the 02-12 reference genome.

InDel primers designed based on the TO1000 reference genome were then applied for further mapping of the *ygl-1* gene. A total of 43 of the 62 pairs of InDel primers designed based on the TO1000 reference genome exhibited polymorphisms according to the F_3 population. The genetic distances of the 16 InDel markers are shown in Table 1 (the sequences of these 16 markers are provided in Supplementary Table S2). The *ygl-1* gene was flanked by the InDel markers T1-36 (18,069,792 kb) and T1-58 (29,537,314 kb), with genetic distances of 0.42 cM and 0.42 cM, respectively. The interval distance between the two markers was 11.47 Mb based on the TO1000 reference genome. In the mapping region, spanning a physical distance of 11.47 Mb with a genetic difference of only 0.84 cM, the recombination rate was almost twenty times lower than the normal level for the cabbage genome (~600 kb/cM), suggesting that recombination suppression existed in this region.

Table 1. Genetic distances of the InDel primers to the *ygl-1* in the two mapping populations.

	YL-1 × 01-20		YL-1 × 11-192
Primers	**Genetic Distance (cM)**	**Primers**	**Genetic Distance (cM)**
T2-3	9.21	T2-3	13.3
T2-5	6.90	-	-
T1-1	6.28	-	-
T1-14	4.39	T1-14	6.5
T1-18	3.97	T1-18	4.4
T1-26	2.51	T1-26	2.3
T1-28	1.46	T1-28	1.5
T1-30	1.05	T1-30	1.3
T1-34	0.63	T1-34	0.3
T1-36	0.42	T1-36	0.00
T1-58	0.42	T1-58	0.7
T2-6	0.42	T2-6	1.04
T2-10	0.63	T2-10	1.04
T2-14	0.63	T2-14	1.04
T2-16	3.14	T2-16	2.61
T2-18	5.02	T2-18	6.02

Another BC_2P_1 population, constructed with YL-1 and 11–192, was used to further identify recombination suppression. The *ygl-1* gene was flanked by InDel markers T1-34 (17,301,717 kb) and T1-58 (29,537,314 kb), with genetic distances of 0.3 cM and 0.7 cM, respectively. This result further demonstrated the existence of a recombination-suppressed region in the *ygl-1* mapping interval.

In a previous study [7], we showed that the region between markers the BCYM585 (14,547,932 bp) and BCYM825 (24,060,605 bp) exhibits recombination suppression. In Figure 1, the sequence of BCYM585 was aligned to an unanchored scaffold (Scaffold00751), and the sequence of BCYM825 was aligned to a physical distance of 19,230,187 bp based on the TO1000 reference genome. Part II was aligned between 21,177,688 bp and 29,307,981 bp based on the TO1000 reference genome. These results showed that the recombination-suppressed region observed between the markers T1-36 (18,069,792 kb) and T1-58 (29,537,314 kb) in this study was consistent with the recombination-suppressed region between the markers BCYM585 and BCYM825 identified in our previous study [7].

4.2. BSR Analysis, DEGs between the Yellow-Green-Leaf and Normal-Leaf Pools

BSR was applied to obtain DEGs using the F_3 segregated population constructed with YL-1 and 01-20 and the BC_2 population constructed with YL-1 and 11-192. A total of 339,481,468 reads were generated from the four cDNA libraries. Among these reads, 82,143,852 were obtained from BC_normal, 91,405,984 from BC_yellow, 86,447,180 from F_normal, and 79,484,452 from F_yellow. The GC contents of the sequences of the four libraries were all approximately 47%, and all Q30% scores (reads with average quality scores >30) were >90%, indicating that the accuracy and quality of the sequencing data were sufficient for further analysis. The sequenced reads were aligned to the *B. oleracea* genome reference (TO1000) (http://plants.ensembl.org/Brassica_oleracea/Info/Index) (accessed on 5 May 2017). An overview of the sequencing process is shown in Supplementary Table S3. The density distribution and boxplot of all the genes exhibited similar patterns among the four samples, indicating that no bias occurred in the construction of the cDNA libraries (Supplementary Figure S2).

The number of DEGs identified between the yellow-green-leaf and normal-leaf samples is shown in Table 2 (Supplementary Figure S3). In the yellow-green-leaf pools, there were approximately 20% fewer down-regulated genes than up-regulated genes. In total, 5730 and 4118 (4924 on average) DEGs were detected between the yellow-green-leaf and normal-leaf pools for the F_3 and BC_2 populations. As shown in the Venn diagram presented in Figure 2, 1884 common DEGs were shared between the DEGs identified in BC_normal vs. BC_yellow and the DEGs identified in F_normal vs. F_yellow, representing approximately half of the total number of DEGs in either population. Cross-comparison showed that only 1112 DEGs (Supplementary Table S4) were common between yellow-leaf and normal-leaf bulks. Thus, BSR analysis using four pools greatly reduced the number of DEGs from 4924 to 1112.

Table 2. Numbers of DEGs between the yellow-leaf and normal-leaf samples.

	No. of DEGs	No. of Up-Regulated DEGs	Percentage (%)	No. of Down-Regulated DEGs	Percentage (%)
BC_normal vs. BC_yellow	4118	2384	58	1734	42
BC_normal vs. F_yellow	8009	4894	60	3315	40
F_normal vs. F_yellow	5730	3226	56	2504	44
F_normal vs. BC_yellow	5405	2844	53	2561	47

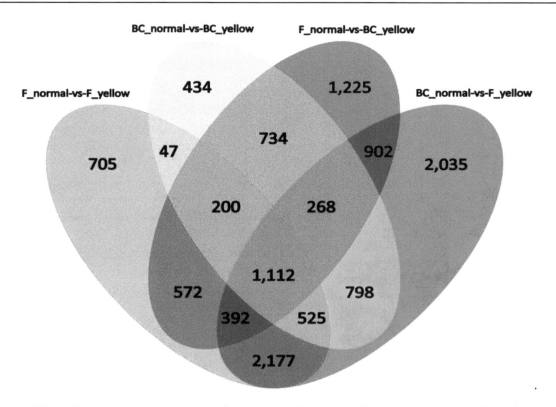

Figure 2. Venn diagram showing the numbers of overlapping and nonoverlapping DEGs (FDR < 0.01 and fold change > 2.0 or < −2.0) in the indicated segments from normal-leaf samples and yellow-leaf samples.

These 1112 DEGs were assigned into three Gene Ontology (GO) classes: biological process, cellular component, and molecular function. Thirty of the most significantly enriched of GO terms are shown in Figure 3, including "carbohydrate binding", "sequence-specific DNA binding transcription factor activity", "receptor activity", "brassinosteroid sulfotransferase activity", "unfolded protein binding" and "protein phosphatase inhibitor activity" under GO molecular functions and "endoplasmic reticulum lumen", "plant-type cell wall", "cytoplasm", "vacuolar membrane", "apoplast", and "nucleus" under GO cellular components. Seventeen biological function or functional groups were enriched in the GO biological process category. In certain biological functions, genes play roles by interacting with each other, and KEGG pathway analysis helps provide an in-depth understanding of the biological functions of genes. A total of 1112 DEGs were annotated in the KEGG database, and 117 KEGG pathways were assigned. These 117 pathways were divided into three levels. Level one included "genetic information processing", "metabolism", "cellular processes", "organismal systems", and "environmental information processing." The nineteen terms in level two are shown in Figure 4.

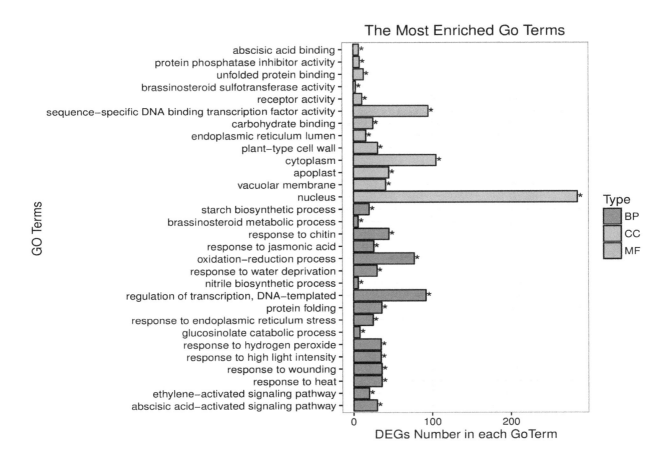

Figure 3. The thirty top GO assignments of 1112 DEGs. Blue: molecular function, green: cellular component, and red: biological process. The Y-axis represents the GO Term; the X-axis represents the number of DEGs for each GO Term. "*" indicates significant enrichment of the GO Term.

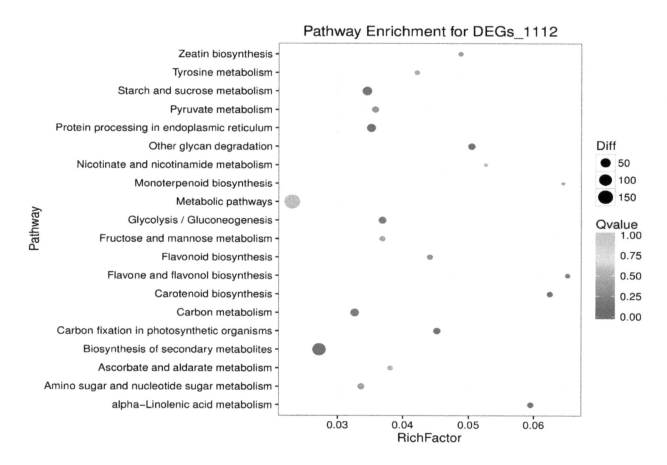

Figure 4. The top-20 enriched KEGG pathways of the 1112 DEGs. The Y-axis represents the pathway term; the X-axis represents the rich factor. The sizes of the points represent different DEG numbers, such that the bigger the point, the greater the DEG number. The colors represent different *Q*-values.

4.3. DEGs Involved in B. oleracea Chlorophyll Synthesis

The chlorophyll a, chlorophyll b, and total chlorophyll contents of the yellow-green-leaf mutant YL-1 were significantly lower than those of wild-type plants over the entire growth period [10]. Among the 1112 identified DEGs, 18 DEGs related to chlorophyll were clustered, which are shown in Supplementary Figure S4, including nine down- and nine up-regulated DEGs. These 18 DEGs were distributed among different chromosomes. Among the nine chromosomes, there were more DEGs on C01, C03, and C06 than on the other chromosomes (Supplementary Figure S5). In the 11.47 Mb recombination suppression region, two genes *Bo1g088040* (homologous gene *AT1G58290*, *HEMA1*) and *Bo1g098190* (homologous gene *AT1G61520*, *LCA3*) were related to chlorophyll according to the annotations, but there were not DEGs among these four pools by transcriptomics analysis and semi-quantitative PCR. Besides, no sequence variation was detected in the CDS region of these two genes of YL-1, compared with the sequences of 01-20, 11-192, and reference genome TO1000.

DEGs located in the *ygl-1* mapping interval with recombination suppression were selected for further analysis. In the BC_normal vs. BC_yellow comparison, 82 DEGs were found in the 11.47 Mb genomic region, with 45 DEGs being down-regulated and 37 being up-regulated. In the F_normal vs. F_yellow comparison, 105 DEGs were found in the 11.47 Mb genomic region, with 47 DEGs being down-regulated and 58 being up-regulated. Among these four pools, 43 common DEGs were present, with 20 DEGs being down-regulated and 23 being up-regulated (Supplementary Table S5). According to the annotations, five of these genes were related to chloroplasts (Table 3), including the down-regulated genes *Bo1g087310*, *Bo1g094360*, and *Bo1g098630* and the up-regulated genes *Bo1g059170* and *Bo1g098440*. These five genes were applied in qRT-PCR and RT-PCR analyses of the three parents (01-20, YL-1, 11-192). The relative normalized expression of these five genes is shown in Figure 5. The primers of qRT-PCR were supplied on Supplementary Table S6. Based on the relative normalized expression, it can be observed that the expression of *Bo1g059170*, *Bo1g087310*, and *Bo1g094360* genes was consistent with the results of BSR, whereas the relative expression of the *Bo1g098440* and *Bo1g098630* genes differed from the results of BSR. We inferred that these two genes' transcription levels were irrelevant to the yellow-green-leaf trait. In the other three genes that related to chloroplasts, *Bo1g087310* (homologous gene *AT1G56340*, Calreticulins-1) plays important roles in calciumion binding, plant growth, and plant height [33]. *Bo1g059170* (homologous gene *AT3G51420*) is involved in strictosidine synthase activity and plant defense [34], and *Bo1g094360* (homologous gene *AT3G08840*) functions in D-alanine-D-alanine ligase activity (Table 3) [35]. Hence, we inferred that these three candidate genes (*Bo1g094360*, *Bo1g087310*, and *Bo1g059170*) may be responsible for the development of the yellow-green-leaf mutant phenotype.

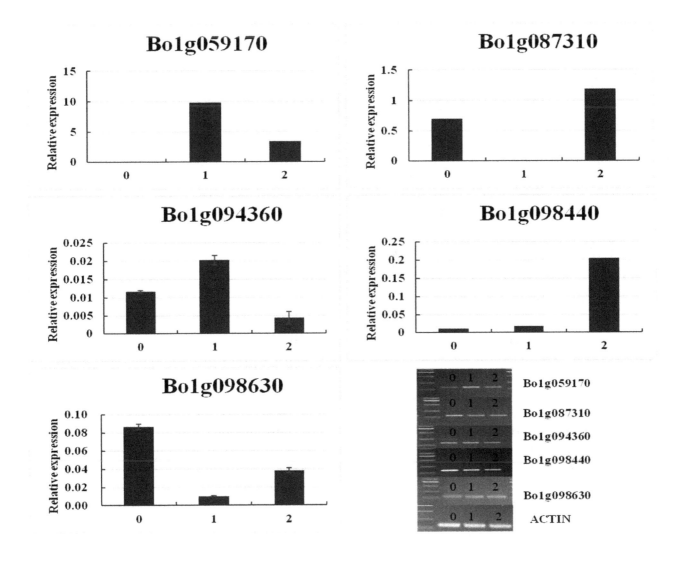

Figure 5. qRT-PCR and RT-PCR validation of transcripts of five DEGs associated with the yellow-green-leaf mutant. 0: the parent 01-20, 1: Mutant YL-1, 2: the parent 11-192.

Table 3. DEGs related to chloroplasts in the recombination-suppressed region.

Gene ID [a]	Physical Distance (TO1000)	F Normal [b]	F_Yellow [b]	BC_Normal [b]	BC_Yellow [b]	Diff [c]	A.T. Annotation [d]
Bo1g087310	C1:25381300-25383803	1837.98	156.85	1920.64	287.42	Down	Calreticulins-1, response to oxidative stress, response to cadmiumion, response to salt stress, calciumion homeostasis; D-alanine-D-alanine ligase activity
Bo1g094360	C1:27829353-27834745	48.65	10.53	29.70	2.04	Down	GPT2: glucose-6-phosphate/ phosphate translocator 2
Bo1g098630	C1:29261755-29263303	4002.89	475.81	1119.81	125.36	Down	
Bo1g059170	C1:18110687-18112080	167.35	828.45	277.80	858.19	Up	SSL4: strictosidine synthase-like 4
Bo1g098440	C1:29037892-29038492	129.41	285.27	120.70	427.05	Up	Protein of unknown function, DUF538

[a] Five *B. oleracea* DEGs related to chloroplasts (reference genome TO1000). [b] Expression levels in the four samples. [c] Differential regulation: up-regulation and down-regulation. [d] GO annotations for seven Bo to AT best-hit genes obtained from The Arabidopsis Information Resource (TAIR).

5. Discussion

5.1. Efficiency of BSR in DEG Detection

BSA (an efficient method for rapidly identifying markers linked to mutant phenotypes) combined with RNA-seq has been performed to map important agronomic traits at the transcription level in some species, such as catfish [23], onion [36] maize [37], Chinese cabbage [38], Chinese wheat cultivar [39], polyploid wheat [40], etc. Using BSR, Kim et al. [35] identified the candidate gene, AcPMS1, which is involved in DNA mismatch repair, for the fertility restoration of cytoplasmic male sterility in onions. Ramirez-Gonzalez et al. [24] mapped *Yr15* to a 0.77-cM interval in hexaploid wheat using a segregated F_2 population through BSR. In the present study, RNA-seq analysis of four bulks detected only 1112 common DEGs between the four pools (4924 on average), which can reduce the number of genes related to the phenotype. Therefore, BSR was further demonstrated to be an efficient method for analyzing the genes associated with the yellow-green-leaf mutant phenotype.

5.2. DEGs Analysis Associated with the Yellow-Green-Leaf in a Recombination-Suppressed Region via RNA-Seq

In recent years, the fine mapping of important agronomic traits in *Brassica* has developed rapidly [41–43]. Some yellow leaf color genes have been mapped in *Brassica* crops. A mutation responsible for chlorophyll deficiency in *Brassica juncea* was mapped between amplified fragment length polymorphism (AFLP) markers EA4TG4 and EA7MC1, with genetic distances of 33.6 and 21.5 cM, respectively [44]. In *B. napus*, Wang et al. [45] mapped the *CDE1* locus to a 0.9 cM interval of chromosome C08, and Zhu et al. [3] mapped a chlorophyll-deficient mutant between the markers BnY5 and CB10534, which are closely linked to the chlorophyll deficiency gene *BnaC.YGL*, with genetic distances of 3.0 and 3.2 cM on C06, respectively. Gene mapping for the above leaf color mutant was based on normal recombination in the segregated population. Recombination suppression was reported in many species, such as tomato [46], barley [47], petunia [48], *Populus* [49], hexaploid wheat [50], and buffelgrass [51]. In this study, we identified a large recombination suppression region spanning ~11 Mb on C01. However, recombination rate of *Brassica oleracea* C01 in previous studies seemed to be normal. The genetic map was constructed based on *Brassica oleracea* re-sequencing data; the C01 linkage groups spanned 97.59 cM, with an average distance of 1.15 cM between neighboring loci; and no recombination suppression was found [52]. Lv et al. (2016) [53] constructed a high-density genetic map while describing a comprehensive QTL analysis of key agronomic traits of cabbage. On C01, twelve markers existed between the markers Indel481 (17,365,179 bp) and

Indel14 (28,513,070 bp), which showed recombination was observed to be normal at the 17.3–28.5 Mb. In the present study, recombination suppression was observed at C01: 18,069,792–29,537,314 bp in the mapping of *ygl-1* gene using the population constructed from YL-1 and 01-20. Moreover, a recombination-suppressed region was identified in the same area while mapping *ygl-1* using another population constructed from YL-1 and 11-192. These two populations have one same parent YL-1. Therefore, we speculated that the suppression of recombination may be due to the YL-1 mutant.

In the recombination-suppressed region, it is difficult to identify candidate genes using fine mapping. Some research has revealed genes related to the phenotype by RNA-seq, such as *Fhb1* in wheat [54] and *BPH15* in rice [55]. In the *ygl-1* gene-mapping interval, a total of 10478 SNPs and Indels, with 455 genes, were identified in the recombination-suppressed region, including 78 genes related to chloroplasts. Comparison of the two bulk RNA-seq groups showed that only 43 genes were common DEGs, only five of which were related to chloroplasts. Furthermore, three of these five genes' expression by qRT-PCR were consistent with the results of BSR. Therefore, BSA combined with RNA-seq was able to greatly reduce the number of DEGs, demonstrating that this method is an effective alternative for identifying candidate genes in a recombination-suppressed region.

5.3. Assembly Error in the Reference Genome

Brassica oleracea reference genome sequencing was completed in 2014 [25,56]. However, the 02-12 reference genome assemblies have been woefully incomplete, and some assembly errors have been identified in recent studies. For example, Lee et al. [47] revised 27 v-blocks, 10 s-blocks, and several other blocks in the 02-12 reference genome assembly during the mapping of clubroot resistance QTLs through genotyping-by-sequencing. The purple leaf gene (*BoPr*) in the ornamental kale was mapped on an unanchored scaffold by Liu et al. (2017) [57]. In a previous study [7], we identified possible assembly errors in the 02-12 reference genome. According to the comparison of marker positions with the TO1000 reference, the physical locations of Part II and Part III in the 02-12 reference genome likely represent assembly errors (Figure 1). The makers' order of linkage map was consistent with the physical map order of TO1000 reference genome. All the results showed that the TO1000 reference genome is reliable. These results will contribute to the improvement of the cabbage genome.

6. Conclusions

In conclusion, we mapped the yellow-green-leaf gene *ygl-1* on a recombination-suppressed genomic region by two populations. Bulk segregant RNA-seq (BSR) was applied to identify differentially expressed genes using two segregate populations. BSR analysis with four pools greatly reduced the number of common DEGs from 4924 to 1112. Eighteen DEGs related to chlorophyll were clustered. In the *ygl-1* gene mapping region with suppressed recombination, 43 common DEGs were identified. Five of the genes were related to chloroplasts; the *Bo1g098440* and *Bo1g098630* genes were excluded by qRT-PCR. Hence, *Bo1g059170*, *Bo1g087310*, and *Bo1g094360* in the mapping interval may be tightly associated with the development of the yellow-green-leaf mutant phenotype. Further studies on these genes may reveal the molecular mechanism of yellow-green-leaf formation in *B. oleracea*.

Author Contributions: X.L. and H.Y. developed the F_2 and BC. populations and wrote and revised the manuscript. H.Y., F.H., Z.L., and X.L. isolated the samples, performed the marker assays, and analyzed the marker data. F.H., Z.F., L.Y., M.Z., H.L., Y.L., Z.L., and Y.Z. conceived the study and critically reviewed the manuscript. All authors read and approved the final manuscript.

Acknowledgments: This work was performed in the Key Laboratory of Biology and Genetic Improvement of Horticultural Crops, Ministry of Agriculture, Beijing 100081, People's Republic of China. The work reported here was performed in the Key Laboratory of Biology and Genetic Improvement of Horticultural Crops, Ministry of Agriculture, Beijing 100081, China.

References

1.   Carol, P.; Stevenson, D. Mutations in the *Arabidopsis* gene IMMUTANS cause a variegated phenotype by inactivating a chloroplast terminal oxidase assoeiated with phytoene desaturation. *Plant Cell.* **1999**, *11*, 57–68. [CrossRef] [PubMed]

2.   Svensson, J.T.; Crosatti, C. Transcriptome analysis of cold acclimation in barley albina and xantha mutants. *Plant Physiol.* **2006**, *141*, 257–270. [CrossRef] [PubMed]

3.   Zhu, L.; Zeng, X. Genetic characterisation and fine mapping of a chlorophyll-deficient mutant (*BnaC.ygl*) in Brassica napus. *Mol. Breed.* **2014**, *34*, 603–614. [CrossRef]

4.   Chen, H.; Cheng, Z. A knockdown mutation of YELLOW—GREEN LEAF2, blocks chlorophyll biosynthesis in rice. *Plant Cell Rep.* **2013**, *32*, 1855–1867. [CrossRef] [PubMed]

5.   Li, C.; Hu, Y. Mutation of FdC2 gene encoding a ferredoxin-like protein with C-terminal extension causes yellow-green-leaf phenotype in rice. *Plant Sci.* **2015**, *238*, 127–134. [CrossRef] [PubMed]

6.   Ma, X.; Sun, X. Map-based cloning and characterization of the novel yellow-green-leaf gene *ys83* in rice (*Oryza sativa*). *Plant Physiol. Biochem.* **2016**, *111*, 1–9. [CrossRef] [PubMed]

7.   Liu, X.; Yang, C. Genetics and fine mapping of a yellow-green-leaf gene (*ygl-1*) in cabbage (*Brassica oleracea* var. capitata L.). *Mol. Breed.* **2016**, *36*, 1–8. [CrossRef]

8.   Whitaker, T.W. Genetic and Chlorophyll Studies of a Yellow-Green Mutant in Muskmelon. *Plant Physiol.* **1952**, *27*, 263–268. [CrossRef] [PubMed]

9.   Zhong, X.M.; Sun, S.F. Research on photosynthetic physiology of a yellow-green mutant line in maize. *Photosynthetica* **2015**, *53*, 1–8. [CrossRef]

10. Yang, C.; Zhang, Y.Y. Photosynthetic Physiological Characteristics and Chloroplast Ultrastructure of Yellow Leaf Mutant YL-1 in Cabbage. *Acta Hortic. Sin.* **2014**, *41*, 1133–1144.

11. Ladygin, V.G. Spectral features and structure of chloroplasts under an early block of chlorophyll synthesis. *Biophysics* **2006**, *51*, 635–644. [CrossRef]

12. Runge, S.; Cleve, B.V.; Lebedev, N.; Armstrong, G.; Apel, K. Isolation and classification of chlorophyll-deficient xantha mutants of *Arabidopsis thaliana*. *Planta* **1995**, *197*, 490–500. [CrossRef] [PubMed]

13. Chantret, N.; Sourdille, P. Location and mapping of the powdery mildew resistance gene MIRE and detection of a resistance QTL by bulked segregant analysis (BSA) with microsatellites in wheat. *Theor. Appl. Genet.* **2000**, *100*, 1217–1224. [CrossRef]

14. Futschik, A.; Schlötterer, C. The next generation of molecular markers from massively parallel sequencing of pooled DNA samples. *Genetics* **2010**, *186*, 207–218. [CrossRef] [PubMed]

15. Zeng, F.; Yi, B.; Tu, J.; Fu, T. Identification of AFLP and SCAR markers linked to the male fertility restorer gene of pol, CMS (*Brassica napus* L.). *Euphytica* **2009**, *165*, 363–369. [CrossRef]

16. Wang, Y.; Thomas, C.E.; Dean, R.A. Genetic mapping of a Fusarium wilt resistance gene (*Fom-2*) in melon (*Cucumis melo* L.). *Mol. Breed.* **2000**, *6*, 379–389. [CrossRef]

17. Subudhi, P.K.; Borkakati, R.P.; Virmani, S.S.; Huang, N. Molecular mapping of a thermosensitive genetic male sterility gene in rice using bulked segregant analysis. *Genome* **1997**, *40*, 188–194. [CrossRef] [PubMed]

18. Cheema, K.K.; Grewal, N.K. A novel bacterial blight resistance gene from Oryza nivara mapped to 38 kb region on chromosome 4L and transferred to *Oryza sativa* L. *Gen. Res.* **2008**, *90*, 397–407. [CrossRef] [PubMed]

19. Mackay, I.J.; Caligari, P.D. Efficiencies of F_2 and backcross generations for bulked segregant analysis using dominant markers. *Crop Sci.* **2000**, *40*, 626–630. [CrossRef]

20. Zhang, G.; Guo, G.W. Deep RNA sequencing at single base-pair resolution reveals high complexity of the rice transcriptome. *Genome Res.* **2010**, *20*, 646–654. [CrossRef] [PubMed]

21. Song, H.K.; Hong, S.E. Deep RNA Sequencing Reveals Novel Cardiac Transcriptomic Signatures for Physiological and Pathological Hypertrophy. *PLoS ONE* **2012**, *7*, e35552. [CrossRef] [PubMed]

22. Jarvie, T.; Harkins, T. Transcriptome sequencing with the Genome Sequencer FLX system. *Nat. Methods* **2008**, *5*. [CrossRef]

23. Wang, R.; Sun, L. Bulk segregant RNA-seq reveals expression and positional candidate genes and allele-specific expression for disease resistance against enteric septicemia of catfish. *BMC Genomics* **2013**, *14*, 929–939. [CrossRef] [PubMed]

24. Ramirez-Gonzalez, R.H.; Segovia, V. RNA-Seq bulked segregant analysis enables the identification of high-resolution genetic markers for breeding in hexaploid wheat. *Plant Biotech. J.* **2014**, *13*, 613–624. [CrossRef] [PubMed]

25. Parkin, I.A.; Koh, C. Transcriptome and methylome profiling reveals relics of genome dominance in the mesopolyploid *Brassica oleracea*. *Genome Biol.* **2014**, *15*, R77. [CrossRef] [PubMed]

26. Kim, D.; Langmead, B.; Salzberg, S.L. HISAT: A fast spliced aligner with low memory requirements. *Nat. Methods* **2015**, *12*, 357–359. [CrossRef] [PubMed]

27. Anders, S.; Huber, W. Differential expression analysis for sequence count data. *Genome Biol.* **2010**, *11*, R106. [CrossRef] [PubMed]

28. Krzywinski, M.; Schein, J. Circos: An information aesthetic for comparative genomics. *Genome Res.* **2009**, *19*, 1639–1645. [CrossRef] [PubMed]

29. Ashburner, M.; Ashburner, M. Gene ontology: Tool for the unification of biology. *Nat. Genet.* **2000**, *25*, 25–29. [CrossRef] [PubMed]

30. Kanehisa, M.; Goto, S. The KEGG resource for deciphering the genome. *Nucleic Acids Res.* **2004**, *32*, D277–D280. [CrossRef] [PubMed]

31. Livak, K.J.; Schmittgen, T.D. Analysis of relative gene expression data using real-time quantitative PCR and the $2^{-\Delta\Delta CT}$ method. *Methods* **2001**, *25*, 402–408. [CrossRef] [PubMed]

32. Guo, J.; Zhang, Y. Transcriptome sequencing and de novo analysis of a recessive genic male sterile line in cabbage (*Brassica oleracea* L. var. capitata). *Mol. Breed.* **2016**, *36*, 117–119. [CrossRef]

33. Piippo, M.; Allahverdiyeva, Y. Chloroplast-mediated regulation of nuclear genes in Arabidopsis thaliana in the absence of light stress. *Physiol. Gen.* **2006**, *25*, 142–152. [CrossRef] [PubMed]

34. Sohani, M.M.; Schenk, P.M.; Schultz, C.J.; Schmidt, O. Phylogenetic and transcriptional analysis of a strictosidine synthase-like gene family in *Arabidopsis thaliana*, reveals involvement in plant defence responses. *Plant Biol.* **2009**, *11*, 105–117. [CrossRef] [PubMed]

35. Jyothi, T.; Duan, H.; Liu, L.; Schuler, M.A. Bicistronic and fused monocistronic transcripts are derived from adjacent loci in the Arabidopsis genome. *RNA* **2005**, *11*, 128–138.

36. Kim, S.; Kim, C.W.; Park, M.; Choi, D. Identification of candidate genes associated with fertility restoration of cytoplasmic male-sterility in onion (*Allium cepa* L.) using a combination of bulked segregant analysis and RNA-seq. *Theor. Appl. Genet.* **2015**, *128*, 2289–2299. [CrossRef] [PubMed]

37. Liu, C.; Zhou, Q.; Dong, L.; Wang, H.; Liu, F.; Weng, J.; Li, X.; Xie, C. Genetic architecture of the maize kernel row number revealed by combining QTL mapping using a high-density genetic map and bulked segregant RNA sequencing. *BMC Genomics* **2016**, *17*, 915. [CrossRef] [PubMed]

38. Huang, Z.; Peng, G.; Liu, X.; Deora, A.; Falk, K.C.; Gossen, B.D.; McDonald, M.R.; Yu, F. Fine Mapping of a Clubroot Resistance Gene in Chinese Cabbage Using SNP Markers Identified from Bulked Segregant RNA Sequencing. *Front. Plant Sci.* **2017**, *8*, 1448–1459. [CrossRef] [PubMed]

39. Wang, Y.; Xie, J.; Zhang, H. Mapping stripe rust resistance gene YrZH22, in Chinese wheat cultivar Zhoumai 22 by bulked segregant RNA-Seq (BSR-Seq) and comparative genomics analyses. *Theor. Appl. Genet.* **2017**, *130*, 2191–2201. [CrossRef] [PubMed]

40. Trick, M.; Adamski, N.M.; Mugford, S.G.; Jiang, C.C.; Febrer, M.; Uauy, C. Combining SNP discovery from next-generation sequencing data with bulked segregant analysis (BSA) to fine-map genes in polyploid wheat. *BMC Plant Biol.* **2012**, *12*, 14–19. [CrossRef] [PubMed]

41. Lei, S.; Yao, X. Towards map-based cloning: Fine mapping of a recessive genic male-sterile gene (*BnMs2*) in *Brassica napus* L. and syntenic region identification based on the *Arabidopsis thaliana* genome sequences. *Theor. Appl. Genet.* **2007**, *115*, 643–651. [CrossRef] [PubMed]

42. Shimizu, M.; Pu, Z.J. Map-based cloning of a candidate gene conferring Fusarium yellows resistance in *Brassica oleracea*. *Theor. Appl. Genet.* **2015**, *128*, 119–130. [CrossRef] [PubMed]

43. Liang, J.; Ma, Y. Map-based cloning of the dominant genic male sterile *Ms-cd1* gene in cabbage (*Brassica oleracea*). *Theor. Appl. Genet.* **2016**, *5*, 1–9. [CrossRef] [PubMed]

44. Tian, Y.; Huang, Q. Inheritance of chlorophyll-deficient mutant *L638-y* in *Brassica juncea* L. and molecular markers for chlorophyll deficient gene *gr1*. *J. Northwest A F Univ.* **2012**, *12*, 17–19.

74   Plant Improvement: Molecular Breeding and Genetic Perspectives

45. Wang, Y.; He, Y. Fine mapping of a dominant gene conferring chlorophyll-deficiency in *Brassica napus*. *Sci. Rep.* **2016**, *6*, 314–319. [CrossRef] [PubMed]

46. Sherman, J.D.; Stack, S.M. Two-dimensional spreads of synaptonemal complexes from Solanaceous plants. VI. Highresolution recombination nodule map for tomato (*Lycopersicon esculentum*). *Genetics* **1995**, *141*, 683–708. [PubMed]

47. Wei, F.; Gobelman-Werner, K. The Mla (powdery mildew) resistance cluster is associated with three NBS-LRR gene families and suppressed recombination within a 240-kb DNA interval on chromosome 5S (1HS) of barley. *Genetics* **1999**, *153*, 1929–1948. [PubMed]

48. Ten, H.R.; Robbins, T.P. Localization of T-DNA Insertions in Petunia by Fluorescence in Situ Hybridization: Physical Evidence for Suppression of Recombination. *Plant Cell* **1996**, *8*, 823–830.

49. Stirling, B.; Newcombe, G. Suppressed recombination around the MXC3 locus, a major gene for resistance to poplar leaf rust. *Theor. Appl. Genet.* **2001**, *103*, 1129–1137. [CrossRef]

50. Neu, C.; Stein, N.; Keller, B. Genetic mapping of the Lr20-Pm1 resistance locus reveals suppressed recombination on chromosome arm 7AL in hexaploid wheat. *Genome* **2002**, *45*, 737–744. [CrossRef] [PubMed]

51. Jessup, R.W.; Burson, B.L. Disomic Inheritance, Suppressed Recombination, and Allelic Interactions Govern Apospory in Buffelgrass as Revealed by Genome Mapping. *Crop Sci.* **2002**, *42*, 1688–1694. [CrossRef]

52. Lee, J.; Izzah, N.K. Genotyping-by-sequencing map permits identification of clubroot resistance QTLs and revision of the reference genome assembly in cabbage (*Brassica oleracea* L.). *DNA Res.* **2015**, *14*, S113. [CrossRef] [PubMed]

53. Lv, H.; Wang, Q. Whole-Genome Mapping Reveals Novel QTL Clusters Associated with Main Agronomic Traits of Cabbage (*Brassica oleracea* var. *capitata* L.). *Front. Plant Sci.* **2016**, *7*, 989–999. [CrossRef] [PubMed]

54. Schweiger, W.; Schweiger, W.; Steiner, B.; Vautrin, S.; Nussbaumer, T.; Siegwart, G.; Zamini, M.; Jungreithmeier, F.; Gratl, V.; Lemmens, M.; et al. Suppressed recombination and unique candidate genes in the divergent haplotype encoding *Fhb1*, a major Fusarium head blight resistance locus in wheat. *Theor. Appl. Genet.* **2016**, *129*, 1607–1623. [CrossRef] [PubMed]

55. Lv, W.; Du, B. BAC and RNA sequencing reveal the brown planthopper resistance gene *BPH15*, in a recombination cold spot that mediates a unique defense mechanism. *BMC Genomics* **2014**, *15*, 674–679. [CrossRef] [PubMed]

56. Liu, S.; Liu, Y.; Yang, X.; Tong, C.; Edwards, D.; Parkin, I.A. The *Brassica oleracea* genome reveals the asymmetrical evolution of polyploid genomes. *Nat. Commun.* **2014**. [CrossRef] [PubMed]

57. Liu, X.P.; Gao, B.Z.; Han, F.Q.; Fang, Z.Y.; Yang, L.M.; Zhuang, M.; Lv, H.H.; Liu, Y.M.; Li, Z.S.; Cai, C.C.; et al. Genetics and fine mapping of a purple leaf gene, BoPr, in ornamental kale (*Brassica oleracea* L. var. acephala). *BMC Genomics* **2017**, *18*, 230–239. [CrossRef] [PubMed]

5

Chrysanthemum *DgWRKY2* Gene Enhances Tolerance to Salt Stress in Transgenic Chrysanthemum

Ling He, Yin-Huan Wu, Qian Zhao, Bei Wang, Qing-Lin Liu * and Lei Zhang

Department of Ornamental Horticulture, Sichuan Agricultural University, 211 Huimin Road, Wenjiang District, Chengdu 611130, Sichuan, China; heling@stu.sicau.edu.cn (L.H.); s20141825@sicau.edu.cn (Y.-H.W.); s20167109@stu.sicau.edu.cn (Q.Z.); s20167108@stu.sicau.edu.cn (B.W.); 14069@sicau.edu.cn (L.Z.)
* Correspondence: 13854@sicau.edu.cn

Abstract: WRKY transcription factors (TFs) play a vital part in coping with different stresses. In this study, *DgWRKY2* was isolated from *Dendranthema grandiflorum*. The gene encodes a 325 amino acid protein, belonging to the group II WRKY family, and contains one typical WRKY domain (WRKYGQK) and a zinc finger motif (C-X4-5-C-X22-23-H-X1-H). Overexpression of *DgWRKY2* in chrysanthemum enhanced tolerance to high-salt stress compared to the wild type (WT). In addition, the activities of antioxidant enzymes (superoxide dismutase (SOD), peroxidase (POD), catalase (*CAT*)), proline content, soluble sugar content, soluble protein content, and chlorophyll content of transgenic chrysanthemum, as well as the survival rate of the transgenic lines, were on average higher than that of the WT. On the contrary, hydrogen peroxide (H_2O_2), superoxide anion ($O_2{}^-$), and malondialdehyde (MDA) accumulation decreased compared to WT. Expression of the stress-related genes *DgCAT*, *DgAPX*, *DgZnSOD*, *DgP5CS*, *DgDREB1A*, and *DgDREB2A* was increased in the *DgWRKY2* transgenic chrysanthemum compared with their expression in the WT. In conclusion, our results indicate that *DgWRKY2* confers salt tolerance to transgenic chrysanthemum by enhancing antioxidant and osmotic adjustment. Therefore, this study suggests that *DgWRKY2* could be used as a reserve gene for salt-tolerant plant breeding.

Keywords: transgenic chrysanthemum; WRKY transcription factor; salt stress; gene expression; *DgWRKY2*

1. Introduction

High-salt stress is one of the most important factors that seriously affects and inhibits the growth and yield of plants [1]. Environmental stresses affect plant growth, causing plants to evolve mechanisms to face these challenges [2]. Under salt stress, transcription factors (TFs) can regulate the expression of multiple stress-related genes, which enhance tolerance to salt compared with the activity of a functional gene [3]. These genes are involved in the salt stress response in plants, forming a complex regulatory network [4]. Therefore, by using transcription factors, the plants' resistance can be improved.

WRKYs are a massive TF family, dominating the genetic transcription of plants. WRKY was named after the highly conserved sequence motif WRKYGQK. The WRKY proteins are divided into 3 types: class I contains two conserved WRKY domains and a zinc finger structure C-X4-5-C-X22-23-H-X1-H; class II contains one conserved WRKY domain and the same zinc finger structure, and most of the WRKY proteins found to date are this type; there is one conserved domain in class III, the zinc finger structure C-X7-C-X22-23-H-X1-C [5]. Overexpression of genes is a commonly used method to study gene function. Many studies had shown that WRKY TFs played a vital role in the physiological processes of plants [6–9]. It has also been proved that overexpression of some WRKY genes successfully increase plant tolerance to abiotic stress. *TaWRKY93* may increase salinity tolerance by enhancing osmotic adjustment, maintaining membrane stability, and increasing transcription of stress-related genes [10]. During salt treatment, *NbWRKY79* enhanced the tolerance of the transgenic plants to oxidant stress. Therefore, it increased the salt tolerance of *Nicotiana benthamiana* [11]. *RtWRKY1* conferred tolerance to salt stress in transgenic *Arabidopsis* by regulating plant growth, osmotic balance, Na$^+$/K$^+$ homeostasis, and the antioxidant system [12]. *VvWRKY30* increased salt resistance by regulating reactive oxygen species (ROS)-scavenging activity and the accumulation of osmoticum [13].

Physiological traits are important indicative indexes of botanical abiotic resistance. Plants produce ROS in the body under environmental pressures, including accumulation of superoxide anions (O$_2^-$), hydroxyl ions (OH$^-$), hydroxyl radicals (-OH), hydrogen peroxide (H$_2$O$_2$), and other types. These species not only lead to membrane lipid peroxidation of plant cells, affecting the redox state of the protein, but also cause oxidative damage to nucleic acids [14]. The plant antioxidant defense system consists of a variety of enzymes (superoxide dismutase (SOD), peroxidase (POD), catalase (*CAT*), ascorbate peroxidase (*APX*), etc.), which act as active oxygen scavengers in plants [15]. Much evidence has shown that the production and removal of ROS are closely related to the mechanism of salt tolerance [16,17]. As penetrating agents, soluble sugar, soluble protein, and proline maintain osmotic balance together. Under salt stress, transcription factors may participate in the regulation of the expression of many salt tolerance-related functional genes, so as to obtain stronger stress resistance than can be imparted by functional genes. The key genes encoding antioxidant enzymes (*Cu/ZnSOD*, *CAT*, *APX*, etc.) can increase the efficiency of ROS elimination in plant cells, so much so that the plant's tolerance to abiotic stresses is improved [18–20]. The proline synthase gene (*P5CS*) can effectively increase the tolerance of transgenic plants to osmotic stress [21]. The DREB (dehydration-responsive element binding proteins) transcription factor can specifically bind to the DRE *cis*-acting element or the core sequence with the DRE element (CCGAC), regulate the expression of stress-related genes, and mediate the transmission of abiotic stress signals [22–25].

Chrysanthemums are cut flower with high economic benefits and appreciable value, but it is sensitive to salinity, which can cause slow growth, plant chlorosis, and even death [26]. In a previous study, we obtained a database of the chrysanthemum transcriptome in response to salinity conditions by using high-throughput sequencing [27]. A large number of salt-induced transcripts were found in the data, especially from the WRKY family. Previously, we identified four WRKY genes (*DgWRKY1*, *DgWRKY3*, *DgWRKY4*, and *DgWRKY5*) and demonstrated that they can increase the salt tolerance of tobacco or chrysanthemum [28–31]. In order to analyze the WRKY family in chrysanthemum from multiple angles to complement our information, the salt stress-related gene *DgWRKY2* was isolated from chrysanthemum. This study investigated the importance of *DgWRKY2* as a transcription regulator under salt stress.

2. Results

2.1. Isolation and Characterization of DgWRKY2

DgWRKY2 contained a complete open reading frame of 1107 bp, which encoded a protein of 325 amino acids with a calculated molecular mass of 36.55 kDa. The theoretical isoelectric point is PI = 6.66 (Figure 1). Multi-sequence alignment analysis of the amino acid sequences of *DgWRKY2* and eight other genes showed that *DgWRKY2* contains a WRKY domain and a zinc finger structure (C-X4-5-C-X22-23-H-X1-H). It was further confirmed that the cloned cDNA sequence was a WRKY transcription factor II family member (Figure 2). Phylogenetic analysis showed that *DgWRKY2* is most closely related to *AtWRKY28* from *Arabidopsis thaliana* (Figure 3).

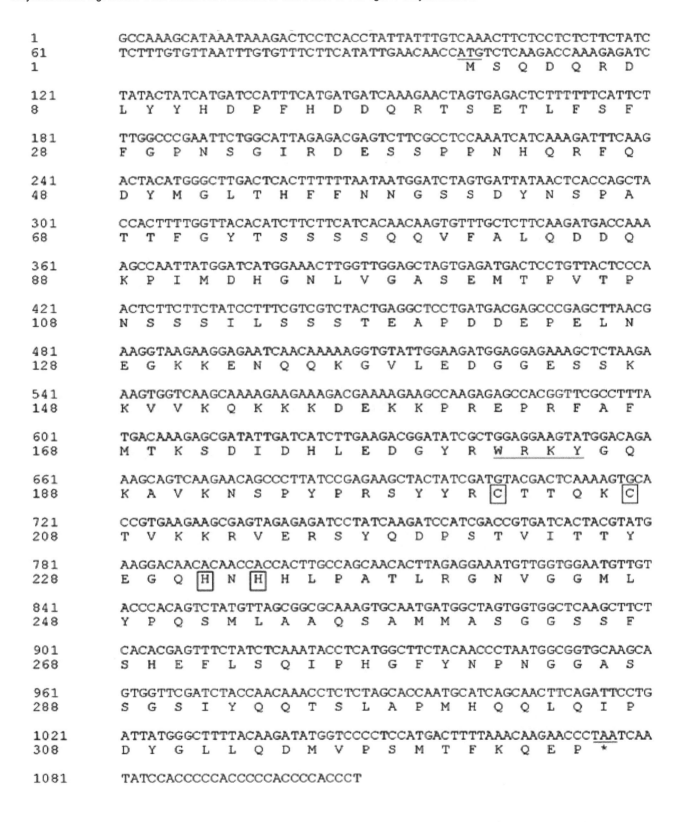

Figure 1. The nucleotide sequence and the deduced amino acid sequence of *DgWRKY2*. The WRKY domain is underlined. The cysteine and histidine in the zinc-finger motifs are boxed.

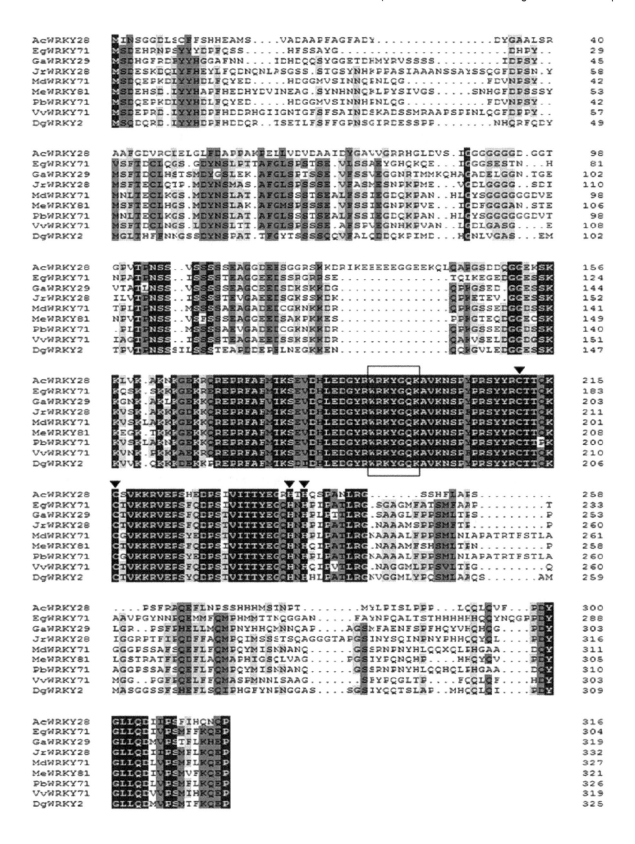

Figure 2. Comparison between the amino acid sequences deduced for the *DgWRKY2* gene. Amino acid residues conserved in all sequences are shaded in black, and those conserved in four sequences are shaded in light gray. The completely conserved WRKYGQK amino acids are boxed. The cysteine and histidine in zinc finger motifs are indicated by arrowheads (▼).

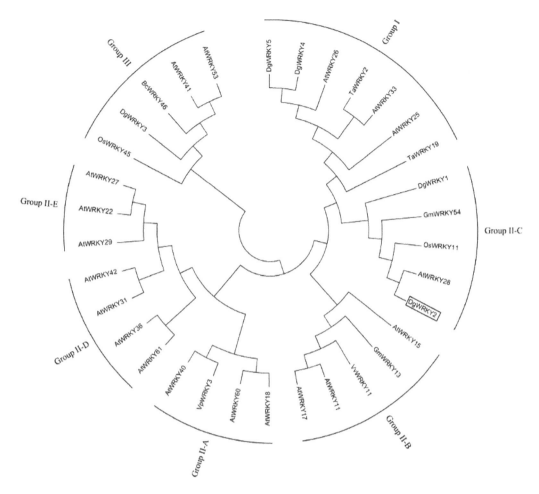

Figure 3. Phylogenetic tree analysis of the WRKY protein in different plants. The phylogenetic tree was drawn using the MEGA 5.0 program with the neighbor-joining method. *DgWRKY2* is boxed. The plant WRKY proteins used for the phylogenetic tree are as follows: *DgWRKY1* (KC153303), *DgWRKY3* (KC292215), *DgWRKY4*, *DgWRKY5* from *Dendranthema grandiflorum*; *AtWRKY11* (NP_849559), *AtWRKY15* (NP_179913.1), *AtWRKY17* (NP_565574.1), *AtWRKY18* (NP_567882), *AtWRKY22* (AEE81999), *AtWRKY25* (NP_180584), *AtWRKY26* (AAK28309), *AtWRKY27* (NP_568777), *AtWRKY28* (NP_193551), *AtWRKY29* (AEE84774), *AtWRKY31* (NP_567644), *AtWRKY33* (NP_181381), *AtWRKY36* (NP_564976), *AtWRKY40* (NP_178199), *AtWRKY41* (NP_192845), *AtWRKY42* (NP_192354), *AtWRKY53* (NP_194112), *AtWRKY60* (NP_180072), *AtWRKY61* (NP_173320) from *Arabidopsis thaliana*. *TaWRKY2* (EU665425), *TaWRKY19* (EU665430) from *Triticicum aestivum*. *GmWRKY13* (DQ322694), *GmWRKY54* (DQ322698) from *Glycine max*. *OsWRKY11* (AK108745), *OsWRKY45* (AY870611) from *Oryza sativa*. *VvWRKY11* (EC935078) from *Vitis vinifera*. *VpWRKY3* (JF500755) from *Vitis pseudoreticulata*. *BcWRKY46* (HM585284) from *Brassica campestris*.

2.2. Salt-Tolerance Analysis of DgWRKY2 Transgenic Chrysanthemum

To determine whether *DgWRKY2* overexpression enhanced salt tolerance, chrysanthemum transgenic lines with overexpressed *DgWRKY2* were produced by *Agrobacterium*-mediated transformation. *DgWRKY2* transcription levels in up in five transgenic lines (OE-3, OE-11, OE-17, OE-21 and OE-24) were detected by qRT-PCR (Figure 4A). We compared the salt stress tolerance between OE-17 and OE-21 transgenic chrysanthemum and the WT. Under normal growth conditions, the phenotypic differences were not significant. The growth rate was consistent. By contrast, under salt stress, wilting and yellowing of leaves of the WT plants were evident (Figure 4B). After the recovery period (2 weeks), the survival rate in the WT was 40.23%, while the survival rates in transgenic lines

OE-17 and OE-21 were 79.07% and 82.60%, respectively (Figure 4C). The survival rate of transgenic chrysanthemums was significantly higher than that of the WT.

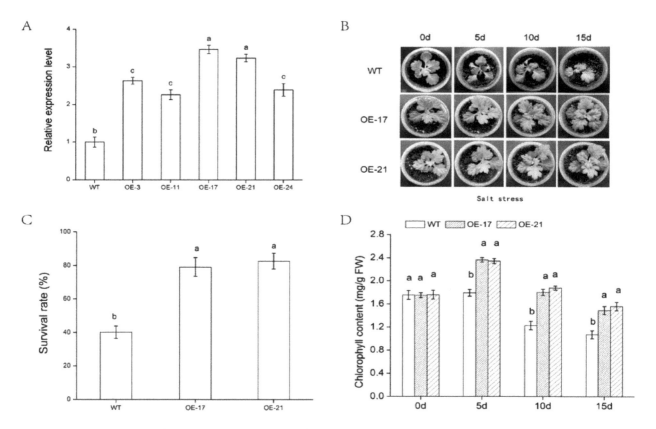

Figure 4. Expression of chrysanthemum in different strains under salt stress. (**A**) Relative expression level of *DgWRKY2* in transgenic chrysanthemums. The different normal letters indicate a significant difference at the 0.05 level among different strain lines, the same below; (**B**) comparison of transgenic plants and wild type plants after different periods under salt stress; (**C**) chrysanthemum survival statistics after recovery; (**D**) chlorophyll contents of chrysanthemum leave under salt stress.

2.3. Analysis of Chlorophyll Content and under Salt Stress

Salt stress significantly inhibited plant photosynthesis [32]. The content of chlorophyll in the leaves of the WT decreased obviously at the 10th day, while reaching the minimum value at the 15th day. However, the chlorophyll content from the transgenic chrysanthemum lines OE-17 and OE-21 increased significantly, by 35% and 33% at the 5th day, and decreased gradually later on. In general, the decrease of chlorophyll content in transgenic chrysanthemum is lower than that of the WT (Figure 4D).

2.4. Accumulation of H_2O_2, O_2^-, and MDA in DgWRKY2 Transgenic Chrysanthemum under Salt Stress

Reactive oxygen species in plant cells have a strong toxic effect. In order to study the effect of transgenic lines on the scavenging of reactive oxygen species, H_2O_2 and O_2^- in different lines were investigated with DAB and NBT staining. Under normal circumstances, there was no significant difference in H_2O_2 and O_2^- between the WT and two transgenic lines. After treatment with salt stress, the H_2O_2 content in each line increased significantly (Figure 5A,B). The contents of O_2^- showed an upward trend with the increase of stress time (Figure 5C,D), but it was not as obvious as that of H_2O_2. Under salt stress, despite the rising trend, the accumulation of H_2O_2 and O_2^- in the transgenic lines was much lower than that of the WT. These results indicate that the overexpression of *DgWRKY2* might regulate the activity of antioxidant protective enzymes, conferring greater tolerance to salt stress in transgenic plants. Similarly, under salt stress, the MDA accumulation level of overexpressed lines was apparently lower than that of the WT (Figure 5E). In all, these results provided strong evidence that the accumulation of ROS in *DgWRKY2* overexpression chrysanthemum was lower than that of WT under salt stress. Thus, *DgWRKY2* overexpression reduced the ROS level and alleviated the oxidant damage under salt stress.

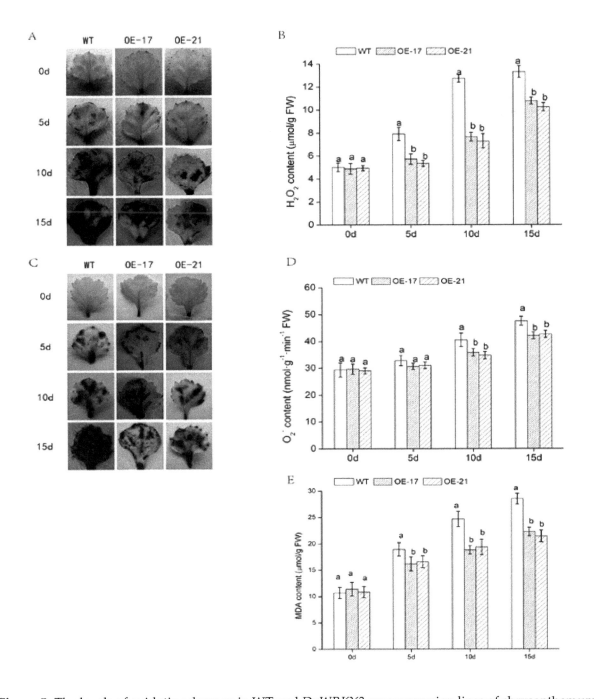

Figure 5. The levels of oxidative damage in WT and *DgWRKY2* overexpression lines of chrysanthemum were analyzed. (**A**) Diaminobenzidine (DAB) staining of chrysanthemum leaves during salt stress treatment; (**B**) changes in H_2O_2 content under salt stress; (**C**) nitroblue tetrazolium (NBT) staining of chrysanthemum leaves during salt stress treatment; (**D**) changes in O_2^- content under salt stress; (**E**) changes in malondialdehyde (MDA) content of chrysanthemum leaves under salt stress. Data represent means and standard errors of three replicates. The different letters above the columns indicate significant differences ($p < 0.05$) according to Duncan's multiple range test.

2.5. Physiological Changes in DgWRKY2 Transgenic Chrysanthemum

Antioxidant enzymes play an important part in botanical stress tolerance. We observed activities of SOD, POD, and *CAT* in the leaves of *DgWRKY2* lines and WT plants at different stages of treatment. Under normal growth conditions, the activities of these three enzymes had no obvious differences in any of the lines. Under salt treatment conditions, there was an increase in the WT and overexpressed lines. Moreover, compared with WT, these increases were extraordinarily greater in the overexpressed lines (Figure 6A–C). As a result, overexpression of *DgWRKY2* increases the antioxidant enzyme activity of transgenic chrysanthemum to counteract injury from ROS. Thus, this reduced oxidative damage.

Osmotic adjustment is one of the most basic characteristics of plant salt tolerance, while proline is the most widely distributed compatible penetrant [33,34]. Under salt stress, we measured the proline content of transgenic lines and the WT in order to understand the osmoregulation ability of transgenic plants (Figure 7A). There was little difference in proline content between transgenic lines and WT under normal circumstances. By contrast, under salt stress, there was a remarkable increase in proline content for both. Nevertheless, the accumulation of proline in the transgenic lines was significantly higher than that of the WT under salt stress. These results indicate that *DgWRKY2* upregulated the accumulation of proline in the transgenic lines under salt stress.

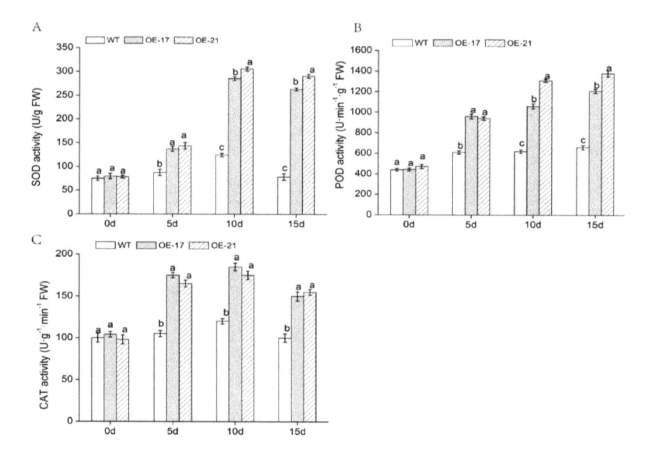

Figure 6. Changes in antioxidant enzyme activities of chrysanthemum leaves under salt stress. (**A**) Superoxide dismutase (SOD) activity under salt stress; (**B**) peroxidase (POD) activity under salt stress; (**C**) catalase (*CAT*) activity under salt stress. Data represent means and standard errors of three replicates. The different letters above the columns indicate significant differences ($p < 0.05$) according to Duncan's multiple range test.

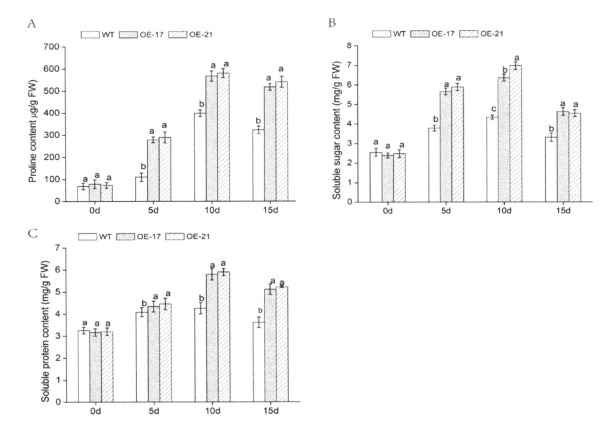

Figure 7. Changes in contents of osmotic adjustment substances of chrysanthemum leaves under salt stress. (**A**) Proline content under salt stress; (**B**) soluble sugar content under salt stress; (**C**) soluble protein content under salt stress. The different letters above the columns indicate significant differences ($p < 0.05$) according to Duncan's multiple range test.

Soluble proteins keep cells appropriately permeable and protect cells from dehydration, while stabilizing and protecting the structure and function of biological macromolecules [35]. We observed the content of soluble protein and of soluble sugar of these three lines under salt stress. In this environment, soluble protein and soluble sugar content of overexpressed lines increased significantly compared with the WT. (Figure 7B,C). The above data suggest that overexpression of *DgWRKY2* enhanced the osmoregulation ability of transgenic chrysanthemum while it increased its salt tolerance.

2.6. Expression of Abiotic Stress-Related Genes in DgWRKY2 Transformed Chrysanthemum

In order to reveal the signal regulatory network of transgenic lines in the stress resistance process, we measured the expression of several functional genes involved in signal transduction pathways by qRT-PCR. Under standard circumstances, there was little difference in the expression of abiotic stress-response genes. When exposed to salt stress, the expression level of the gene encoding ROS-scavenging enzymes (*CAT*, *APX*, and *Cu/ZnSOD*) in the transgenic lines was much higher than in the WT (Figure 8A–C). Additionally, *P5CS*, a gene related to proline synthase, showed an expression level with a similar trend (Figure 8D). Furthermore, other genes, such as *DREB1A* and *DREB2A*, that are closely related to plant responses to environmental stresses, were all significantly upregulated in the overexpressed lines compared to the WT under salinity conditions (Figure 8E,F). Our data suggest that *DgWRKY2* overexpression could reduce osmotic pressure by clearing excess ROS and accumulating proline, thereby promoting salt tolerance.

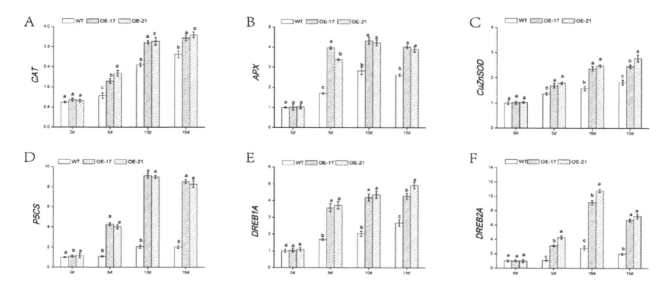

Figure 8. Expression of stress-related genes in wild type (WT) and overexpressed lines. (**A**) Expression analysis of *Cu/ZnSOD* under salt stress; (**B**) expression analysis of *CAT* under salt stress; (**C**) expression analysis of ascorbate peroxidase (*APX*) under salt stress; (**D**) expression analysis of *P5CS* in chrysanthemum under salt stress; (**E**) expression analysis of *DREB1A* under salt stress; (**F**) expression analysis of *DREB2A* under salt stress. Data represent means and standard errors of three replicates. The different letters above the columns indicate significant differences ($p < 0.05$) according to Duncan's multiple range test.

3. Discussion

To date, the WRKY gene has been cloned from *Arabidopsis thaliana* [36], wheat [37,38], rice [39], soybean [40], chrysanthemum [28,29], birch [41], and other plants. It was confirmed that the WRKY gene is related to plant stress resistance. We isolated a new WRKY transcription factor—*DgWRKY2*—from chrysanthemum, and found it to be induced by salt stress. The deduced amino acid sequence of the *DgWRKY2* gene from this study contains one WRKY domain (WRKYGQK) and a zinc finger structure (C-X4-5-C-X22-23-H-X1-H), which could be considered part of the group II WRKY family.

The same group of WRKY proteins might have similar capabilities. Previous studies have shown that *GmWRKY54* might improve the salt and cold tolerance of plants through the regulation of *DREB2A* and STZ/Zat10 [40]. *OsWRKY11* overexpression increased rice drought tolerance [42]. The expression of *AtWRKY28* changed significantly under NaCl stress, indicating that *AtWRKY28* had much to do with adaptation to environmental stress [43]. In our previous study, an overexpression *DgWRKY1* tobacco line was more tolerant to salt stress than the WT [28]. *DgWRKY2* belongs to group II with *GmWRKY54*, *OsWRKY11*, *AtWRKY28*, and *DgWRKY1*, thus, we hypothesized that *DgWRKY2* has a positive effect on salt tolerance. In addition, the previous studies demonstrated that *DgWRKY3*, *DgWRKY4*, and *DgWRKY5* also played a positive regulatory role on salt stress [29–31]. *DgWRKY1* and *DgWRKY3* were only studied for their role in salt tolerance in tobacco, and the salt tolerance in chrysanthemum has yet to be studied. Previous studies have confirmed that *DgWRKY4* and *DgWRKY5* belong to the group III, and *DgWRKY2* in this study belongs to group II. The results of these studies showed that *DgWRKY4* and *DgWRKY5* imparted stronger salt tolerance than *DgWRKY2*. This is partly due to different groups playing different roles in the stress regulatory network. Additional work is needed to understand the mechanisms.

In this study, the *DgWRKY2* overexpression transgenic chrysanthemum was compared with the WT from physiological and biochemical aspects, and the function of *DgWRKY2* overexpression was verified. Chlorophyll content in chrysanthemum leaves continued to decrease in the late stage of salt stress. We speculated that ROS inhibited the photosynthesis of chrysanthemum [44]. However, chlorophyll content in the overexpressed lines was higher than that of the WT at respectively different salt stress stages. Increased ROS activity causes a great deal of physiological and metabolic changes in plants, enabling them to cope with environmental stress. In *CmWRKY17*-overexpressing plants, *CmWRKY17* altered the salinity sensitivity via regulation of ROS levels [45]. *NbWRKY79* was involved with the regulation of SOD, POD, *CAT*, and *APX* activities, which resulted in the suppression of ROS accumulation so that the plant could endure less oxidative damage under salt conditions [11]. *MsWRKY11* might reduce ROS levels and thus increase salt tolerance in soybean [46]. The activity of antioxidant enzymes SOD, POD, and *CAT* in *DgWRKY2* overexpression lines increased, and the activity of the enzymes was significantly higher than that of the WT at each stage of salt treatment. Moreover, the content of H_2O_2 and O_2^- in transgenic chrysanthemum leaves was also lower than that of WT. The above results indicate that *DgWRKY2* overexpression could enhance plant antioxidant capacity by increasing the activities of SOD, POD, and *CAT*, thereby enhancing the salt tolerance of transgenic chrysanthemum.

Accumulation of MDA content can lead to membrane lipid peroxidation of plant cells, causing changes in the cell membrane structure and permeability, reducing cell function [47]. In contrast, proline prevents membrane lipid peroxidation, maintains normal cellular structure, and maintains a stable cell osmotic pressure [48]. Under salt treatment, compared with the WT, MDA content of the overexpressed lines was lower, but proline content was higher. The contents of soluble sugar and soluble protein in *DgWRKY2* overexpression lines were higher than those of WT. The results suggest that *DgWRKY2* might increase its salt tolerance by regulating the osmotic pressure of plant cells.

The expression of antioxidant genes (*Cu/ZnSOD*, *CAT*, and *APX*) was upregulated under salinity, which is consistent with physiological results. Under salt stress, the expression of antioxidant enzyme genes was significantly higher in *RtWRKY1*-overexpressed *Arabidopsis* than in the wild type [12]. The *P5CS* gene is associated with a proline-synthesizing enzyme in plants. When the expression of the *P5CS* gene was induced by environmental stress, the proline content in plants increased. Under salt stress, the expression of genes related to proline biosynthesis was upregulated in *VvWRKY30* transgenic lines compared with their expression in the WT [13]. These results show that transgenic plants exhibited increased expression levels of *P5CS* under stress conditions. The DREB gene belongs to the *AP2/EREBP* transcription factor family. These TFs are closely related to the response of plants to the environment [49,50]. In this study, *DREB1A* was upregulated to a greater extent in overexpressed lines than in WT, and *DREB2A* first increased and later decreased. Previous studies indicated that *OsDREB2A* might participate in abiotic stress by directly binding the DREB element to regulate the expression of downstream genes. Overexpression of *OsDREB2A* in soybean might be used to improve its tolerance to salt stress [51]. Cong found that overexpression of the *OjDREB* gene improved salt tolerance in tobacco plant [52]. These results suggest that enhanced salt tolerance was associated with the induction of downstream stress-related gene expression in *DgWRKY2* transgenic plants.

4. Materials and Methods

4.1. Plant Materials

The experimental material used for treatment is a wild-type chrysanthemum: *Dendranthema grandiforum*—'Jinba'. All plant materials were provided by Sichuan Agricultural University, Chengdu, China. Chrysanthemum seedlings grew on MS culture medium (200 μL m^{-2} s^{-1}, 16 h photoperiod, 25 °C/22 °C day/night temperature, and 70% relative humidity) for 20 days. Then, 20-day-old seedlings were planted in basins filled with a 1:1 mixture of peat and perlite, incubated for 3 days, and watered once daily (70% of field capacity). Seedlings at the six-leaf stage were harvested, frozen in liquid nitrogen immediately, and stored at −80 °C for RNA extraction.

4.2. Cloning of DgWRKY2 and Sequence Analysis

The RNA extraction of chrysanthemum leaves was performed by TRNzol reagent (Mylab, Beijing, China). The full-length cDNA of the *DgWRKY2* sequence was obtained by PCR (polymerase chain reaction) utilizing gene-specific primers (Table 1). The RACE reactions were carried out according to the manufacturer's protocol (Invitrogen RACE cDNA amplification kit, Clontech, Mountain View, CA, USA). The fragment generated was cloned into pEASY-T1 Cloning Kit (Transgene Biotech, Beijing, China) and sequenced.

Table 1. Primers and their sequences in experiment.

Primer	Sequence (5′-3′)
DgWRKY2	F: ATTTGTCAAACTTCTCCTCTCTTCT
	R: GTGGGGGTGGGGGTGGATA
EF1a	F: TTTTGGTATCTGGTCCTGGAG
	R: CCATTCAAGCGACAGACTCA
Cu/Zn SOD	F: CCATTGTTGACAAGCAGATTCCACTCA
	R: ATCATCAGGATCAGCATGGACGACTAC
CAT	F: TACAAGCAACGCCCTTCAA
	R: GACCTCTGTTCCCAACAGTCA
APX	F: GTTGGCTGGTGTTGTTGCT
	R: GATGGTCGTTTCCCTTAGTTG
P5CS	F: TTGGAGCAGAGGTTGGAAT
	R: GCAGGTCTTTGTGGGTGTAG
DREB1A	F: CGGTTTTGGCTATGAGGGGT
	R: TTCTTCTGCCAGCGTCACAT
DREB2A	F: GATCGTGGCTGAGAGACTCG
	R: TACCCCACGTTCTTTGCCTC

The sequence of *DgWRKY2* was analyzed by the National Center for Biotechnology Information (NCBI, http://www.ncbi.nlm.nih.gov/gorf/gorf.html) to obtain its open reading frame (ORF). Identification of protein domains and significant sites was performed with Motifscan (http://myhits.isb-sib.ch/cgi-bin/motif_scan). The phylogenetic tree was drawn with the MEGA 5.0 program (Sudhir Kumar, Arizona State University, Tempe, AZ, USA) using the neighbor-joining method.

4.3. Generation of Transgenic Chrysanthemum

The pEASY-WRKY2 cloning vector was constructed by TA cloning technology (The complementarity between the vector 3′-T overhangs and PCR product 3′-A overhangs allows direct ligation of Taq-amplified PCR products into the T-vector). The plasmid containing the pEASY-WRKY2 and pBI121 expression vectors were double digested with *SacI* and *XbaI* to construct the *pBI121-DgWRKY2* expression vector. The fused construction of *pBI121-DgWRKY2* was transformed into the leaf disk of chrysanthemum by *Agrobacterium tumefaciens* (strain LBA4404) [53]. Callus induction from chrysanthemum was used to form seedlings [54]. The obtained *DgWRKY2* transgenic chrysanthemum lines (OE-17 and OE-21) were employed in subsequent experiments. The transgenic lines OE-17 and OE-21 were expanded for subsequent replication experiments.

4.4. Expression of DgWRKY2 under Salt Treatment

The method of RNA extraction is the same as above. Then RNA was used for first-strand cDNA synthesis with reverse transcriptase (TransScript II All-in-one First-Strand cDNA Synthesis SuperMix for PCR, Transgene, Beijing, China) according to the manufacturer's protocol. Quantitative real-time PCR (qRT-PCR) was performed by SsoFast EvaGreen supermix (Bio-Rad, Hercules, CA, USA) and Bio-Rad CFX96TM detection system. The gene elongation factor 1α (*EF1α*) was used as a reference for quantitative expression analysis. A final 20 μL qPCR reaction mixture contained: 10 μL SsoFast

EvaGreen supermix, 2 µL diluted cDNA sample, and 300 nM primers. Then, the reactions were incubated following the standard process: 1 cycle of 95 °C for 30 s, 40 cycles of 15 s at 95 °C and 30 s at 60 °C, and a single melting cycle from 65 to 95 °C. To avoid experimental errors, each reaction was repeated at least three times. To avoid variables and statistic error, a negative control group was set up, in which water supplanted the above solution. Relative expression levels were calculated by the $2^{-\Delta\Delta Ct}$ method [55].

4.5. Salt Treatment of Transgenic Chrysanthemum and Stress Tolerance Assays

Two overexpressed lines (OE-17 and OE-21) and the WT of chrysanthemum, all 20 days old, were sown into a 1:1 mixture of peat and perlite, then cultured in a light incubator (200 µL m^{-2} s^{-1}, 16 h photoperiod, 25 °C/22 °C day/night temperature, and 70% relative humidity). Soil-grown chrysanthemum seedlings at the six-leaf stage were irrigated with an increasing concentration of NaCl solution: 100 mm for 1–5 days (d), 200 mm for 6–10 days, and 400 mm for 11–15 days, using Chen as a reference [56]. Under salt stress, leaves 4–5 were harvested at 0, 5, 10, and 15 days for both physiological and molecular experiments. After a 2-week recovery, the surviving plants were collected to calculate the survival rate.

4.6. Determination of Physiological Indexes of Transgenic Chrysanthemum under Salt Stress

Activities of superoxide dismutase (SOD), peroxidase (POD), and catalase (CAT) were measured according to Li [57]. Malondialdehyde (MDA) content in chrysanthemum was measured according to Zhang [58]. Accumulation of proline, soluble sugar, and soluble protein was measured according to Sun [59]. The chlorophyll content was detected according to Jin [60].

4.7. Histochemical Detection of Reactive Oxygen Species (ROS)

Nitroblue tetrazolium (NBT) and diaminobenzidine (DAB) staining was measured according to Shi [61]. The standard steps were as follows: chrysanthemum leaves were completely immersed in 10 mm phosphate buffer (pH = 7.8) containing 1 mg/mL NBT or DAB at room temperature. The leaves were not placed in 95% ethanol for decolorization until the spots appeared. After that, the sample was observed, and photos of the sample were taken. Finally, H_2O_2 and O_2^- concentration were determined by detection kits (Nanjing Jiancheng Bioengineering Institute, Nanjing, China).

4.8. Expression of Salt Stress Response Genes in Dgwrky2 Transgenic Chrysanthemum

To evaluate the expression of abiotic stress-related genes, RNA from the WT and transgenic lines was extracted for reverse transcription. Transgenic chrysanthemum stress-responsive gene expression was detected by qRT-PCR. The abiotic stress-response genes monitored were Cu/ZnSOD, CAT, APX, P5CS, DREB1A, and DREB2A. All relevant primers used in the study are listed in Table 1.

4.9. Statistical Analysis

All experiments were performed three times for biological repetition to avoid all types of error. All data were analyzed by SPSS version 24.0 (International Business Machines Corporation, Armonk, NY, USA). A one-way analysis of variance, Tukey's multiple range test ($p < 0.05$), was employed to identify the treatment means to avoid static errors.

5. Conclusions

In summary, this study demonstrated that DgWRKY2 could positively regulate salt stress tolerance. To alleviate the damage of salt stress to plants, DgWRKY2 overexpression improved expression of stress-related genes, resulting in relatively enhanced photosynthetic capacity, greatly increased activities of antioxidant enzymes, and high accumulation of proline, soluble sugar, and soluble protein. This indicates that DgWRKY2 may enhance the sensitivity to salinity by enabling antioxidant and osmotic adjustment capabilities. Overall, this study identified *DgWRKY2* as a potential genetic resource for plant salt tolerance. Not only did *DgWRKY2* play an important role in supplementing and perfecting chrysanthemum-tolerant germplasm resources, but it could also be used as a reserved gene for salt-tolerant plant breeding.

Author Contributions: L.H., Y.-H.W. and Q.-L.L. conceived and designed the experiments. L.H., Y.-H.W., B.W., Q.-L.L. and Q.Z. performed the experiments. L.Z. and L.H. analyzed the data. L.H. wrote the paper. All authors read and approved the manuscript.

Acknowledgments: This study was supported by Sichuan Agricultural University Ornamental Horticulture lab. We would like to acknowledge the contribution of Qing-Lin Liu for the provision of experimental materials and instruments.

References

1.   Jamil, A.; Riaz, S.; Ashraf, M.; Foolad, M.R. Gene Expression Profiling of Plants under Salt Stress. *Crit. Rev. Plant Sci.* **2011**, *30*, 435–458. [CrossRef]

2.   Munns, R. Comparative physiology of salt and water stress. *Plant Cell Environ.* **2002**, *25*, 239–250. [CrossRef] [PubMed]

3.   Bing, L.; Zhao, B.C.; Shen, Y.Z.; Huang, Z.J.; Ge, R.C. Progress of Study on Salt Tolerance and Salt Tolerant Related Genes in Plant. *J. Hebei Norm. Univ.* **2008**, *2*, 243–248.

4.   Tuteja, N. Mechanisms of high salinity tolerance in plants. *Methods Enzymol.* **2007**, *428*, 419–438. [CrossRef] [PubMed]

5.   Eulgem, T.; Rushton, P.J.; Robatzek, S.; Somssich, I.E. The WRKY superfamily of plant transcription factors. *Trends Plant Sci.* **2000**, *5*, 199–206. [CrossRef]

6.   Zhang, Y.; Yu, H.; Yang, X.; Li, Q.; Ling, J.; Wang, H.; Gu, X.; Huang, S.; Jiang, W. CsWRKY46, a WRKY transcription factor from cucumber, confers cold resistance in transgenic-plant by regulating a set of cold-stress responsive genes in an ABA-dependent manner. *Plant Physiol. Biochem.* **2016**, *108*, 478. [CrossRef] [PubMed]

7.   Bakshi, M.; Oelmüller, R. WRKY transcription factors: Jack of many trades in plants. *Plant Signal. Behav.* **2014**, *9*, e27700. [CrossRef] [PubMed]

8.   Tripathi, P.; Rabara, R.C.; Rushton, P.J. A systems biology perspective on the role of WRKY transcription factors in drought responses in plants. *Planta* **2014**, *239*, 255–266. [CrossRef] [PubMed]

9.   Guo, Y.; Cai, Z.; Gan, S. Transcriptome of Arabidopsis leaf senescence. *Plant Cell Environ.* **2004**, *27*, 521–549. [CrossRef]

10.   Qin, Y.; Tian, Y.; Liu, X. A wheat salinity-induced WRKY transcription factor TaWRKY93 confers multiple abiotic stress tolerance in Arabidopsis thaliana. *Biochem. Biophys. Res. Commun.* **2015**, *464*, 428. [CrossRef] [PubMed]

11.   Nam, T.N.; Le, H.T.; Mai, D.S.; Tuan, N.V. Overexpression of NbWRKY79, enhances salt stress tolerance in Nicotiana benthamiana. *Acta Physiol. Plant.* **2017**, *39*, 121. [CrossRef]

12.   Du, C.; Zhao, P.; Zhang, H.; Li, N.; Zheng, L.; Wang, Y. The Reaumuria trigyna transcription factor RtWRKY1 confers tolerance to salt stress in transgenic Arabidopsis. *J. Plant Physiol.* **2017**, *215*, 48–58. [CrossRef] [PubMed]

13.   Zhu, D.; Hou, L.; Xiao, P.; Guo, Y.; Deyholos, M.K.; Liu, X. VvWRKY30, a grape WRKY transcription factor, plays a positive regulatory role under salinity stress. *Plant Sci.* **2018**. [CrossRef]

14.   Powers, S.K.; Lennon, S.L.; Quindry, J.; Mehta, J.L. Exercise and cardioprotection. *Curr. Opin. Cardiol.* **2002**, *17*, 495–502. [CrossRef] [PubMed]

15.   Jiang, M.; Zhang, J. Water stress-induced abscisic acid accumulation triggers the increased generation of reactive oxygen species and up-regulates the activities of antioxidant enzymes in maize leaves. *J. Exp. Bot.* **2002**, *53*, 2401–2410. [CrossRef] [PubMed]

16. Meloni, D.A.; Oliva, M.A.; Martinez, C.A.; Cambraia, J. Photosynthesis and activity of superoxide dismutase, peroxidase and glutathione reductase in cotton under salt stress. *Environ. Exp. Bot.* **2003**, *49*, 69–76. [CrossRef]

17. Moradi, F.; Ismail, A.M. Responses of photosynthesis, chlorophyll fluorescence and ROS-scavenging systems to salt stress during seedling and reproductive stages in rice. *Ann. Bot.* **2007**, *99*, 1161–1173. [CrossRef] [PubMed]

18. Negi, N.P.; Sharma, V.; Sarin, N.B. Pyramiding of Two Antioxidant Enzymes CuZnSOD and cAPX from Salt Tolerant Cell Lines of Arachis hypogeae Confers Drought Stress Tolerance in Nicotiana tabacum. *Indian J. Agric. Biochem.* **2017**, *30*, 141. [CrossRef]

19. Hui, Y.; Qiang, L.; Park, S.C.; Wang, X.; Liu, Y.J.; Zhang, Y.G.; Tang, W.; Kou, M.; Ma, D.F. Overexpression of CuZnSOD, and APX, enhance salt stress tolerance in sweet potato. *Plant Physiol. Biochem.* **2016**, *109*, 20–27. [CrossRef]

20. Yang, Z.; Zhou, Y.; Ge, L.; Li, G.; Liu, Q.; Xu, Y.; Jiang, L.; Yang, Y.; School of Agriculture, Jiangxi Agricultural University; School of Sciences, Jiangxi Agricultural University. Expression of Cucumber CsCAT3 Gene under Stress and Its Salt Tolerance in Transgenic Arabidopsis thaliana. *Mol. Plant Breed.* **2018**.

21. Guerzoni, J.T.S.; Belintani, N.G.; Moreira, R.M.P.; Hoshino, A.A.; Domingues, D.S.; Filho, J.C.B.; Vieira, L.G.E. Stress-induced Δ1-pyrroline-5-carboxylate synthetase (P5CS) gene confers tolerance to salt stress in transgenic sugarcane. *Acta Physiol. Plant.* **2014**, *36*, 2309–2319. [CrossRef]

22. Wang, W.; Vinocur, B.; Altman, A. Plant responses to drought, salinity and extreme temperatures: Towards genetic engineering for stress tolerance. *Planta* **2003**, *218*, 1–14. [CrossRef] [PubMed]

23. Qin, F.; Kakimoto, M.; Sakuma, Y.; Maruyama, K.; Osakabe, Y.; Tran, L.S.; Shinozaki, K.; Yamaguchi-Shinozaki, K. Regulation and functional analysis of ZmDREB2A in response to drought and heat stresses in *Zea mays* L. *Plant J.* **2010**, *50*, 54–69. [CrossRef] [PubMed]

24. Zhou, M.L.; Ma, J.T.; Zhao, Y.M.; Wei, Y.H.; Tang, Y.X.; Wu, Y.M. Improvement of drought and salt tolerance in Arabidopsis and Lotus corniculatus by overexpression of a novel DREB transcription factor from Populus euphratica. *Gene* **2012**, *506*, 10–17. [CrossRef] [PubMed]

25. Ma, J.T.; Yin, C.C.; Guo, Q.Q.; Zhou, M.L.; Wang, Z.L.; Wu, Y.M. A novel DREB transcription factor from Halimodendron halodendron, leads to enhance drought and salt tolerance in Arabidopsis. *Biol. Plant.* **2014**, *59*, 74–82. [CrossRef]

26. Akça, Y.; Samsunlu, E. The effect of salt stress on growth, chlorophyll content, proline and nutrient accumulation, and k/na ratio in walnut. *Am. Bank.* **2012**, *1999*, 1513–1520.

27. Wu, Y.H.; Wang, T.; Wang, K.; Liang, Q.Y.; Bai, Z.Y.; Liu, Q.L.; Jiang, B.B.; Zhang, L. Comparative Analysis of the Chrysanthemum Leaf Transcript Profiling in Response to Salt Stress. *PLoS ONE* **2016**, *11*, e0159721. [CrossRef] [PubMed]

28. Liu, Q.L.; Xu, K.D.; Pan, Y.Z.; Jiang, B.B.; Liu, G.L.; Jia, Y.; Zhang, H.Q. Functional Analysis of a Novel Chrysanthemum WRKY Transcription Factor Gene Involved in Salt Tolerance. *Plant Mol. Biol. Rep.* **2014**, *32*, 282–289. [CrossRef]

29. Liu, Q.L.; Zhong, M.; Li, S.; Pan, Y.Z.; Jiang, B.B.; Jia, Y.; Zhang, H.Q. Overexpression of a chrysanthemum transcription factor gene, DgWRKY3, intobacco enhances tolerance to salt stress. *Plant Physiol. Biochem.* **2013**, *69*, 27–33. [CrossRef] [PubMed]

30. Wang, K.; Wu, Y.H.; Tian, X.Q.; Bai, Z.Y.; Liang, Q.Y.; Liu, Q.L.; Pan, Y.Z.; Zhang, L.; Jiang, B.B. Overexpression of DgWRKY4 Enhances Salt Tolerance in Chrysanthemum Seedlings. *Front. Plant Sci.* **2017**, *8*, 1592. [CrossRef] [PubMed]

31. Liang, Q.Y.; Wu, Y.H.; Wang, K.; Bai, Z.Y.; Liu, Q.L.; Pan, Y.Z.; Zhang, L.; Jiang, B.B. Chrysanthemum WRKY gene DgWRKY5 enhances tolerance to salt stress in transgenic chrysanthemum. *Sci. Rep.* **2017**, *7*, 4799. [CrossRef] [PubMed]

32. Diao, M.; Ma, L.; Wang, J.; Cui, J.; Fu, A.; Liu, H. Selenium Promotes the Growth and Photosynthesis of Tomato Seedlings Under Salt Stress by Enhancing Chloroplast Antioxidant Defense System. *J. Plant Growth Regul.* **2014**, *33*, 671–682. [CrossRef]

33. Ben, K.R.; Abdelly, C.; Savouré, A. Proline, a multifunctional amino-acid involved in plant adaptation to environmental constraints. *Biol. Aujourdhui* **2012**, *206*, 291. [CrossRef]

34. Chaleff, R.S. Further characterization of picloram tolerant mutance of *Nicotinana tabacum*. *Theor. Appl. Genet.* **1980**, *58*, 91–95. [CrossRef]

35. Wang, F.; Liu, P.; Zhu, J. Effect of magnesium (Mg) on contents of free proline, soluble sugar and protein in soybean leaves. *J. Henan Agric. Sci.* **2004**, *6*, 35–38.

36. Fu, Q.T.; Yu, D.Q. Expression profiles of AtWRKY25, AtWRKY26 and AtWRKY33 under abiotic stresses. *Hereditas* **2010**, *32*, 848–856. [CrossRef] [PubMed]

37. Qin, Y.X. Salt-Tolerant Drought-Tolerant Wheat Gene TaWRKY79 and Application Thereof. CN 102703465A, 3 October 2012.

38. Tian, Y.C.; Qin, Y.X. Wheat Salt-Tolerant and Drought-Resistant Gene TaWRKY80 and Application Thereof. CN 102703466B, 3 July 2013.

39. Wang, H.; Hao, J.; Chen, X.; Hao, Z.; Wang, X.; Lou, Y.; Peng, Y.; Guo, Z. Overexpression of rice WRKY89 enhances ultraviolet B tolerance and disease resistance in rice plants. *Plant Mol. Biol.* **2007**, *65*, 799–815. [CrossRef] [PubMed]

40. Zhou, Q.Y.; Tian, A.G.; Zou, H.F.; Xie, Z.M.; Lei, G.; Huang, J.; Wang, C.M.; Wang, H.W.; Zhang, J.S.; Chen, S.Y. Soybean WRKY-type transcription factor genes, GmWRKY13, GmWRKY21, and GmWRKY54, confer differential tolerance to abiotic stresses in transgenic Arabidopsis plants. *Plant Biotechnol. J.* **2008**, *6*, 486–503. [CrossRef] [PubMed]

41. Wang, F.; Hou, X.; Tang, J.; Wang, Z.; Wang, S.; Jiang, F.; Li, Y. A novel cold-inducible gene from Pak-choi (*Brassica campestris*, ssp. *chinensis*), BcWRKY46, enhances the cold, salt and dehydration stress tolerance in transgenic tobacco. *Mol. Biol. Rep.* **2012**, *39*, 4553. [CrossRef]

42. Song, Y.; Jing, S.J.; Yu, D.Q. Overexpression of the stress-induced OsWRKY08 improves osmotic stress tolerance in Arabidopsis. *Chin. Sci. Bull.* **2009**, *54*, 4671–4678. [CrossRef]

43. Zhong, G.M.; Wu, L.T.; Wang, J.M.; Yang, Y.; Li, X.F. Subcellular localization and expression analysis of transcription factor AtWRKY28 under biotic stresses. *J. Agric. Sci. Technol.* **2012**, *14*, 57–63.

44. Zhao, Y.; Zhou, Y.; Jiang, H.; Li, X.; Gan, D.; Peng, X.; Zhu, S.; Cheng, B. Systematic Analysis of Sequences and Expression Patterns of Drought-Responsive Members of the HD-Zip Gene Family in Maize. *PLoS ONE* **2011**, *6*, e28488. [CrossRef] [PubMed]

45. Raghavendra, A.S.; Padmasree, K.; Saradadevi, K. Interdependence of photosynthesis and respiration in plant cells: Interactions between chloroplasts and mitochondria. *Plant Sci.* **1994**, *97*, 1–14. [CrossRef]

46. Li, P.; Song, A.; Gao, C.; Wang, L.; Wang, Y.; Sun, J.; Jiang, J.; Chen, F.; Chen, S. Chrysanthemum WRKY gene CmWRKY17, negatively regulates salt stress tolerance in transgenic chrysanthemum and Arabidopsis plants. *Plant Cell Rep.* **2015**, *34*, 1365–1378. [CrossRef] [PubMed]

47. Wang, Y.; Jiang, L.; Chen, J.; Tao, L.; An, Y.; Cai, H.; Guo, C. Overexpression of the alfalfa WRKY11 gene enhances salt tolerance in soybean. *PLoS ONE* **2018**, *13*, e0192382. [CrossRef] [PubMed]

48. Skórzyńskapolit, E. Lipid peroxidation in plant cells, its physiological role and changes under heavy metal stress. *Acta Soc. Bot. Pol.* **2007**, *76*, 49–54. [CrossRef]

49. Jain, M.; Mathur, G.; Koul, S.; Sarin, N. Ameliorative effects of proline on salt stress-induced lipid peroxidation in cell lines of groundnut (*Arachis hypogaea* L.). *Plant Cell Rep.* **2001**, *20*, 463–468. [CrossRef]

50. Tang, M.; Liu, X.; Deng, H.; Shen, S. Over-expression of JcDREB, a putative AP2/EREBP domain-containing transcription factor gene in woody biodiesel plant Jatropha curcas, enhances salt and freezing tolerance in transgenic Arabidopsis thaliana. *Plant Sci. Int. J. Exp. Plant Biol.* **2011**, *181*, 623. [CrossRef] [PubMed]

51. Sakuma, Y.; Maruyama, K.; Osakabe, Y.; Qin, F.; Seki, M.; Shinozaki, K. Functional analysis of an Arabidopsis transcription factor, DREB2A, involved in drought-responsive gene expression. *Plant Cell* **2006**, *18*, 1292–1309. [CrossRef] [PubMed]

52. Mallikarjuna, G.; Mallikarjuna, K.; Reddy, M.K.; Kaul, T. Expression of OsDREB2A, transcription factor confers enhanced dehydration and salt stress tolerance in rice (*Oryza sativa*, L.). *Biotechnol. Lett.* **2011**, *33*, 1689–1697. [CrossRef] [PubMed]

53. An, G.; Watson, B.D.; Chiang, C.C. Transformation of Tobacco, Tomato, Potato, and Arabidopsis thaliana Using a Binary Ti Vector System. *Plant Physiol.* **1986**, *81*, 301–305. [CrossRef] [PubMed]

54. Xue, J.P.; Yu, M.; Zhang, A.M. Studies on callus induced from leaves and plantlets regeneration of the traditional Chinese medicine Chrysanthemum morifolium. *China J. Chin. Mater. Med.* **2003**, *28*, 213–216.

55. Schmittgen, T.D. Analysis of relative gene expression data using real-time quantitative PCR and the 2(-Delta Delta C(T)) Method. *Methods* **2001**, *25*, 402–408. [CrossRef]

56. Chen, L.; Chen, Y.; Jiang, J.; Chen, S.; Chen, F.; Guan, Z.; Fang, W. The constitutive expression of Chrysanthemum dichrum ICE1 in Chrysanthemum grandiflorum improves the level of low temperature, salinity and drought tolerance. *Plant Cell Rep.* **2012**, *31*, 1747–1758. [CrossRef] [PubMed]
57. Li, H.S. *Principles and Techniques of Plant Physiological and Biochemical Experiment*, 3rd ed.; Higher Education Press: Beijing, China, 2015; pp. 182–184.
58. Zhang, L.; Tian, L.H.; Zhao, J.F.; Song, Y.; Zhang, C.J.; Guo, Y. Identification of an apoplastic protein involved in the initial phase of salt stress response in rice root by two-dimensional electrophoresis. *Plant Physiol. Plant Signal. Behav.* **2009**, *149*, 916–928. [CrossRef] [PubMed]
59. Sun, H.J.; Wang, S.F.; Chen, Y.T. Effects of salt stress on growth and physiological index of 6 tree species. *For. Res.* **2009**, *22*, 315–324.
60. Jin, Y.; Donglin, L.I.; Ding, Y.; Wang, L. Effects of salt stress on photosynthetic characteristics and chlorophyll content of Sapium sebiferum seedlings. *J. Nanjing For. Univ.* **2011**, *35*, 29–33.
61. Shi, J.; Fu, X.Z.; Peng, T.; Huang, X.S.; Fan, Q.J.; Liu, J.H. Spermine pretreatment confers dehydration tolerance of citrus in vitro plants via modulation of antioxidative capacity and stomatal response. *Tree Physiol.* **2010**, *30*, 914–922. [CrossRef] [PubMed]

6

Molecular Genetics and Breeding for Nutrient use Efficiency in Rice

Jauhar Ali [1,*,†], **Zilhas Ahmed Jewel** [1,†], **Anumalla Mahender** [1,†], **Annamalai Anandan** [2], **Jose Hernandez** [3] **and Zhikang Li** [4]

[1]   Rice Breeding Platform, International Rice Research Institute (IRRI), Los Baños, Laguna 4031, Philippines; jeweluplb@gmail.com (Z.A.J.); m.anumalla@irri.org (A.M.)
[2]   ICAR-National Rice Research Institute, Cuttack, Odisha 753006, India; anandanau@yahoo.com
[3]   Institute of Crop Science, College of Agriculture and Food Science, University of the Philippines Los Baños, Laguna 4031, Philippines; joehernandez56@gmail.com
[4]   Institute of Crop Sciences, Chinese Academy of Agricultural Science, Beijing 100081, China; zhkli1953@126.com
*   Correspondence: J.Ali@irri.org
†   These authors contributed equally to this work.

Abstract: In the coming decades, rice production needs to be carried out sustainably to keep the balance between profitability margins and essential resource input costs. Many fertilizers, such as N, depend primarily on fossil fuels, whereas P comes from rock phosphates. How long these reserves will last and sustain agriculture remains to be seen. Therefore, current agricultural food production under such conditions remains an enormous and colossal challenge. Researchers have been trying to identify nutrient use-efficient varieties over the past few decades with limited success. The concept of nutrient use efficiency is being revisited to understand the molecular genetic basis, while much of it is not entirely understood yet. However, significant achievements have recently been observed at the molecular level in nitrogen and phosphorus use efficiency. Breeding teams are trying to incorporate these valuable QTLs and genes into their rice breeding programs. In this review, we seek to identify the achievements and the progress made so far in the fields of genetics, molecular breeding and biotechnology, especially for nutrient use efficiency in rice.

Keywords: NPK fertilizers; agronomic traits; molecular markers; quantitative trait loci

1. Introduction

Global rice production increased by three-fold over the past three decades despite rice production constraints and rising input costs. Rice is a nutritionally important cereal crop and staple food of Asia. There is an urgent need for developing high-yielding, nutritious, resource use-efficient and multi-stress-tolerant rice varieties to keep up with the tremendous human population growth, especially in Asia, where rice remains the primary source of caloric intake. The yields of rice grain had seen remarkable improvement during the green revolution and post-green revolution. This increase in yield was primarily achieved through high-input-responsive varieties requiring more chemical fertilizers and pesticides and under an ample supply of irrigation water. This kind of approach that predominated over the past three to four decades now stands exhausted amidst our hope to raise productivity per se sustainably. We are now finding that yields are fast approaching a theoretical limit set by the crop's efficiency in harnessing applied inputs. In exploratory managed experimental plots, N fertilizer retrieval in a single year averaged 65% for maize, 57% for wheat and 46% for rice [1,2]. Alterations in the scale of farming operations and management practices such as tillage, seeding, weed and pest control, irrigation and harvesting usually resulted in on-farm variation (lower nutrient use efficiency) and did not accurately reflect the efficiencies obtained in the experimental plot. N recovery efficiency on average ranges from 20–30% for farmer-managed fields under rainfed conditions, from 30–55% under irrigated conditions [3,4] and rarely exceeds 50%.

Over the years, the rice varieties bred did not improve in nutrient absorption and were not developed to maximize nutrient absorption, but they have the capacity to use less than 50% of the applied nutrients. Breeding rice cultivars with improved nutrient use efficiency (NuUE) is becoming a prerequisite for lowering production costs. Such cultivars with NuUE protects the environment by reducing fertilizer application, decreasing the rate of nutrient application losses to ecosystems, decreasing input costs and improving rice yield with a guarantee for sustainability in agriculture while maintaining soil and ground water quality. On the other hand, improvement of NuUE is an essential prerequisite for expanding crop production into marginal lands with low nutrient availability. In light of high energy costs and progressively unpredictable resources, future agricultural systems with concern for improving yield productivity need to be more fruitful and efficient, especially considering fertilizer and irrigation water. In this context, the identification and development of rice varieties with superior grain yield under low input conditions have therefore become a high breeding priority [5]. Even though significant genotypic differences in nitrogen use efficiency exist in rice, genetic selection for this trait has not been carried out systematically [6–8]. This may be primarily because of the complexity involved in the overall phenotype and its evaluation and the non-availability of genetic tools to use. However, with the recent use of high-throughput single nucleotide polymorphism (SNP) markers with ease and high precision, this area of research needs improvement for better understanding [9–11].

Genetic and physiological traits often change with the interaction with environmental variables. Plants are efficient in the absorption and use of nutrients in controlled environments. Therefore, there is a need for a systematic breeding program to develop cultivars with high NuUE and water use efficiency (WUE) [12,13]. The traits involved, particularly nutrient absorption, transport, use and mobilization, should be identified to enhance NuUE and coupled with best management practices for sustainable agriculture.

Use of the wild species of *Oryza* and native landraces becomes imperative for exploiting the untapped reservoir of useful QTLs and genes, especially to broaden the genetic basis of rice and to enrich existing varieties [14,15]. Genetic selection and plant breeding techniques helped to develop rice varieties that are resistant to pests, diseases and adverse environmental conditions such as drought, submergence and salinity. However, for improving NuUE in rice crop, a proper genetic selection approach is necessary. Superior N-efficient genotypes are required as evidenced from the low recovery of N fertilizer, associated economic and environmental concerns and the lack of adoption of more efficient N management strategies [16,17]. Nitrogen use efficiency (NUE) mostly depends on interactions and the use of the nutrient in a proper way, water availability, light intensity, disease pressure and genotype, which could also be improved through appropriate genetic manipulation [6]. Plant ability to absorb and use nutrients under various environmental and ecological conditions is largely influenced by the genetic makeup and physiological components [12]. There are two major approaches to understand NuUE. First, the nutrient deficiency stress triggers a response of plants to it, which may lead to the identification of the processes affecting it. It would help us to understand how to sustain plants under low nutrient inputs. The second approach would be to exploit genetic variability (both natural and induced) through innovative molecular breeding schemes.

Molecular linkage genetic maps and quantitative trait locus (QTL) mapping technologies are helpful for estimating the number and position of the loci governing genetic variation using different types of segregating and fixed populations. Characterizing these loci to their map positions in the genome, as well as their phenotypic effects and epistatic interactions with other QTLs and loci [18–21] has enabled us to explore the genetic loci associated with complex traits such as drought, salinity, disease, NuUE and insect resistance in crop plants [18,22–29]. The rapid advancement in genome sequencing technologies and marker-aided breeding approaches has resulted in a change in breeding methods, providing new opportunities [5]. Association mapping is a method used to identify genes and QTLs underlying quantitatively inherited variation based on a diverse set of fixed lines. It allows the discovery of QTLs/genes using historical phenotypic data and eventually leads to identifying gene functions, under used alleles and allele combinations that can be useful for crop improvement [30,31]. Genome-wide association mapping depends on the strength of linkage disequilibrium (LD) across a diverse population besides identifying the relationships between markers and traits of agronomic and evolutionary interest [32,33].

Understanding the genetic basis of agronomic, physiological and morphological traits in rice is critical for developing new and improved rice varieties. Rice breeders can use this information to select parental lines for hybridization and screen segregating populations (Figure 1). Recently, researchers have been gaining access to the enormous online wealth of genomic and plant breeding resources, including high-quality genome sequences [34–36], dense SNP maps [37–39], extensive germplasm collections and public databases of genomic information [35,36,39–41]. In this review, we have attempted to gather all the necessary information on QTLs related to N, P and K for the benefit of breeders involved in developing rice varieties with NuUE for sustainable agriculture.

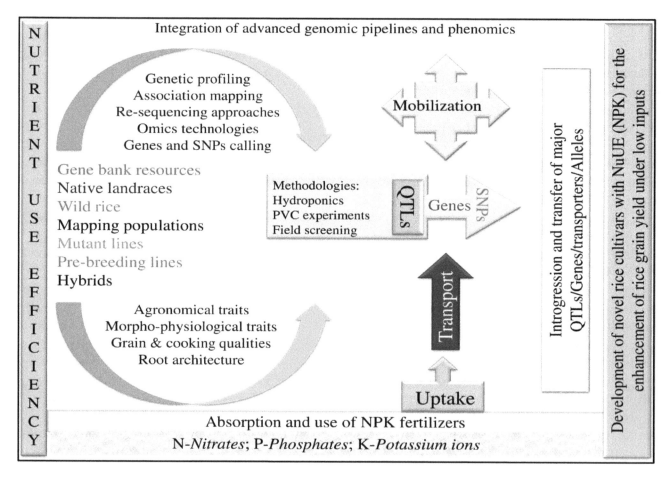

Figure 1. Integrated breeding and genomic approaches for improvement of rice cultivars superior in nutrient use efficiency (NuUE).

2. Screening Protocols and Breeding Efforts for Traits Related to Nutrient Use Efficiency

The literature is replete with NuUE screening protocols, especially for varieties, and very few are available for the systematic breeding of varieties with NuUE. Most of these NuUE studies use minus plots for different nutrients under study [42]. Research plots in institutions practice using omission or minus plots for any given target nutrient under study. Furthermore, researchers have always used natural sites with nutrient deficiencies for screening for any given nutrient such as the Pangil and Tiaong locations in the Philippines for P and Zn_ deficiency conditions, respectively.

2.1. Phosphorus

Deficiency of phosphorus is widespread in tropical and temperate acid soils. Screening and breeding for low phosphorus-tolerant (LPT) genotypes are some of the primary criteria for improving the use efficiency of P fertilizers. Worldwide, one-third of cultivable lands lack P in the soil to meet the requirement for ideal plant growth and development [43]. To avoid these stressful conditions, P is applied widely as an artificial fertilizer for improving grain yield for the burgeoning global population. The inconsistent use of fertilizers severely reduces income, and extreme conditions may cause environmental pollution [44]. Therefore, to overcome this crisis, the identification and improvement of P-efficient rice genotypes adapted to low-P soils would be a favorable solution for the enhancement of grain yield [45]. Developing P-efficient genotypes started with breeders involved in developing upland rice genotypes in an inadvertent manner. On the other hand, the mega-variety of India Swarna is a widely adaptable and popular variety among farmers perhaps because among its necessary traits is P responsiveness, as it possesses the *Pup1* QTL. Therefore, breeders should give more emphasis to developing lines tolerant of P_ deficiency with high grain yield. Krishnamurthy et al. [46] identified six LPT genotypes as Rasi, IET5854, IET14554, PRH122, IET15328 and IET17467, based on grain yield in field experiments at the Directorate of Rice Research in Hyderabad, India. Fageria et al. [47] reported seven lines (CAN 5164, CAN 4097, CAN 5170, IR3646-8-1-2, CAN 4137, A8-391 and IAC-47) at the National Rice and Bean Research Center of Embrapa in Brazil. In 2015, Saito et al. [48] found two varieties (Mudgo and DJ123) based on aboveground biomass at two locations. The development of such genotypes from diverse rice collections and mapping populations, along with cautious screening methodologies, is essential at the laboratory level to reduce the necessity for large-scale field evaluations. Several researchers used hydroponic nutrient solution and field experiments with different doses of P fertilizer to characterize rice varieties. This identified promising traits involved in tolerance of low P [48–58].

For grain yield and response to a graded level of applied phosphorus in low soil fertility conditions, Krishnamurthy et al. [50]) evaluated 28 pre-release promising rice varieties and hybrids at the Directorate of Rice Research farm in Hyderabad. They followed the protocol of 0–60 kg P_2O_5 ha^{-1} (i.e., 0, 10, 20, 30 40, 50 and 60 kg P_2O_5 ha^{-1}) for the P application rate. Among the 28 rice varieties, four distinct patterns were identified in response to grain yield. Eight rice varieties at 0–10 kg P_2O_5 ha^{-1} and six varieties at 20–30 kg P_2O_5 ha^{-1} exhibited higher grain yield, while five varieties recorded higher grain yield in responses at higher P rates of 50–60 kg P_2O_5 ha^{-1}. Out of the 28 varieties, three lines (IET 17190, Sumati and Rajavadlu) did not show any significant change in grain yield at 0–10 or 50–60 kg P_2O_5 ha^{-1}, indicating the existence of genetic variability for P-use efficiency. Chin et al. [58] suggested a soil-based screening method as the most favorable approach for identifying genotypes with tolerance of P_ deficiency. Aluwihare et al. [53] experimented with Ultisol soils, without any application of fertilizer for four decades at Rice Research and Development Institute (RRDI), Sri Lanka, and this also confirmed the absence of P [58]. At P0 and P30 (30 mg/kg P_2O_5) conditions, during early vegetative, late vegetative and flowering stages, plant height (PH), number of tillers (NT), SDW (shoot dry weight), SPC (shoot P concentration), SPU (shoot P uptake) and PUE (P use efficiency) were found to be the major indicators for P_ deficiency tolerance (PDT). Among the total genotypes, 13 were considered as highly tolerant, 13 as moderate and 4 as sensitive to P_ deficiency based on SDW and P use efficiency under P0 conditions. Cancellier et al. [59] and Fageria et al. [60] elucidated that plant height is a vital morphological trait for PDT screening as it significantly correlates with dry weight and yield. Panigrahy et al. [61] identified four low P-tolerant and four susceptible mutants by screening 300-ethane methane sulfonate (EMS)-induced (Nagina 22 [N22]) mutants under low-P field conditions.

However, experimentations at the gene expression level were carried out in controlled test tube, Petri plate or potted conditions with different rates of nutrients, which often included the zero condition (control) for less than a month's duration [13,62,63]. Li et al. [64] carried out expression profile studies using a DNA chip by subjecting rice at 6, 24 and 72 h under low-P stress and compared to a control treatment under normal P conditions. The study showed that genes directly involved in phosphorus absorption and use did not change significantly in transcription in rice shoots, relating to the inadequate low-P treatment. At 72 h under low phosphorus limitation, rice shoots did not develop severe phosphorus stress [65].

Specific genotypes known for their susceptibility to nutrient deficiency stress are useful for selection purposes, especially for different target nutrients. P_ deficiency tolerance was identified in a rice population derived from a cross between P-inefficient *japonica* cultivar "Nipponbare" and P-efficient *indica* landrace "Kasalath" [65].

On the other hand, several traits were studied to understand the phenotyping behavior of plants for precision screening and to progress in breeding activities. Root dry weight (RDW) is an important feature for evaluating the selection index for low-P tolerance in rice. Li et al. [49] reported that, at the seedling stage, dry weight had a significant genotypic variation (19.60%) in both standard and low-P conditions. TDW correlated with RRDW (relative root dry weight), RPH (relative plant height), RPUP (relative total P uptake), RSPA (relative shoot P accumulation), RPUE (relative P use efficiency) and RPC (relative P concentration) at $p < 0.01$. Several key morphological and physiological traits such as plant height, number of tillers, shoot root length, relative shoot and root dry weight and leaf age and root-attributed traits such as root diameter, root hair number and number of roots were used for screening and identifying tolerant genotypes under P_ deficiency conditions [61,66–70]. Increasing the productivity of grain yield under P_ deficiency conditions, increasing P taken up from the soil and improving the dry matter of internal use of P help to enhance the number of panicles and grain productivity [53,71]. Relative tiller dry weight (RTW), shoot dry weight and plant dry weight used as better screening criteria for identifying genotypes tolerant of low-P stress, especially RTW being sensitive, proved to be a reliable screening test. In recent days, image analysis has been becoming popular in high-throughput screening. Chen et al. [72] established an accurate, fast and operable method for diagnosing the crop nutrition status of NPK deficiencies in the color and shape of leaf parameters using a static scanning technology (SST) and hierarchical method in a pot experiment.

2.2. Nitrogen

Nitrogen fertilizer is an essential element for many aspects to improve grain yield, grain quality, flowering time and root development for extracting water and other nutrient elements from the soil [73,74]. On the other hand, the application of N is not uniform in all geographic regions of nations worldwide [75]. Several morphological and agronomic factors were found to influence the deficiency or high rates of N. Higher rates of N fertilizer consumption repeatedly led to environmental pollution and decreased nitrogen use efficiency (NUE) [76]. Therefore, the immediate focus should be to exploit the available variability in the use efficiency of rice cultivars through classical plant breeding methods and advanced biotechnological approaches to increase NUE in rice. Numerous research efforts have been conducted with different rates of N fertilizer in field experiments and hydroponic nutrient solution, and this was correlated with N use-efficient genotypes and higher grain yield (GY) parameters [77–79]. Chaturvedi [80] conducted a field experiment with different treatments of N fertilizer at the Agricultural Research Station in Chhattisgarh, India. Using an application of sulfur-containing nitrogenous fertilizer (Super Net) has significantly increased the grain yield and grain nitrogen content in hybrid rice variety Proagro 6207. Manzoor et al. [81] directed an experiment with nine different N rates (i.e., 0, 50, 75, 100, 125, 150, 175, 200 and 225 kg ha^{-1}) at the Rice Research Institute in Lahore, Pakistan, with Super basmati. Interestingly, at 200 kg N ha^{-1} and above, yield-attributed traits declined, and higher grain yield, number of grains per panicle, 1000-grain weight, number of tillers and panicle length significantly improved at 175 kg N ha^{-1}.

Likewise, Swamy et al. [82] evaluated ten rice genotypes under recommended rates of nitrogen (100 kg N ha^{-1}) and deficient N as no external nitrogen (i.e., N0) in a treatment grown in field conditions at Indian Institute of Rice Research (IIRR), Hyderabad. They found that 14% of root length (RL) decreased significantly under N_ deficiency. Haque and Haque [83] detected higher grain yield (5.36 t ha^{-1}) in 60 kg N ha^{-1}, and the highest NUE (344.50 kg grain kg^{-1} N) was recorded for BU dhan 1 at six different N rates (0, 20, 40, 60, 80 and 100 kg N ha^{-1}); they found an intermediate rate of N as economical and environment-friendly.

Employing a hydroponic experiment, Nguyen et al. [74] determined the effect of N supply in low and excess NH_4NO_3 concentration in Yoshida nutrient solution using three rice cultivars: IR64 (*Oryza sativa* ssp. *indica*), Azucena (*O. sativa* ssp. *japonica*) and TOG7105 (*O. glaberrima*). The rate of absorption of NUE (aNUE) and agronomic NUE (agNUE) decreased significantly, although at a gradual pace as the N supply increased, and physiological NUE (pNUE) declined progressively upon lowering the N supply.

To minimize N application and to use available N more efficiently, agronomic practices still need to be standardized. Nitrogen use efficiency is a complex trait and is associated with different components such as pNUE, aNUE, agNUE [77,78,84] and alteration in morpho-agronomic and physiological traits such as plant height, tiller number, grain yield, dry weight of shoots and roots, spikelet number, number of filled grains per panicle, 1000-grain weight, the leaf color chart (LCC) and chloroplasts [25,26,74,81,83,85–91] in rice. Alteration of the main traits was influenced by the response of N fertilizers, which may enhance the availability of N, which can lead to higher photo-assimilates and dry matter accumulation [80,92]. Therefore, considering the absorption, physiological and agronomic NUEs associated with morpho-agronomic traits will help to attain the balance between high grain yield and the eco-friendly nature of farm systems, which would be useful in developing crops with superior NUE.

2.3. Potassium

The availability of K in the soil is insufficient in developing countries, and it plays a significant role in crop grain yield and quality [93]. From 2012–2016, K fertilizer consumption globally increased from 28.6 Mt (K_2O) to 33.2 Mt (K_2O) [94]. Notably, East and South Asia are promising agricultural areas consuming 44.9% of the world K fertilizer, which is not adequate for improving grain yield under deficiency of K. The price of K fertilizers increased rapidly from 2003 (USD 165 per ton) to 2013 (USD 595 per ton) [94]. Therefore, the identification of K use efficiency (KUE) in rice is essential and needs to be used in developing genotypes with higher grain yield for K-deficient conditions. Dobermann et al. [95] mentioned that, as compared with other cereal crops, rice acquires 56–112 kg of K from soils in each harvest of yield of 4–8 t ha^{-1}, and yearly K demand for irrigated rice would be 9–15 × 106 tons by 2025. In physiological aspects, K is involved in many functions related to regulating osmotic potential, transporting assimilates, root development for uptaking water and nutrients, reducing the frequency of diseases, drought tolerance and photosynthetic activity [96–99]. Under different rates of K fertilizer (0, 25, 50, 75 and 100 kg ha^{-1}), Mehdi et al. [100] evaluated the response of rice cultivars in saline-sodic soil during 2005 and achieved the highest paddy yield (3.24 t ha^{-1}) and straw yield (3.92 t ha^{-1}) at 100 kg K_2O ha^{-1}. Similarly, Fageria et al. [101] elucidated lowland rice grain yield varying from 5.88–6.24 t ha^{-1} with an application of 125 kg ha^{-1} in different years. Analysis of six upland rice genotypes evaluated in a greenhouse under natural soil of 200 mg K kg^{-1} revealed that K uptake in shoot and grain and the K use efficiency ratio (KUER) were significantly and positively associated with grain yield [101], whereas, compared with grain, K concentration and uptake were higher in shoots. Arif et al. [102] conducted a pot experiment with three genotypes in a rain-protected wire house at the University of Agriculture in Faisalabad using hydroponic nutrient solution with different K rates of 0, 30, 60, 90 and 120 kg ha^{-1}, respectively. Among the three genotypes, IR6 (low KUE), Super basmati (medium KUE), genotype 99509 (high KUE), the highest thousand grain weight (TGW) (IR6), grain yield (g pot^{-1}) (Super basmati, 99509), number of panicles and tillers per pot (Super basmati) were recorded at optimum rates of 60 kg ha^{-1}. Earlier reports revealed that a higher rate of K influences increases in yield-attributed traits [103–106]. The increase in yield with an optimum rate of K plays a crucial role in increased N use and increasing chlorophyll synthesis and translocation of assimilates to reproductive parts [107]. Recently, Islam et al. [108] compared the application of K fertilizer between 40 and 80 kg ha^{-1} in a randomized complete block design. The significant ($p < 0.05$) increases in grain and straw yield in the treatment with K application rates of 40 and 80 kg ha^{-1} were 54% and 68% in the dry season and 39% and 45% in the wet season from 2003–2010 in field experiments at the Bangladesh Rice Research Institute farm. Hence, improving uptake, transport and translocation of K efficiency in shoots and rice grain is possible for identifying superior genotypes to further enhance grain yield by proper management practices.

3. Identification and Use of QTLs Related to Nutrient Use Efficiency

Developing rice varieties with multiple tolerance is possible provided large-effect QTLs/genes are available and exploited with innovative molecular breeding approaches. The number of reported QTLs is unwaveringly increasing day by day, but still, very few are applied in breeding programs. Obtaining more data that validate QTLs/genes in different genetic backgrounds and environments is a prerequisite for their large-scale application. In rice, there is an attempt to bring a few large-effect QTLs that confer tolerance of submergence, drought, salinity and P deficiency together through molecular marker-assisted breeding. *Pup1* is the best model for exploiting the NuUE QTLs currently being used, for which molecular markers are now available and evaluated in different genetic backgrounds under field conditions [5].

3.1. QTLs Related to Nitrogen Use Efficiency

Among the essential nutrient elements, nitrogen is the most important one for rice growth in natural ecosystems. The green revolution, which was a breakthrough in agricultural production to secure human nutrition in the past century, depended mainly on fertilizer application and high-yielding modern varieties [109–113]. In this context, nitrogen use-efficient crop varieties are of great concern. Further, genes and QTLs related to agronomy for NUE are presented in Tables 1 and 2. Deeper understanding of the molecular basis of NUE would enable us to provide valuable information for crop improvement through biotechnological approaches. Recent advances in genomics and proteomics approaches such as subtractive hybridization, differential display and microarray techniques are transforming our approach to identify the candidate genes that play a crucial role in the regulation of NUE [4,7,114–118]. In addition, marker-trait association for NUE through quantitative real-time polymerase chain reaction (RT-PCR) technology is being used [119–121]. The identification of potential candidate genes/proteins will serve as biomarkers in the regulation of NUE for screening genotypes for their nitrogen responsiveness. This will help to optimize nitrogen inputs in agriculture.

The modern rice varieties were all selected earlier for higher N uptake to obtain maximum grain yields. Conversely, the biggest problem with the increased N supply often leads to a decrease in N use efficiency. This is mainly due to high N uptake before flowering, but is also due to low N uptake during the reproductive growth phase and incomplete N translocation from vegetative plant parts to the grains [15,178]. Sustainable agriculture requires developing crop varieties with high yield potential and less dependency on heavy applications of N and P fertilizer. Similar to P, N has no systematic breeding program and screening protocol. The genotypes were screened either in nutrient minus fields or under solution culture.

In recent years, heavy nitrogen fertilization during panicle development has been popular in China to improve population dynamics and increase grain yield [179]. Panicle fertilization was adopted to increase grain yield and N recovery efficiency at IRRI [180]. Nitrogen use efficiency positively correlates with photosynthetic characteristics. The measures for promoting photosynthetic function and delaying senescence of leaves may indirectly enhance N absorption and use of rice and ultimately increase NUE. Some research efforts had been devoted to developing genotypes that use N more efficiently. This highly complicated objective requires an in-depth understanding of the genetic basis ot N assimilation and N use at different developmental stages. The QTLs underlying related traits toward the late developmental stage in rice at two different nitrogen rates were investigated using a population of chromosome segment substitution lines (CSSLs) derived from a cross between Teqing and Lemont. A total of 31 QTLs referencing five traits, especially plant height, panicle number per plant, chlorophyll content, shoot dry weight and grain yield per plant, were detected. Under the normal nitrogen (150 kg/h^{-1} N fertilizer) rate, three QTLs were identified for each trait, and the under low nitrogen (0N) rate, five, four, five and two QTLs were detected for plant height, panicle number per plant, chlorophyll content and shoot dry weight, respectively. Most of the QTLs were located on chromosomes 2, 3, 7, 11 and 12 [166].

Table 1. Rice genes/QTLs governing key agronomic traits, the protein encoded, level of allele expression and their possible use in breeding programs.

S. No.	Traits	Name of QTL	Encoded Protein	Nature of Allele Suitable for Use in Breeding Programs	References
1	Grain number	Gn1a	Cytokinin oxidase	Low expression	[122]
2	Grain number and strong culm	dep1	PEBP-like domain protein	Loss of function	[123]
3	Grain number	WFP	OsSPL14	High expression	[124]
4	Grain number, low tiller number, and strong culm	Ipa	OsSPL14	High and ectopic expression	[125]
5	Grain size	gs3	Transmembrane protein	Loss of function	[126]
6	Grain size and filling	gw2	RING-type ubiquitin E3 ligase	Loss of function	[127]
7	Grain size	qSW5/GW5	Unknown	Loss of function	[128]
8	Grain filling	GIF1	Cell wall invertase	Restricted expression in the ovular vascular trace	[129]
9	Heading date	Hd1	CONSTANS-like protein	Loss-of-function allele leads to late heading	[130]
10	Heading date	Hd6	Subunit of protein kinase	Loss–of-function allele leads to early heading	[131]
11	Heading date	Hd3a	FT-like	Low expression leads to late heading	[132–134]
12	Heading date	Ehd1	B-type response regulator	Loss-of-function allele leads to late heading	[135]
13	Grain number, plant height and heading date	Ghd7	CCT domain protein	Functional allele	[136]
14	Days to heading	DTH8	CCT domain protein	Functional allele	[137]
15	Plant height	sd1	Gibberellin 20 oxidase	Loss of function	[138]
16	Lodging resistance	SCM2	F-box protein	High expression	[139]
17	Disease resistance	pi21	Proline-rich protein	Loss of function	[140]
18	Disease resistance	Pb1	CC-NBS-LRR protein	Functional allele	[141]
19	Salt tolerance	SKC1	HKT-type transporter	Gain of function	[142]
20	Cold tolerance	qLTG3-1	GRP and LTP domain	Functional allele	[143]
21	Submerge tolerance	Sub1A	ERF-related factor	Gain of function	[144]
22	Internode elongation under submergence conditions	SK2	ERF-related factor	Gain of function	[145]
23	Cadmium accumulation	OsHMA3	Putative heavy metal transporter	Functional allele	[146]
24	Seed shattering	sh4	Myb3 transcription factor	Loss of function	[147]
25	Seed shattering	qSH1	BEL1-like homeobox protein	Low expression in abscission layer between panicle and spikelet	[148]
26	Prostrate growth	PROG1	Zinc finger transcription factor	Loss of function	[149,150]
27	Disease resistance	RHBV	NS3 protein	Favorable gene or QTL alleles	[151]
28	Phosphorus uptake	Pup1	OsPupK46-2	High expression	[57]
29	Deep rooting	DRO1	Auxin signaling pathway	Functional allele	[152]

Table 2. Quantitative trait loci identified for traits related to nitrogen, phosphorus and potassium use efficiency in rice.

Entry		Phosphorus				
S. No.	Traits	Population	Cross	No. of QTLs		Reference
				M	E	
1	Phosphorus uptake, plant dry weight, tiller number; phosphorus use efficiency	NILs	*Nipponbare/Kasalath*	8	-	[65]
2	Relative tillering ability, relative shoot dry weight, relative root dry weight	RILs	*IR20/IR55178*	4	-	[153]
3	Phosphorus uptake, tiller number	NIL	*Nipponbare/Kasalath*	1 (Pup)	-	[154]
4	Root elongation, shoot dry weight, relative phosphorus content, relative Fe content	F_8	*Gimbozu/Kasalath*	6	-	[155]
5	Relative root length, relative shoot length, relative shoot dry weight, relative root dry weight	BILs	*OM2395/AS996*	1	-	[156]
6	Root elongation under phosphorus deficiency	CSSLs	*Nipponbare/Kasalath CSSL29*	1	-	[157]
7	Plant height, maximum root length, root number, root volume, root fresh weight, root dry weight, shoot dry weight, total dry weight, root/shoot dry weight ratio	ILs	*Yuefa/IRAT109*	24	29	[63]
8	Relative root length, relative root dry weight, relative shoot dry weight, relative total dry weight, relative root-shoot ratio of dry weight	BC_2F_4	*Shuhui 527/Minghui 86*	48	-	[158]
9	Total aboveground biomass, harvest index, P use efficiency for grain yield based on P accumulation in grains, P harvest index, P translocation, P translocation efficiency, P total aboveground P uptake, P use efficiency for biomass accumulation, P use efficiency for grain yield, P use efficiency for straw dry weight based on P accumulation in straw	RILs	*Zhenshan 97/Minghui 63*	36	-	[159]
10	Root dry weight, relative shoot dry weight, relative total dry weight	DHs	*ZYQ8/JX17*	6	-	[160]
		Nitrogen				
1	Plant height	DHs	*IR64/Azucena*	10	-	[161]
2	Rubisco, total leaf nitrogen, soluble protein content	BILs	*Nipponbare/Kasalath*	15	-	[162]
3	N uptake (NUP), grain yield, biomass yield, N use efficiency (NUE)	CSSLs	*9311/Nipponbare*	13	-	[118]
4	Toot system architecture, NDT, and morphological and physiological traits	CSSLs	*Curinga/IRGC105491*	13	-	[163]
5	Twelve physiological and agronomic traits	RILs	*IR64/Azucena*	63	-	[27]
6	Glutamine synthetase, glutamate synthase	BILs	*Nipponbare/Kasalath*	13	-	[164]
7	Glutamine synthetase, panicle number per plant, panicle weight	NILs	*Koshihikari/Kasalath*	1	-	[164]
8	Total grain nitrogen, total shoot nitrogen, nitrogen uptake, nitrogen use efficiency, nitrogen translocation efficiency	F_3	*Basmati370/ASD16*	43	-	[165]
9	Root dry weight, shoot dry weight, biomass	RILs	*Zhenshan97/Minghui 63*	52	103	[166]
10	Plant height, panicle number per plant, chlorophyll content, shoot dry weight	CSSLs	*Teqing/Lemont*	31	-	[167]
11	Total grain number, total leaf nitrogen, total shoot nitrogen, nitrogen uptake, specific leaf nitrogen	RILs	*IR69093-4-3-2/IR72*	32	-	[168]
12	Root length, root thickness, root biomass, biomass, etc.	RILs	*Bala/Azucena*	17	-	[169]
13	Relative root dry weight, spikelet number per panicle, spikelet fertility, 1000-grain weight	ILs	*Shuhui 527 × Minghui 86*	48		[170]
14	Total grain number, total leaf nitrogen, total shoot nitrogen, physiological nitrogen-use efficiency, biomass	RILs	*Dasanbyeo/TR22183*	20	58	[170]
15	Total plant nitrogen, nitrogen-use efficiency	DHs	*IR64/Azucena*	16	-	[171]

Table 2. Cont.

Entry		Phosphorus				
S. No.	Traits	Population	Cross	No. of QTLs		Reference
				M	E	
16	Total plant nitrogen, nitrogen dry matter production efficiency, nitrogen grain production efficiency, total grain number	RIL	Dasanbyeo/TR22183	28	23	[172]
17	Grain yield per plant, biomass, harvest index, etc.	RILs	IR64/INRC10192	46	-	[173]
18	Plant height, root dry weight, shoot dry weight, chlorophyll content, root length, biomass	RILs	R9308/Xieqingzao B	7	-	[161]
19	Grain yield per plant, grain number per panicle	RILs	Zhenshan 97/HR5	19	11	[174]
20	Number of panicles per plant, number of spikelets per panicle, number of filled grains per panicle, grain density per panicle	RILs	Xieqingzao B/Zhonghui 9308	52	-	[175]
21	Nitrogen deficiency tolerance and nitrogen-use efficiency	RILs	Zhenshan 97 and Minghui 63	12		[176]
Potassium						
1	Plant height, tiller number, shoot and root oven-dry weight	DHs	IR64/Azucena.	4	-	[177]

M = main-effect QTLs; E = epistatic QTLs.

Based on the use of two N supply levels, 5 mg N L^{-1} for low N and 40 mg N L^{-1} [167] for high N, QTLs for plant height in rice were mapped onto the Restriction Fragment Length Polymorphism (RFLP) linkage map of a doubled-haploid population derived from a cross between IR64 and Azucena. Two QTLs, one on chromosome 1 and the other on chromosome 8, were detected at high N levels (40 mg N L^{-1}) in soil-based nutrient solution culture experiments. Furthermore, a total of eight QTLs were identified at low N level and located on chromosomes 1, 2, 3, 4, 5 and 6, whereas the QTL flanked by molecular markers RZ730 and RZ801 on chromosome 1 was identified in all experimental conditions. The hypothesis suggests that the genotype showing higher N efficiency under low N level may carry the gene(s) for higher N efficiency. This study demonstrated that the effects of low N stress on plant height lessened. In the present study, the female parent IR64 was found to have a relatively higher N efficiency than the male parent Azucena under low N levels due to its lesser decline in plant height than Azucena. Furthermore, some of the QTLs associated with plant height were detected only at low N levels and might have some relationship with N efficiency [162]. QTL analysis was related to N and P tolerance traits such as root length at the seedling stage, productive panicles, seed setting ratio and yield. A few QTLs out of these were found to be located on similar chromosomal sections that showed the genes associated with the N or P metabolism pathway [181,182]. QTLs for rice panicle number and grain yield were detected under low nitrogen (N0) and low phosphorus (P0) conditions and helped to analyze the genetic basis of tolerance of soil nutrient deficiency. A total of 125 CSSLs with relatively few introgression segments were derived from japonica cultivar Nipponbare within the genetic background of indica cultivar 93–11. These were screened using an augmented design in field experiments with regular fertilization (NF), low nitrogen (N0) and low phosphorus (P0) treatments. Grain yield and panicle number per plant were measured for each CSSL, and their relative values based on regular fertilization treatment considered as the measurement for tolerance of the nutrient deficiency. Both regular fertilization and low phosphorus treatments showed adverse effects on grain yield and panicle number. The different responses observed among the CSSLs refer to the deficiency of nitrogen or phosphorus. The relative traits had a significantly negative correlation with the traits in the regular fertilizer treatment. Cultivar 93–11 showed higher tolerance of low-nutrient stresses than Nipponbare. The negative allelic effects of 38 QTLs were contributed by Nipponbare under nitrogen and phosphorus deficiency stresses. Out of these, 26 QTLs were responsible for yield and panicle number, and the remaining 12 QTLs specified the relative traits. Five QTLs were identified in common under both stresses. Moreover, 81% of the QTLs were specifically detected only in low nitrogen (N0) or phosphorus (P0) conditions. These different QTLs suggest that the response to limiting nitrogen and phosphorus conditions was regulated by various sets of genes in rice [168].

The application of N fertilizer is of particular importance for cultivating high-yielding rice. However, heavy nitrogen fertilizer uses with high loss of nitrogen in rice-growing areas have led to low N recovery rates and environmental pollution. Grain yields are used as an indicator of NUE since it is difficult to evaluate the amount of plant-available N from the soil or any source of N inputs, including fertilizer application and N fixation [183]. Genotypes with high NUE are those cultivars that produce high grain yields with the application of N, while those that do not yield well are genotypes with low NUE. Cultivars with high NUE have the ability to take up N and efficiently use it to produce grains [184]. The relative weight of root, shoot and plant under two different N treatments could reveal the cultivars showing tolerance of low N stress. The QTLs identified for relative performance were distinctive from those for root, shoot and plant weight detected under the two N treatment conditions [182].

The study of Tong et al. [174] revealed a correlation with path analysis indicating that spikelet fertility percentage had the most significant contribution to grain yield per plant at the 300-and 150-kg urea ha^{-1} rates, but filled grains per panicle contributed a strong positive relationship with grain yield per plant at the N0 level. Six of 15 QTLs identified with main effects were detected for each trait except SFP. Clusters of main-effect QTLs associated with several key traits were observed on chromosomes 1, 2, 3, 5, 7 and 10, respectively. The main-effect QTLs (qGYPP-4b and qGNPP-12) were identified at the N0 rate only, which explained 10.9% and 10.2% of the total phenotypic variation explained (PVE). The identification of genomic regions associated with yield and its components at different nitrogen rates will be useful in marker-assisted selection for improving the NUE of rice. The NUE-related trait in rice is so complex that different results were obtained in previous publications because of various experimental conditions, methods and materials. The main-effect QTL (M-QTL), epistatic QTL (E-QTL) and QTL × environment (Q × E) interactions of six traits were investigated using a fully-saturated simple sequence repeat (SSR) linkage map. Obara et al. [185] found a QTL region associated with panicle number and panicle weight on chromosome 2 that contains a regulator gene (GS1) for glutamine synthetase activity. The selected rice plants based on this QTL region showed superiority in tillering ability, panicle number and total panicle weight under low N rates.

Several researchers identified main-effect QTLs on chromosome 3 [171], chromosome 6 [186] and chromosomes 2 and 9 [170] by using doubled haploids and Recombinant Inbred Lines (DHs and RILs) populations.

Among these QTLs, one QTL was identified as being associated with the number of grains per panicle under low N rate, and it was located in a similar region to the Pup1 locus on chromosome 12, thus encouraging the use of Pup1 materials for testing low-N tolerance [5]. Recently, in a hydroponic experiment with CSSLs, Zhou et al. [118] identified a total 23 QTLs, with seven QTLs for N uptake (NUP) located on different chromosomes (2, 3, 6, 8, 10 and 11), with phenotypic variation (PV) ranging from 3.16–13.99%. Six QTLs for N use efficiency were located on chromosomes 2, 4, 6 and 10 and had explained PV ranging from 3.76–12.34%, respectively. The remaining 10 QTLs were responding to grain yield (GY) and biomass yield (BY). With the results of correlation analysis, Zhou et al. [118] suggested that both NUP and NUE had large effects on grain yield. Previous reports of Dong et al. [187,188] showed the NUP trait more closely associated with grain yield than NUE. NUE and NUP trait-linked QTLs are highly useful for improving grain yield under low-input conditions.

3.2. Phosphorus Use Efficiency and Related QTLs

Phosphorus is one of the essential macro-nutrients required for plant growth and development. Low availability of phosphorus in a variety of soils, especially in the tropics, often limits rice grain yields [189], along with the lack of available P sources locally in many countries. The higher importation and transportation costs of P fertilizers frequently prevent resource-poor farmers, especially in developing countries, from applying P to their deficient farmlands. Thus, developing rice cultivars with improved tolerance of P deficiency may therefore be a cost-effective solution to this problem. Rose and Wissuwa [45], optimistic that breeding for poor soil with high P uptake and high PUE needs to be developed and to maximize crop grain yield in such low-input systems, noticed that continuous cropping of poor soil is often related to poverty. It is also important to breed efficient crops. A combination of both P uptake and P internal nutrient efficiency is equally desirable for high-input systems, whereas it would facilitate a reduction in fertilizer rates without yield compensation. Dobermann and Fairhurst [190] reported in rice that P fertilizer use efficiency is only ~25%, which suggests considerable scope for improvement.

Several researchers have identified genes and QTLs governing agronomic traits related to nutrient use efficiency, and these are shown in Tables 1 and 2 and are represented in Figure 2 with the respective NPK QTLs located on 12 chromosomes associated with morpho-physiological traits under low-input conditions. The *Pup1* gene responsible for phosphorus uptake was identified and characterized by Chin et al. [57] (Table 1). Quantitative trait loci for P deficiency tolerance were identified in a rice population derived from a cross between P-inefficient *japonica* cultivar Nipponbare and P-efficient *indica* landrace Kasalath [65]. Tolerance of P deficiency was primarily caused by genotypic differences in P uptake; internal PUE had a negligible effect, and even phosphorus content changed slightly within 72 h in the shoots under low phosphorus stress, but phosphorus content decreased rapidly at 24 h in the roots [62].

Several studies were carried out to understand the genetics of tolerance of phosphorus deficiency in crops, and they identified several QTLs associated with it [54–66,154,156,191]. Su et al. [192] reported that 39 QTLs were associated with panicle number and weight of dry matter, chosen as the indices of P deficiency tolerance in wheat (*Triticum aestivum* L.).

The QTLs related to root traits, panicle number and seed set percentage were reported in rice [66,153,156]. Yield component traits such as panicle number and seed-setting percentage could be used as selection indices for P deficiency tolerance in rice [192]. However, only a few reports are available for the QTL mapping of grain yield and its components for P_ deficiency tolerance.

A significant QTL for P uptake was mapped to a 13.2-cM interval on the long arm of chromosome 12 flanked by markers C443-G2140. The position was estimated to be at 54.5-cM, a 3-cM distance from marker C443. Additional minor QTLs were found on chromosomes 2, 6 and 10 [155]. However, the first evidence supporting the presence of a significant QTL for P_ deficiency tolerance came from a study by Ni et al. [154].

A doubled-haploid population was derived from a cross between P_ deficiency-tolerant *japonica* rice IRAT109 and P deficiency-sensitive *japonica* rice Yuefu [193]. A total of 116 lines were evaluated for yield per plant and its component traits under P deficiency and normal conditions. There were significant differences in seed-setting percentage, panicle number per plant and yield per plant for the doubled haploid DH population between the two conditions, whereas there was no significant difference in 1000-grain weight and grain number per panicle. The results indicated that seed-setting percentage, panicle number per plant and yield per plant were easily influenced by P_ deficiency. Restricted fragment length polymorphism (RFLP) and simple sequence repeat (SSR) markers were used to cover 1535-cM of the rice genome to discover a total of 17 QTLs for plant yield and its components (1000-grain weight, seed-setting %, panicle number per plant, grain number per panicle) under P deficiency conditions. These QTLs explained from 2.65–20.78% of the phenotypic variance, with 12 QTLs showing higher than 10%. For 1000-grain weight, one QTL was detected, which had an logarithm of the odds LOD score of 5.13 and high contribution of PV (14.38%). Five QTLs were linked with seed-setting percentage, and three QTLs were linked with panicle number per plant [193]. Out of these five, three SP QTLs (*qSP2*, *qSP5* and *qSP11*) contributed more than 10%, and the three QTLs for panicle number per plant had high general contributions of more than 17%. Two QTLs (*qPN10* and *qPN12*) had an opposite additive effect. For grain number per panicle, four QTLs were detected, two of which (*qGN6* and *qGN7*) had high general contributions and positive effects. Four additive QTLs were found on chromosomes 2, 3, 6 and 7, which explained 4.77–13.55% of the phenotypic variance, for yield per plant. Three of them, *qYP3*, *qYP6* and *qYP7*, had high general contributions of more than 10% [194].

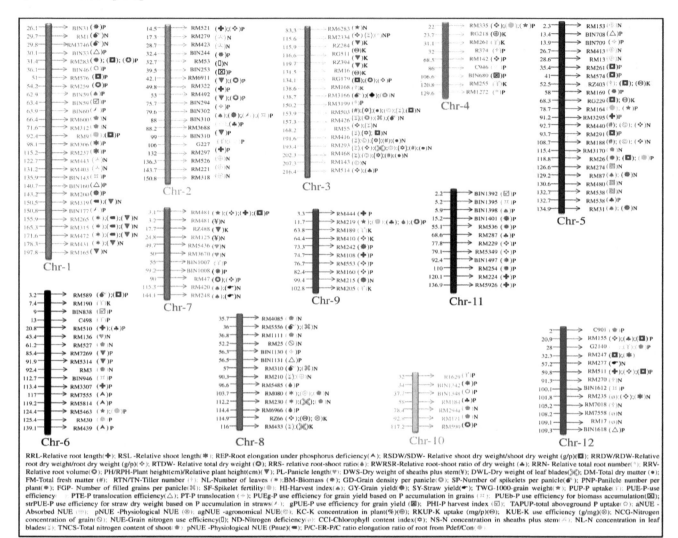

Figure 2. Diagram of 12 chromosomes with reported nutrient use efficiency (NuUE)-NPK QTLs linked to markers associated with the respective traits were identified through marker assisted selection (MAS) breeding approaches in a low-NPK environment using diverse mapping populations of rice.

3.3. Potassium Use Efficiency and Related QTLs

Among the essential elements, potassium is necessary for plant growth. It is the activator of many enzymes in plants and the osmotic regulator of cell solute potential, and it plays a significant role in plant growth and metabolism. In rice, increased application of K fertilizer significantly improves grain and milling quality, such as increasing the percentages of brown rice, milled rice and head milled rice; reducing chalkiness; and enhancing grain protein content [194]. Fageria et al. [101] reported on K uptake and the use efficiency of upland rice under Brazilian conditions. They conducted a greenhouse experiment with the K rate as zero (natural soil level) and 200 mg K kg⁻¹ of soil with the objective of evaluating the influence of K on grain yield, K uptake and their use efficiency, especially for six upland rice genotypes grown on a Brazilian Oxisol. Shoot dry weight and grain yield were significantly influenced by K rate and genotype treatments. The potassium concentration in the shoot was about six-fold greater than that of the grain, across two K rates and six genotypes. However, the K use efficiency ratio (KUER) was about 6.5-times higher in the grain than in the shoot, over two K rates and six genotypes. Potassium uptake in shoot and grain and KUER were significantly and positively associated with grain yield. Besides these, soil Ca, K, base saturation, acidity saturation, Ca saturation, K saturation, Ca/K ratio and Mg/K ratio showed a significant influence on the K application rate.

A greenhouse experiment was conducted at four levels of saline water irrigation (tap water and 2, 4 and 6 dS m^{-1} of salinity) and four different methods of K application (spraying with distilled water as the control, application of potassium on soil, potassium spraying and application of potassium on soil plus spraying). The purpose was to study the efficiency of potassium spraying and use in the soil and their effect on yield and its components under salinity stress. The results showed that grain yield, number of shoots, 100-seed weight, tiller number, dry root weight and K uptake in seeds and shoots decreased significantly with increasing salinity. The best method of K application was soil intake plus spraying [195]. In an investigation of a DH population consisting of 123 lines derived from *indica* variety IR64 and *japonica* variety Azucena under a hydroponic experiment, Wu et al. [177] identified three QTLs associated with shoot and root dry weight under K-deficient conditions. These same three QTLs were also influencing the effect on K content in the plant (KC), K uptake and K use efficiency. The QTLs individually had PVE ranging from 8–15% and were positioned on chromosomes 2, 3, 5 and 8 in K_ deficiency conditions.

4. Effect of Nutrient Use Efficiency across Medium- and Long-Duration Rice

Singh et al. [6] assessed the variability in grain yield and N use of 10 medium-duration (119 ± 4 days after seeding) and 10 long-duration (130 ± 4 DAS) genotypes. These genotypes showed varying rates of acquisition and use of soil and fertilizer N. Significant diversity within genotypes was found in grain yield and N uptake, efficiency and partitioning parameters (physiological N use efficiency, agronomic N use efficiency, apparent recovery, partial factor productivity (PFP) of applied N, N productivity index and N harvest index). The N use-efficient genotypes were IR54790-B-B-38, BG380-2 and BG90-2 (medium duration) and IR3932-182-2-3-3-2, IR54853-B-B-318 and IR29723-88-2-3-3 (long duration), producing high grain yields at both low and high rates of N, whereas inefficient genotypes produced low grain yields at low N rates, but responded well to N application. Increases in grain yields were highly correlated with N uptake. The grain yield-N uptake relationship for individual genotypes indicated significant differences in slope and the grain yield obtained with soil N (GY0). Significant differences in GY0 were due to genotypic variation in N uptake and efficiency of use. The N harvest index was related to both N uptake and use efficiency. The N productivity index, which integrated both GY0 and PFP of applied N, provided a better ranking of rice genotypes. The performance levels of efficient and inefficient genotypes over a range of soil and fertilizer N supply were consistent across three seasons of trials.

5. QTLs for Both Low Nitrogen and Phosphorus Stress

Eight QTLs explained panicle number per plant under the three treatments. Five of the QTLs were identified under the low-nitrogen treatment, and three were identified under the low-phosphorus treatment. The alleles from Nipponbare at all the QTLs_ had adverse effects on panicle number (decreasing it by 42.6–62.9%). No common QTLs were identified for panicle number under both low-N and low-P stresses. A total of 18 QTLs for yield per plant were detected in three treatments [175]. Located on chromosome 4, a QTL (*Qyd-4c*) was identified in all treatments with relatively higher phenotypic variance explained (58.2%, 55.2% and 88.1%) under normal, low-N and low-P conditions, respectively. The authors detected another four QTLs (*Qyd-3a*, *Qyd-4a*, *Qyd-7a* and *Qyd-10*) in two treatments. The rest of the 13 QTLs were identified in only low-nitrogen or low-phosphorus treatments. Regarding relative yield, two and three QTLs were identified in different N and P treatments, respectively, of which *Qryd-7a* was a common QTL, suggesting that the CSSL containing the *Qryd-7a* locus was sensitive to both N and P_ deficiency stresses [127,172]. QTL *Qyd-4a* was located in the same chromosomal region as the QTL for dry weight of seedling root [167]. The authors conjectured this substitution region to be associated with root response to nutrient stresses, probably containing genes for regulating nutrient absorption and consequently affecting yield per plant in rice. Root elongation gets hit by either N or P_ deficiency [126,167,172], resulting in various nutrition assimilation in plants. Several QTLs from this study correspond to known genes in the N or P metabolic pathway. For example, *Qyd-2b* for N_ deficiency tolerance was located near the gene encoding cytosolic glutamine synthetase (*GS1*), and *Qyd-3b* and *Qpn-3* were nearby the genes for glutamate dehydrogenase (*GDH2*) [182]. Furthermore, *Qyd-12* was detected only under low-P conditions, and it co-localized with a significant QTL (*Pup1*) on chromosome 12, which was involved in P absorption [154]. These results indicate that the QTLs specifically detected under single N or P_ deficiency conditions may be involved in different pathways of N and P metabolism. Their tightly linked markers have breeding potential in pyramiding elite QTLs for N and P use efficiency.

Tolerance of low nitrogen stress conditions is a highly desired characteristic for sustainable crop production. The genetic components associated with low N tolerance in rice at the seedling stage, including main QTL effects, epistatic QTL effects and QTL by environment interactions, using a population of 239 RILs derived from a cross between popular Zhenshan 97 and Minghui 63, were studied [182] in solution culture. Root, shoot and plant weight over two N treatments were measured and the relative weight of the two treatments for each trait considered as measurements for low-N tolerance. Four to eight QTLs with main effects were detected for each of the nine traits. Very few QTLs were detected in both low and normal nitrogen conditions, and interestingly, most of the QTLs for the relative measurements were distinct from those for traits under the two nitrogen treatments, indicating very little commonality in the genetic basis of the traits and their relative performance under low and normal nitrogen conditions. In rice, some agronomic traits involving effective tiller number, spikelet fertility percentage and grain yield were studied under low nitrogen stress [166,170,185,196]. Two main-effect QTLs with large contribution rates were detected at the N0 rate. One of them affecting grain number per plant was detected at the interval RM117-RM101 on chromosome 12, accounting for 10.2% of the total phenotypic variance. There was no significant interaction between this M-QTL and environmental factors. This QTL is from the same region as a QTL (*Pup1*) related to phosphorus uptake [156]. Zhao et al. [33] reported that single segment substitution lines (SSSLs) each having a single chromosome segment derived from a donor under the same genetic background as the recipient parent were developed in rice by advanced backcrossing and genome-assisted selection. The QTLs for 22 essential traits were detected in rice with 32 SSSLs by a randomized block design in two to four cropping seasons. However, the QTLs controlling grain weight, grain length, the ratio of grain length to width and heading date were relatively stable. Fifty-nine QTLs were detected and distributed on chromosomes 1, 2, 3, 4, 6, 7, 8, 10 and 11, of which 18 were detected more than twice. Only 30.5% of the QTLs were repeatedly identified across different cropping seasons. Mostly the QTLs governing important agronomic traits showed small additive effects and instability. The stable QTLs usually had larger additive effects and were less affected by environment. With recent successful achievements in the Green Super Rice (GSR) project, efforts were made for highly adaptive rice cultivars with higher grain yield under low-input conditions [13,196–201]. About a 10% yield increase was obtained in elite GSR rice cultivars as compared with the local check variety NSIC Rc222 under multiple abiotic stress tolerance and low-input conditions, without compromising grain yield and quality [200]. Further progress in the genetic regulation of NuUE of GSR cultivars may provide valuable materials to understand the molecular and physiological pathways for the improvement of yield and grain quality under low-input conditions.

6. Agronomic Efficiency and Partial Factor Productivity QTLs

There is a significant increase in grain yield for each kg of fertilizer applied, termed agronomic efficiency (AE). Efficient fertilizer use is defined as maximum returns per unit of fertilizer applied [202]. According to Yadav [202], PFP and AE are useful measures of NUE, as they provide a basis for an integrative index that quantifies total economic output relative to the use of all nutrient resources in the system. Cassman et al. [203] defined PFP and AE to increase by increasing the amount, uptake and use of available nutrients and further by increasing the efficiency of applied nutrients that are taken up by the crop and used to produce grain.

Several researchers have studied AE and PFP in rice and other cereal crops. Dobermann [204] reported cereal crops in terms of AE of 10–30 kg grain kg^{-1} N, where >30 kg grain kg^{-1} is found in a well-managed system or at a low rate of N use or low soil N supply and for PFP 40–70 kg grain kg^{-1} N, with >70 kg^{-1} at low rates of N or in a well-managed efficient system. Wen-xia et al. [205] reported AE in two kinds of rice, one being Jinzao, with AE ranging from 8.02–20.14 kg grain kg^{-1} N, and the second one being Shanyou63, with an AE range of 3.4–18.37 kg grain kg^{-1} N absorbed. Yoshida [206]) estimated AE to be 15–25 kg grain kg^{-1} N, and Cassman et al. [203] reported AE at 15–20 kg grain kg^{-1} N in the dry season in farmers' fields in the Philippines.

Amanullah et al. [207] declared that in maize, PFP for applied N was 36.62 kg grain kg^{-1} N and AE for applied N was 22.49 kg grain kg^{-1} N, using DAP and SSP in the field for the AE of two fertilizer applications, resulting in 13.01 and 13.71 kg grain kg^{-1} P, and PFP resulting in 63.58 and 61.92 kg grain kg^{-1} P. Rao [208] reported AE for applied K in hybrid cotton to be 8.8 kg grain kg^{-1} K, where the application rate of the fertilizer is NPK at 200-150-100 kg ha^{-1}, and for non-hybrid cotton, 5.9 kg grain kg^{-1} K at the same rate of fertilizer application. In hybrid rice, AE for applied P was 5.2 kg grain kg^{-1} P and 11.8 kg grain kg^{-1} K with a fertilizer application rate of NPK of 200-75-200 and 200-150-200 kg ha^{-1}, respectively. The AE for applied P in non-hybrid rice was 2.3 kg grain kg^{-1} P and 4.7 kg grain kg^{-1} P, where the fertilizer rate was the same. Rao [208] in another study showed that only the application of P (N and K as blanket doses) gave AE for non-hybrid rice of 4.2–15.6 kg grain kg^{-1} P and 5.9–11.4 kg grain kg^{-1} P, where the P application rate was 75 and 150 kg ha^{-1} and plant spacing was 12.5 × 10 cm and 10 × 10 cm, respectively.

The application of a unit of fertilizer is economical if the increase in crop yield due to the quantity of fertilizer added is higher than the cost of the fertilizer used. However, if a unit of fertilizer does not increase the grain yield enough to pay for its cost, then its application will not be considered economical and will not be profitable even after a constant increase in grain yield [209]. The application of essential plant nutrients in optimum split dosages and proportion, dispensed to plants in an appropriate method and timing, is the key to increased and sustained crop production.

7. Conclusions

Improving global rice yield productivity under low-input conditions is the main challenge for plant breeders and molecular biologists to develop/improve appropriate rice cultivars. Improving NuUE (nutrient use efficiency) is a key component from an agronomic, economic and environmental viewpoint. Despite the highly complex nature of NuUE in rice, recent trends in molecular marker-assisted selection and advanced biotechnological tools can accelerate the dissecting of the polygenic nature of complex traits. Apart from several breeding and agronomic strategies, balanced N, P and K nutrient elements are required to maintain soil fertility, uptake and transportation from soil to grain to produce higher grain yield with nutrient quality traits. The combined genomic and phenomic studies are valuable to distinguish the QTL and gene responses to NPK acquisition and transportation identified, and very few of them are strongly used with the target trait of interest in plant breeding programs. So far, plenty of QTLs have been identified in diverse genetic backgrounds with significant PVE under different treatment doses of NPK. By using this QTL information, better NuUE genotypes can be developed suitable for resource-poor farmers. Further, by employing these rapid developments, an integrative SNP array with innovative techniques such as Next-generation sequencing (NGS) and Genotyping by sequencing (GBS)technologies, high-density and SNP linkage maps and molecular breeding approaches are feasible solutions for identifying cultivars with superior NuUE by incorporating them into breeding cycles and understanding the molecular genetics and physiological mechanisms of N, P and K status in plants under different fertilizers or deficiency conditions. However, a combined holistic approach requires different aspects of work in the pipeline and omic technologies for its implementation in modern NuUE breeding programs.

Author Contributions: J.A., Z.A.J. and A.M. worked on outlining the contents in the manuscript and prepared the draft article. A.A., A.M., J.A. and Z.L. contributed to the screening methodologies associated with aspects of molecular and genomic regions. J.A. and Z.L. conceived of the basic idea, gave suggestions, corrected the entire article and improved the prospects for breeding programs. All the authors read and approved the review article.

Acknowledgments: The authors would like to thank and acknowledge the Bill & Melinda Gates Foundation (BMGF) for providing a research grant to Z.L. for the Green Super Rice project under ID OPP1130530. We would also like to thank the Department of Agriculture (DA) of the Philippines for providing funds to J.A. under the Next-Gen project.

References

1. Ladha, J.K.; Pathak, H.; Krupnik, T.J.; Six, J.; Kessel, C.V. Efficiency of fertilizer nitrogen in cereal production: Retrospects and prospects. *Adv. Agron.* **2005**, *87*, 85–156.

2. Xu, X.; Liu, X.; He, P. Yield Gap, Indigenous Nutrient Supply and Nutrient Use Efficiency for Maize in China. *PLoS ONE* **2015**, *10*, e0140767. [CrossRef] [PubMed]

3. Roberts, T.L. Improving Nutrient Use Efficiency. *Turk. J. Agric. For.* **2008**, *32*, 177–182.

4. Liu, Z.; Zhu, C.; Jiang, Y.; Tian, Y.; Yu, J.; An, H.; Tang, W.; Sun, J.; Tang, J.; Chen, G.; et al. Association mapping and genetic dissection of nitrogen use efficiency-related traits in rice (*Oryza sativa* L.). *Funct. Integr. Genom.* **2016**, *16*, 323–333. [CrossRef] [PubMed]

5. Vinod, K.K.; Heuer, S. Approaches towards nitrogen- and phosphorus-efficient rice. *AoB Plants* **2012**, *28*, 1–18. [CrossRef] [PubMed]

6. Singh, U.; Ladha, J.K.; Castillo, E.G.; Punzalam, G.; Tirol-Padre, A.; Duqueza, M. Genotypic variation in nitrogen use efficiency in medium and long-duration rice. *Field Crops Res.* **1998**, *58*, 35–53. [CrossRef]

7. Han, M.; Okamoto, M.; Beatty, P.H.; Rothstein, S.J.; Good, A.G. The Genetics of Nitrogen Use Efficiency in Crop Plants. *Annu Rev. Genet.* **2015**, *49*, 269–289. [CrossRef] [PubMed]

8. Van Bueren, E.T.L.; Struik, P.C. Diverse concepts of breeding for nitrogen use efficiency, a review. *Agron. Sustain. Dev.* **2017**, *37*, 50. [CrossRef]

9. Chen, H.; Xie, W.; He, H.; Yu, H.; Chen, W.; Li, J.; Yu, R.; Yao, Y.; Zhang, W.; He, Y.; et al. A high-density SNP genotyping array for rice biology and molecular breeding. *Mol. Plant* **2014**, *7*, 541–553. [CrossRef] [PubMed]

10. Thomson, M.J.; Singh, N.; Dwiyanti, M.S.; Wang, D.R.; Wright, M.H.; Perez, F.A.; DeClerck, G.; Chin, J.H.; Malitic-Layaoen, G.A.; Juanillas, V.M.; et al. Large-scale deployment of a rice 6 K SNP array for genetics and breeding applications. *Rice* **2017**, *10*, 40. [CrossRef] [PubMed]

11. Feng, B.; Chen, K.; Cui, Y.; Wu, Z.; Zheng, T.; Zhu, Y.; Ali, J.; Wang, B.; Xu, J.; Zhang, W.; et al. Genetic Dissection and Simultaneous Improvement of Drought and Low Nitrogen Tolerances by Designed QTL Pyramiding in Rice. *Front. Plant Sci.* **2018**, *9*, 306. [CrossRef] [PubMed]

12. Baligar, V.C.; Fageria, N.K.; Hea, Z.L. Nutrient Use Efficiency in Plants. *Commun. Soil Sci. Plant Anal.* **2001**, *32*, 7–8. [CrossRef]

13. Ali, J.; Xu, J.L.; Gao, Y.M.; Fontanilla, M.A.; Li, Z.K. Green super rice (GSR) technology: An innovative breeding strategy-achievements & advances. In Proceedings of the 12th SABRAO Congress-Plant Breeding towards 2025: Challenges in a Rapidly Changing World, Chiang Mai, Thailand, 13–16 January 2012; pp. 16–17.

14. Kole, C.; Muthamilarasan, M.; Henry, R.; Edwards, D.; Sharma, R.; Abberton, M.; Batley, J.; Bentley, A.; Blakeney, M.; Bryant, J.; et al. Application of genomics-assisted breeding for generation of climate resilient crops: Progress and prospects. *Front. Plant Sci.* **2015**, *6*, 563. [CrossRef] [PubMed]

15. Stein, J.C.; Yu, Y.; Copetti, D.; Zhang, L.; Zhang, C.; Chougule, K.; Gao, D.; Iwata, A.; Goicoechea, J.L.; Wei, S.; et al. Genomes of 13 domesticated and wild rice relatives highlight genetic conservation, turnover and innovation across the genus *Oryza*. *Nat. Genet.* **2018**, *50*, 285–296. [CrossRef] [PubMed]

16. Broadbent, F.E.; De Datta, S.K.; Laureles, E.V. Measurement of nitrogen utilization efficiency in rice genotypes. *Agron. J.* **1987**, *79*, 786–791. [CrossRef]

17. Singh, U.; Cassman, K.G.; Ladha, J.K. *Innovative Nitrogen Management Strategies for Lowland Rice Systems;* Fragile Lives in Fragile Ecosystems, International Rice Research Institute, P.O. Box 933: Manila, Philippines, 1995; pp. 229–254.

18. Xiao, J.H.; Li, J.M.; Yuan, L.P.; Yuan, S.R. Identification of QTLs affecting traits of agronomic importance in a recombinant inbred population derived from a sub-specific rice cross. *Theor. Appl. Genet.* **1996**, *92*, 230–244. [CrossRef] [PubMed]

19. Wisser, R.J.; Sun, Q.; Hulbert, S.H.; Kresovich, S.; Nelson, R.J. Identification and Characterization of Regions of the Rice Genome Associated with Broad-Spectrum, Quantitative Disease Resistance. *Genetics* **2015**, *169*, 2277–2293. [CrossRef] [PubMed]

20. Bocianowski, J. Epistasis interaction of QTL effects as a genetic parameter influencing estimation of the genetic additive effect. *Genet. Mol. Biol.* **2013**, *36*, 93–100. [CrossRef] [PubMed]

21. Zhu, H.; Liu, Z.; Fu, X.; Dai, Z.; Wang, S.; Zhang, G.; Zeng, R.; Liu, G. Detection and characterization of epistasis between QTLs on plant height in rice using single segment substitution lines. *Breed. Sci.* **2015**, *65*, 192–200. [CrossRef] [PubMed]

22. Li, Z.S.; Pinson, R.M.; Stansel, J.W.; Park, D. Identification of two major genes and quantitative trait loci (QTLs) for heading date and plant height in cultivated rice (*Oryza sativa* L.). *Theor. Appl. Genet.* **1995**, *91*, 371–381. [CrossRef] [PubMed]

23. Kang, H.J.; Cho, Y.G.; Tlee, Y.; Eun, M.Y.; Shim, J.U. QTL mapping of genes conferring days to heading, culm length and panicle length based on molecular map of rice (*Oryza sativa* L.). *RDA J. Crop Sci.* **1998**, *40*, 55–61.

24. Yamamoto, T.; Yonemaru, J.; Yano, M. Towards the Understanding of Complex Traits in Rice: Substantially or Superficially? *DNA Res.* **2009**, *16*, 141–154. [CrossRef] [PubMed]

25. Zhao, K.; Tung, C.W.; Eizenga, G.C.; Wright, M.H.; Ali, M.L.; Price, A.H.; Norton, G.J.; Islam, M.R.; Reynolds, A.; Mezey, J.; et al. Genome-wide association mapping reveals a rich genetic architecture of complex traits in *Oryza sativa. Nat. Commun.* **2011**, *2*, 467. [CrossRef] [PubMed]

26. Nongpiur, R.C.; Singla-Pareek, S.L.; Pareek, A. Genomics Approaches for Improving Salinity Stress Tolerance in Crop Plants. *Curr. Genom.* **2016**, *17*, 343–357. [CrossRef] [PubMed]

27. Nguyen, H.T.T.; Dang, D.T.; Pham, C.V.; Bertin, P. QTL mapping for nitrogen use efficiency and related physiological and agronomical traits during the vegetative phase in rice under hydroponics. *Euphytica* **2016**, *212*, 473–500. [CrossRef]

28. Wang, H.; Qin, F. Genome-Wide Association Study Reveals Natural Variations Contributing to Drought Resistance in Crops. *Front. Plant Sci.* **2017**, *30*, 1110. [CrossRef] [PubMed]

29. Yadav, M.K.; Aravindan, S.; Ngangkham, U.; Subudhi, H.N.; Bag, M.K.; Adak, T.; Munda, S.; Samantary, S.; Jena, M. Correction: Use of molecular markers in identification and characterization of resistance to rice blast in India. *PLoS ONE* **2017**, *12*, e0179467. [CrossRef] [PubMed]

30. Flint-Garcia, S.A.; Thuillet, A.C.; Yu, J.; Pressoir, G.; Romero, S.M.; Mitchell, S.E.; Doebley, J.; Kresovich, S.; Goodman, M.M.; Buckler, E.S. Maize association population: A high-resolution platform for quantitative trait locus dissection. *Plant J.* **2005**, *44*, 1054–1064. [CrossRef] [PubMed]

31. Ersoz, E.S.; Yu, J.; Buckler, E.S. Applications of linkage disequilibrium and association mapping in maize. *Mol. Genet. Approach Maize Improv.* **2009**, *63*, 173–195.

32. Clark, R.M.; Schweikert, G.; Toomajian, C.; Ossowski, S.; Zeller, G.; Shinn, P.; Warthmann, N.; Hu, T.T.; Fu, G.; Hinds, D.A.; et al. Common sequence polymorphisms shaping genetic diversity in *Arabidopsis thaliana. Science* **2007**, *317*, 338–342. [CrossRef] [PubMed]

33. Zhao, F.M.; Zhu, H.T.; Ding, X.H.; Zeng, R.Z.; Zhang, Z.L.; Zhang, G. Detection of QTLs for Important Agronomic Traits and Analysis of Their Stabilities Using SSSLs in Rice. *Agric. Sci. China* **2007**, *6*, 769–778. [CrossRef]

34. Goff, S.A.; Ricke, D.; Lan, T.H.; Presting, G.; Wang, R.; Dunn, M.; Glazebrook, J.; Sessions, A.; Oeller, P.; Varma, H.; et al. A draft sequence of the rice genome (*Oryza sativa* L. ssp *japonica*). *Science* **2002**, *296*, 92–100. [CrossRef] [PubMed]

35. Yu, J.; Hu, S.; Wang, J.; Wong, G.K.; Li, S.; Liu, B.; Deng, Y.; Dai, L.; Zhou, Y.; Zhang, X.; et al. A draft sequence of the rice genome (*Oryza sativa* L. ssp *indica*). *Science* **2002**, *296*, 79–92. [CrossRef] [PubMed]

36. Du, H.; Yu, Y.; Ma, Y.; Gao, Q.; Cao, Y.; Chen, Z.; Ma, B.; Qi, M.; Li, Y.; Zhao, X.; et al. Sequencing and de novo assembly of a near complete *indica* rice genome. *Nat. Commun.* **2017**, *8*, 15324. [CrossRef] [PubMed]

37. McNally, K.L.; Childs, K.L.; Bohnert, R.; Davidson, R.M.; Zhao, K.; Ulat, V.J.; Zeller, G.; Clark, R.M.; Hoen, D.R.; Bureau, T.E.; et al. Genome wide SNP variation reveals relationships among landraces and modern varieties of rice. *Proc. Natl. Acad. Sci. USA* **2009**, *106*, 12273–12278. [CrossRef] [PubMed]

38. Huang, X.; Wei, X.; Sang, T.; Zhao, Q.; Feng, Q.; Zhao, Y.; Li, C.; Zhu, C.; Lu, T.; Zhang, Z.; et al. Genome wide association studies of 14 agronomic traits in rice landraces. *Nat. Genet.* **2010**, *42*, 961–967. [CrossRef] [PubMed]

39. Ebana, K.; Yonemaru, J.; Fukuoka, S.; Iwata, H.; Kanamori, K.; Namiki, N.; Nagasaki, H.; Yano, M. Genetic structure revealed by a whole-genome single nucleotide polymorphism survey of 5 diverse accessions of cultivated Asian rice (*Oryza sativa* L.). *Breed. Sci.* **2010**, *60*, 390–397. [CrossRef]

40. Zhao, K.; Aranzana, M.J.; Kim, S.; Lister, C.; Shindo, C.; Tang, C.; Toomajian, C.; Zheng, H.; Dean, C.; Marjoram, P.; et al. An *Arabidopsis* example of association mapping in structured samples. *PLoS Genet.* **2007**, *3*, e4. [CrossRef] [PubMed]

41. Agrama, H.A.; Yan, W.; Jia, M.; Fjellstrom, R.; McClung, A.M. Genetic structure associated with diversity and geographic distribution in the USDA rice world collection. *Nat. Sci.* **2010**, *2*, 247–291. [CrossRef]

42. Ali, J.; Xu, J.; Ismail, A.M.; Fu, B.Y.; Vijaykumar, C.H.M.; Gao, Y.M.; Domingo, J.; Maghirang, R.; Yu, S.B.; Gregorio, G.; et al. Hidden diversity for abiotic and biotic stress tolerances in the primary gene pool of rice revealed by a large backcross breeding program. *Field Crops Res.* **2006**, *97*, 66–76. [CrossRef]

43. MacDonald, G.K.; Bennett, E.M.; Potter, P.A.; Ramankutty, N. Agronomic phosphorus imbalances across the world's croplands. *Proc. Natl. Acad. Sci. USA* **2007**, *108*, 3086–3091. [CrossRef] [PubMed]

44. Cordell, D.; Drangert, J.O.; White, S. The story of phosphorus global food security and food for thought. *Glob. Environ. Chang.* **2009**, *19*, 92–305. [CrossRef]

45. Rose, T.J.; Wissuwa, M. Rethinking internal phosphorus utilization efficiency: A new approach is needed to improve PUE in grain crops. *Adv. Agron.* **2012**, *116*, 185–217.

46. Krishnamurthy, P.; Sreedevi, B.; Ram, T.; Padmavathi, G.; Kumar, R.M.; Rao, P.R; Rani, N.S.; Latha, P.C.; Singh, S.P. Evaluation of rice genotypes for phosphorus use efficiency under soil mineral stress conditions. *Oryza* **2010**, *47*, 29–33.

47. Fageria, N.K.; Morais, O.P.; Baligar, V.C.; Wrigh, R.J. Response of rice cultivars to phosphorus supply on an Oxisol. *Fertilizer Res.* **1988**, *16*, 195–206. [CrossRef]

48. Saito, K.; Vandamme, E.; Segda, Z.; Fofana, M.; Ahouanton, K. A screening protocol for vegetative-stage tolerance to phosphorus deficiency in upland rice. *Crop Sci.* **2015**, *55*, 1223–1229. [CrossRef]

49. Li, Z.K.; Fu, B.Y.; Gao, Y.M.; Xu, J.L.; Ali, J.; Lafitte, H.R.; Jiang, Y.Z.; Rey, J.D.; Vijayakumar, C.H.; Maghirang, R.; et al. Genome-wide ILs and Their Use in Genetic and Molecular Dissection of Complex Phenotypes in Rice (*Oryza sativa* L.). *Plant Mol. Biol.* **2005**, *59*, 33–52. [CrossRef] [PubMed]

50. Panigrahy, M.; Rao, D.N.; Sarla, N. Molecular mechanisms in response to phosphate starvation in rice. *Biotechnol. Adv.* **2009**, *27*, 389–397. [CrossRef] [PubMed]

51. Wissuwa, M.; Kretzschmar, T.; Rose, T.J. From promise to application: Root traits for enhanced nutrient capture in rice breeding. *J. Exp. Bot.* **2016**, *67*, 3605–3615. [CrossRef] [PubMed]

52. Vejchasarn, P.; Lynch, J.P.; Brown, K.M. Genetic Variability in Phosphorus Responses of Rice Root Phenotypes. *Rice* **2016**, *9*, 29. [CrossRef] [PubMed]

53. Aluwihare, Y.C.; Ishan, M.; Chamikara, M.D.M.; Weebadde, C.K.; Sirisena, D.N.; Samarasinghe, W.L.G.; Sooriyapathirana, S.D.S.S. Characterization and Selection of Phosphorus Deficiency Tolerant Rice Genotypes in Sri Lanka. *Rice Sci.* **2016**, *23*, 184–195. [CrossRef]

54. Mahender, A.; Anandan, A.; Pradhan, S.K.; Singh, O.N. Traits-related QTLs and genes and their potential applications in rice improvement under low phosphorus condition. *Arch. Agron. Soil Sci.* **2017**, *64*, 449–464. [CrossRef]

55. Yugandhar, P.; Veronica, N.; Panigrahy, M.; Nageswara Rao, D.; Subrahmanyam, D.; Voleti, S.R.; Mangrauthia, S.K.; Sharma, R.P.; Sarla, N. Comparing Hydroponics, Sand, and Soil Medium to Evaluate Contrasting Rice Nagina 22 Mutants for Tolerance to Phosphorus Deficiency. *Crop Sci.* **2017**, *57*, 1–9. [CrossRef]

56. Chithrameenal, K.; Vellaikumar, S.; Ramalingam, J. Identification of rice (*Oryza sativa* L.) genotypes with high phosphorus use efficiency (PUE) under field and hydroponic conditions. *Indian Res. J. Genet. Biotechnol.* **2017**, *9*, 23–37.

57. Chin, J.H.; Gamuyao, R.; Dalid, C.; Bustamam, M.; Prasetiyono, J.; Moeljopawiro, S.; Wissuwa, M.; Heuer, S. Developing rice with high yield under phosphorus deficiency: *Pup1* sequence to application. *Plant Physiol.* **2011**, *156*, 1202–1216. [CrossRef] [PubMed]

58. Sirisena, D.N.; Wanninayake, W.M.N. Identification of promising rice varieties for low fertile soils in the low country intermediate zone in Sri Lanka. *Ann. Sri Lanka Dep. Agric.* **2014**, *14*, 95–105.

59. Cancellier, E.L.; Brandao, D.R.; Silva, J.; Santos, M.M.; Fidelis, R.R. Phosphorus use efficiency of upland rice cultivars on Cerrado soil. *Ambience* **2012**, *8*, 307–318.

60. Fageria, N.K.; Knupp, A.M.; Moraes, M.F. Phosphorus Nutrition of Lowland Rice in Tropical Lowland Soil. *Commun. Soil Sci. Plant Anal.* **2013**, *44*, 2932–2940. [CrossRef]

61. Panigrahy, M.; Nageswara Rao, D.; Yugandhar, P.; Sravan Raju, N.; Krishnamurthy, P.; Voleti, S.R.; Ashok Reddy, G.; Mohapatra, T.; Robin, A.; Singh, A.K.; et al. Hydroponic experiment for identification of tolerance traits developed by rice Nagina 22 mutants to low-phosphorus in field condition. *Arch. Agron. Soil Sci.* **2014**, *60*, 565–576. [CrossRef]

62. Wu, P.; Ma, L.; Hou, X. Phosphate starvation triggers distinct alterations of genome expression in *Arabidopsis* roots and leaves. *Plant Physiol.* **2003**, *132*, 1260–1271. [CrossRef] [PubMed]

63. Li, J.; Xie, Y.; Dai, A.; Liu, L.; Li, Z. Root and shoot traits responses to phosphorus deficiency and QTL analysis at seedling stage using ILs of rice. *J. Genet. Genom.* **2009**, *36*, 173–183. [CrossRef]

64. Li, L.; Qiu, X.; Li, X.; Wang, S.; Zhang, Q.; Lian, X.M. Transcriptomic analysis of rice responses to low phosphorus stress. *Chin. Sci. Bull.* **2010**, *55*, 251–258. [CrossRef]

65. Wissuwa, M.; Yano, M.; Ae, N. Mapping of QTLs for phosphorus-deficiency tolerance in rice (*Oryza sativa* L.). *Theor. Appl. Genet.* **1998**, *97*, 777–783. [CrossRef]

66. Wissuwa, M.; Ae, N. Further characterization of two QTLs that increase phosphorus uptake of rice (*Oryza sativa* L.) under phosphorus deficiency. *Plant Soil.* **2001**, *237*, 275–286. [CrossRef]

67. Guo, Y.; Lin, W.; Shi, Q.; Liang, Y.; Chen, F.; He, H.; Liang, K. Screening methodology for rice (*Oryza sativa*) genotypes with high phosphorus use efficiency at their seedling stage. *J. Appl. Ecol.* **2002**, *13*, 1587–1591.

68. Yuan, H.; Liu, D. Signaling components involved in plant responses to phosphate starvation. *J. Integr. Plant Biol.* **2008**, *50*, 849–859. [CrossRef] [PubMed]

69. Chin, J.H.; Lu, X.; Haefele, S.M.; Gamuyao, R.; Ismail, A.; Wissuwa, M.; Heuer, S. Development and application of gene-based markers for the major rice QTL *Phosphate uptake 1. Theor. Appl. Genet.* **2010**, *120*, 1073–1086. [CrossRef] [PubMed]

70. Ramaekers, L.; Remans, R.; Rao, I.M.; Blair, M.W.; Vanderleyden, J. Strategies for improving phosphorus acquisition efficiency of crop plants. *Field Crops Res.* **2010**, *117*, 169–176. [CrossRef]

71. Wissuwa, M.; Mazzola, M.; Picard, C. Novel approaches in plant breeding for rhizosphere-related traits. *Plant Soil* **2009**, *321*, 409–430. [CrossRef]

72. Chen, L.; Lin, L.; Cai, G.; Sun, Y.; Huang, T.; Wang, K.; Deng, J. Identification of Nitrogen, Phosphorus, and Potassium Deficiencies in Rice Based on Static Scanning Technology and Hierarchical Identification Method. *PLoS ONE* **2014**, *9*, e113200. [CrossRef] [PubMed]

73. Place, G.A.; Sims, J.L.; Hall, U.L. Effects of nitrogen and phosphorous on the growth yield and cooking characteristics of rice. *Agron. J.* **1970**, *62*, 239–241. [CrossRef]

74. Nguyen, H.T.T.; Pham, C.V.; Bertin, P. The effect of nitrogen concentration on nitrogen use efficiency and related parameters in cultivated rices (*Oryza sativa* L. subsp. *indica* and *japonica* and *O. glaberrima* Steud.) in hydroponics. *Euphytica* **2014**, *198*, 137–151. [CrossRef]

75. Vitousek, P.M.; Naylor, R.; Crews, T.; David, M.B.; Drinkwater, L.E.; Holland, E.; Johnes, P.J.; Katzenberger, J.; Martinelli, L.A.; Matson, P.A.; et al. Agriculture. Nutrient imbalances in agricultural development. *Science* **2009**, *324*, 1519–1520. [CrossRef] [PubMed]

76. Bouwman, A.F.; Boumans, L.J.M.; Batjes, N.H. Emissions of N_2O and NO from fertilised fields: Summary of available measurement data. *Glob. Biogeochem. Cycles* **2002**, *16*, 6-1–6-13. [CrossRef]

77. Samborski, S.; Kozak, M.; Azevedo, R.A. Does nitrogen uptake affect nitrogen uptake efficiency or vice versa? *Acta Physiol. Plant.* **2008**, *30*, 419–420. [CrossRef]

78. Li, Y.; Yang, X.; Ren, B.; Shen, Q.; Guo, S. Why nitrogen use efficiency decreases under high nitrogen supply in rice (*Oryza sativa* L.) seedlings. *J. Plant Growth Regul.* **2012**, *31*, 47–52. [CrossRef]

79. Singh, H.; Verma, A.; Ansari, M.A.; Shukla, A. Physiological response of rice (*Oryza sativa* L.) genotypes to elevated nitrogen applied under field conditions. *Plant Signal. Behav.* **2015**, *9*, e29015. [CrossRef] [PubMed]

80. Chaturvedi, I. Effect of Nitrogen Fertilizers on Growth, Yield and Quality of Hybrid Rice (*Oryza sativa* L.). *J. Cent. Eur. Agric.* **2005**, *6*, 611–618.

81. Manzoor, Z.; Awan, T.H.; Zahid, M.A.; Faiz, F.A. Response of rice crop (Super Basmati) to different nitrogen levels. *J. Anim. Plant Sci.* **2006**, *16*, 1–2.

82. Swamy, K.N.; Kondamudi, R.; Kiran, T.V.; Vijayalakshmi, P.; Rao, Y.V.; Rao, P.R.; Subrahmanyam, D.; Voleti, S.R. Screening for nitrogen use efficiency with their root characteristics in rice (*Oryza* spp.) genotypes. *Ann. Biol. Sci.* **2015**, *3*, 8–11.

83. Haque, M.A.; Haque, M.M. Growth, Yield and Nitrogen Use Efficiency of New Rice Variety under Variable Nitrogen Rates. *Am. J. Plant Sci.* **2016**, *7*, 612–622. [CrossRef]

84. Kumagai, E.; Araki, T.; Kubota, F. Effects of nitrogen supply restriction on gas exchange and photosystem 2 function in flag leaves of a traditional low-yield cultivar and a recently improved high-yield cultivar of rice (*Oryza sativa* L.). *Photosynthetica* **2007**, *45*, 489–495. [CrossRef]

85. Maske, N.S.; Borkar, S.L.; Rajgire, H.J. Effects of Nitrogen Levels on Growth, Yield and Grain Quality of Rice. *J. Soil Crop* **1997**, *7*, 83–86.

86. Peng, S.; Cassman, K.G.; Virmani, S.S.; Sheehy, J.; Khush, G.S. Yield potential trends of tropical rice since the release of IR8 and the challenge of increasing rice yield potential. *Crop Sci.* **1999**, *39*, 1552–1559. [CrossRef]

87. Lawlor, D.W. Carbon and nitrogen assimilation in relation to yield: Mechanisms are the key to understanding production systems. *J. Exp. Bot.* **2002**, *53*, 773–787. [CrossRef] [PubMed]

88. Yang, J.; Peng, S.; Zhang, Z.; Wang, Z.; Visperas, R.M.; Zhu, Q. Grain and dry matter yields and portioning of assimilates in Japonica/Indica hybrid rice. *Crop Sci.* **2002**, *42*, 766–772. [CrossRef]

89. Ahmed, M.; Islam, M.M.; Paul, S.K.; Khulna, B. Effect of Nitrogen on Yield and Other Plant Characters of Local, T. Aman Rice Var. Jatai. *Res. J. Agric. Biol. Sci.* **2005**, *1*, 158–161.

90. Hamaoka, N.; Uchida, Y.; Tomita, M.; Kumagai, E.; Araki, T.; Ueno, O. Genetic variations in dry matter production, nitrogen uptake, and nitrogen use efficiency in the AA genome *Oryza* species grown under different nitrogen conditions. *Plant Prod. Sci.* **2013**, *16*, 107–116. [CrossRef]

91. Yogendra, N.D.; Kumara, B.H.; Chandrashekar, N.; Prakash, N.B.; Anantha, M.S.; Shashidhar, H.E. Real-time nitrogen management in aerobic rice by adopting leaf color chart (LCC) as influenced by silicon. *J. Plant Nutr.* **2017**, *40*, 1277–1286. [CrossRef]

92. Mandal, N.N.; Chaudhry, P.P.; Sinha, D. Nitrogen, phosphorus and potash uptake of wheat (var. Sonalika). *Environ. Ecol.* **1992**, *10*, 297.

93. Wang, Y.; Wu, W.H. Genetic approaches for improvement of the crop potassium acquisition and utilization efficiency. *Curr. Opin. Plant Biol.* **2015**, *25*, 46–52. [CrossRef] [PubMed]

94. FAO. *Current World Fertilizer Trends and Outlook to 2016*; Food and Agriculture Organization of the United Nations: Rome, Italy, 2012.

95. Dobermann, A.; Cassman, K.G.; Mamaril, C.P.; Sheehy, J.E. Management of phosphorus, potassium and sulfur in intensive irrigated lowland rice. *Field Crops Res.* **1998**, *56*, 113–138. [CrossRef]

96. Xiaoe, Y.; Romheld, V.; Marschner, H.; Baligar, V.C.; Martens, D.C. Shoot photosynthesis and root growth of hybrid and conventional rice cultivars as affected by N and K levels in the root zone. *Pedosphere* **1997**, *7*, 35–42.

97. Epstein, E.; Bloom, A.J. *Mineral Nutrition of Plants: Principles and Perspectives*, 2nd ed.; Sinauer Associates: Sunderland, MA, USA, 2005.

98. Fageria, N.K.; Dos Santos, A.B.; Moreira, A.; Moraes, M.F. Potassium soil test calibration for lowland rice on an inceptisol. *Commun. Soil Sci. Plant Anal.* **2010**, *41*, 2595–2601. [CrossRef]

99. Grzebisz, W.; Gransee, A.; Szczepaniak, W. The effects of potassium fertilization on water-use efficiency in crop plants. *J. Pant Nutr. Soil Sci.* **2013**, *176*, 355–374. [CrossRef]

100. Mehdi, S.M.; Sarfraz, M.; Hafeez, M. Response of rice advanced line PB-95 to potassium in saline sodic soil. *Pak. J. Biol. Sci.* **2007**, *10*, 2938–2939.

101. Fageria, N.K.; Dos Santos, A.B.; De Moraes, M.F. Yield, Potassium Uptake, and Use Efficiency in Upland Rice Genotypes. *Commun. Soil Sci. Plant Anal.* **2010**, *41*, 2676–2684. [CrossRef]

102. Arif, M.; Arshad, M.; Asghar, H.N.; Basara, S.M.A. Response of rice (*Oryza sativa*) genotypes varying in K use efficiency to various levels of potassium. *Int. J. Agric. Biol.* **2010**, *12*, 926–930.

103. Kalita, U.; Ojha, N.J.; Talukdar, M.C. Effect of levels and time of potassium application on yield and yield attributes of upland rice. *J. Potassium Res.* **1993**, *11*, 203–206.

104. Dunn, D.; Stevens, G. Rice potassium nutrition research progress (Missouri). *Better Crops* **2005**, *89*, 15–17.

105. Awan, T.H.; Manzoor, Z.; Safdar, M.E.; Ahmad, M. Yield response of rice to dynamic use of potassium in traditional rice growing area of Punjab. *Pak. J. Agric. Sci.* **2007**, *44*, 130–135.

106. Bahmaniar, M.A.; Ranjbar, G.A. Effects of nitrogen and potassium fertilizers on rice (*Oryza sativa* L.) genotypes processing characteristics. *Pak. J. Biol. Sci.* **2007**, *10*, 1829–1834. [PubMed]

107. Sarkar, R.K.; Malik, G.C. Effect of foliar spray of KNO$_3$ and Ca (NO$_3$)^2 on grass pea (*Lathyrus sativus* L.) grown in rice fallows. *Lathyrus Lathyrism Newslett.* **2001**, *2*, 47–48.

108. Islam, A.; Saha, P.K.; Biswas, J.C.; Saleque, M.A. Potassium Fertilization in Intensive Wetland Rice System: Yield, Potassium Use Efficiency and Soil Potassium Status. *Int. J. Agric. Pap.* **2016**, *1*, 7–21.

109. De, D.S.K.; Broadbent, F.E. Development changes related to nitrogen-use efficiency in rice. *Field Crops Res.* **1993**, *34*, 47–56.

110. William, R.R.; Johnson, G.V. Improving nitrogen use efficiency for cereal production. *Agron. J.* **1999**, *91*, 357–363.

111. Gregard, A.; Gelanger, G.; Michaud, R. Nitrogen use efficiency and morphological characteristics of timothy populations selected for low and high forage nitrogen concentrations. *Crop Sci.* **2000**, *40*, 422–429. [CrossRef]

112. Anil, K.; Nidhi, G.; Atul, K.G.; Vikram, S.G. Identification of Biomarker for Determining Genotypic Potential of Nitrogen-Use-Efficiency and Optimization of the Nitrogen Inputs in Crop Plants. *J. Crop Sci. Biotechnol.* **2009**, *12*, 183–194.

113. Pingali, P.L. Green revolution: Impacts, limits, and the path ahead. *Proc. Natl. Acad. Sci. USA* **2012**, *109*, 12302–12308. [CrossRef] [PubMed]

114. Beatty, P.H.; Shrawat, A.K.; Carroll, R.T.; Zhu, T.; Good, A.G. Transcriptome analysis of nitrogen-efficient rice over-expressing alanine aminotransferase. *Plant Biotechnol. J.* **2009**, *7*, 562–576. [CrossRef] [PubMed]

115. Kant, S.; Bi, Y.; Rothstein, S.J. Understanding plant response to nitrogen limitation for the improvement of crop nitrogen use efficiency. *J. Exp. Bot.* **2011**, *62*, 1499–1509. [CrossRef] [PubMed]

116. Kabir, G. Genetic approaches of increasing nutrient use efficiency especially nitrogen in cereal crops—A review. *J. Bio-Sci.* **2014**, *22*, 111–125. [CrossRef]

117. Rose, T.J.; Kretzschmar, T.; Waters, D.L.E.; Balindong, J.L.; Wissuwa, M. Prospects for Genetic Improvement in Internal Nitrogen Use Efficiency in Rice. *Agronomy* **2017**, *7*, 70. [CrossRef]

118. Zhou, Y.; Tao, Y.; Tang, D.; Wang, J.; Zhong, J.; Wang, Y.; Yuan, Q.; Yu, X.; Zhang, Y. Identification of QTL Associated with Nitrogen Uptake and Nitrogen Use Efficiency Using High Throughput Genotyped CSSLs in Rice (*Oryza sativa* L.). *Front. Plant Sci.* **2017**, *8*, 1166. [CrossRef] [PubMed]

119. Sinha, S.K.; Sevanthi, V.A.M.; Chaudhary, S.; Tyagi, P.; Venkadesan, S.; Rani, M.; Mandal, P.K. Transcriptome Analysis of Two Rice Varieties Contrasting for Nitrogen Use Efficiency under Chronic N Starvation Reveals Differences in Chloroplast and Starch Metabolism-Related Genes. *Genes (Basel)* **2018**, *11*, E206. [CrossRef] [PubMed]

120. Duan, Y.H.; Zhang, Y.L.; Ye, L.T.; Fan, X.R.; Xu, G.H.; Shen, Q.R. Responses of Rice Cultivars with Different Nitrogen Use Efficiency to Partial Nitrate Nutrition. *Ann. Bot.* **2007**, *99*, 1153–1160. [CrossRef] [PubMed]

121. Fan, X.; Xie, D.; Chen, J.; Lu, H.; Xu, Y.; Ma, C.; Xu, G. Over-expression of OsPTR6 in rice increased plant growth at different nitrogen supplies but decreased nitrogen use efficiency at high ammonium supply. *Plant Sci.* **2014**, *227*, 1–11. [CrossRef] [PubMed]

122. Ashikari, M.; Sakakibara, H.; Lin, S.; Yamamoto, T.; Takashi, T.; Nishimura, A.; Angeles, E.R.; Qian, Q.; Kitano, H.; Matsuoka, M. Cytokinin oxidase regulates rice grain production. *Science* **2005**, *309*, 741–745. [CrossRef] [PubMed]

123. Huang, X.; Qian, Q.; Liu, Z.; Sun, H.; He, S.; Luo, D.; Xia, G.; Chu, C.; Li, J.; Fu, X. Natural variation at the DEP1 locus enhances grain yield in rice. *Nat. Genet.* **2009**, *41*, 494–497. [CrossRef] [PubMed]

124. Miura, K.; Ikeda, M.; Matsubara, A.; Song, X.J.; Ito, M.; Asano, K.; Matsuoka, M.; Kitano, H.; Ashikari, M. OsSPL14 promotes panicle branching and higher grain productivity in rice. *Nat. Genet.* **2010**, *42*, 545–549. [CrossRef] [PubMed]

125. Jiao, Y.; Wang, Y.; Xue, D.; Wang, J.; Yan, M.; Liu, G.; Dong, G.; Zeng, D.; Lu, Z.; Zhu, X.; Qian, Q.; Li, J. Regulation of OsSPL14 by OsmiR156 defines ideal plant architecture in rice. *Nat. Genet.* **2010**, *42*, 541–544. [CrossRef] [PubMed]

126. Song, X.J.; Huang, W.; Shi, M.; Zhu, M.Z.; Lin, H.X. A QTL for rice grain width and weight encodes a previously unknown RING-type E3 ubiquitin ligase. *Nat. Genet.* **2007**, *39*, 623–630. [CrossRef] [PubMed]

127. Shomura, A.; Izawa, T.; Ebana, K.; Ebitani, T.; Kanegae, H.; Konishi, S.; Yano, M. Deletion in a gene associated with grain size increased yields during rice domestication. *Nat. Genet.* **2008**, *40*, 1023–1028. [CrossRef] [PubMed]

128. Weng, J.; Gu, S.; Wan, X.; Gao, H.; Guo, T.; Su, N.; Lei, C.; Zhang, X.; Cheng, Z.; Guo, X.; et al. Isolation and initial characterization of GW5,a major QTL associated with rice grain width and weight. *Cell Res.* **2008**, *18*, 1199–1209. [CrossRef] [PubMed]

129. Wang, E.; Wang, J.; Zhu, X.; Hao, W.; Wang, L.; Li, Q.; Zhang, L.; He, W.; Lu, B.; Lin, H.; et al. Control of rice grain-filling and yield by a gene with a potential signature of domestication. *Nat. Genet.* **2008**, *40*, 1370–1374. [CrossRef] [PubMed]

130. Yano, M.; Katayose, Y.; Ashikari, M.; Yamanouchi, U.; Monna, L.; Fuse, T.; Baba, T.; Yamamoto, K.; Umehara, Y.; Nagamura, Y.; et al. *Hd1*, a major photoperiod sensitivity quantitative trait locus in rice, is closely related to the Arabidopsis flowering time gene *CONSTANS*. *Plant Cell* **2000**, *12*, 2473–2483. [CrossRef] [PubMed]

131. Takahashi, Y.; Shomura, A.; Sasaki, T.; Yano, M. Hd6, a rice quantitative trait locus involved in photoperiod sensitivity, encodes the subunit of protein kinase CK2. *Proc. Natl. Acad. Sci. USA* **2001**, *98*, 7922–7927. [CrossRef] [PubMed]

132. Kojima, S.; Takahashi, Y.; Kobayashi, Y.; Monna, L.; Sasaki, T.; Araki, T.; Yano, M. Hd3a, a rice ortholog of the Arabidopsis FT gene, promotes transition to flowering downstream of Hd1 under shortday conditions. *Plant Cell Physiol.* **2002**, *43*, 1096–1105. [CrossRef] [PubMed]

133. Izawa, T.; Oikawa, T.; Sugiyama, N.; Tanisaka, T.; Yano, M.; Shimamoto, K. Phytochrome mediates the external light signal to repress FT orthologs in photoperiodic flowering of rice. *Genes Dev.* **2002**, *16*, 2006–2020. [CrossRef] [PubMed]

134. Tamaki, S.; Matsuo, S.; Wong, H.L.; Yokoi, S.; Shimamoto, K. Hd3a protein is a mobile flowering signal in rice. *Science* **2007**, *316*, 1033–1036. [CrossRef] [PubMed]

135. Doi, K.; Izawa, T.; Fuse, T.; Yamanouchi, U.; Kubo, T.; Shimatani, Z.; Yano, M.; Yoshimura, A. Ehd1, a B-type response regulator in rice, confers short-day promotion of flowering and controls FT-like gene expression independently of Hd1. *Genes Dev.* **2004**, *18*, 926–936. [CrossRef] [PubMed]

136. Xue, W.; Xing, Y.; Weng, X.; Zhao, Y.; Tang, W.; Wang, L.; Zhou, H.; Yu, S.; Xu, C.; Li, X.; Zhang, Q. Natural variation in Ghd7 is an important regulator of heading date and yield potential in rice. *Nat. Genet.* **2008**, *40*, 761–767. [CrossRef] [PubMed]

137. Wei, X.; Xu, J.; Guo, H.; Jiang, L.; Chen, S.; Yu, C.; Zhou, Z.; Hu, P.; Zhai, H.; Wan, J. DTH8 suppresses flowering in rice, influencing plant height and yield potential simultaneously. *Plant Physiol.* **2010**, *153*, 1747–1758. [CrossRef] [PubMed]

138. Sasaki, A.; Ashikari, M.; Ueguchi-Tanaka, M.; Itoh, H.; Nishimura, A.; Swapan, D.; Ishiyama, K.; Saito, T.; Kobayashi, M.; Khush, G.S.; et al. Green revolution: A mutant gibberellin-synthesis gene in rice. *Nature* **2002**, *416*, 701–702. [CrossRef] [PubMed]

139. Ookawa, T.; Hobo, T.; Yano, M.; Murata, K.; Ando, T.; Miura, H.; Asano, K.; Ochiai, Y.; Ikeda, M.; Nishitani, R.; et al. New approach for rice improvement using a pleiotropic *QTL* gene for lodging resistance and yield. *Nat. Commun.* **2010**, *1*, 132. [CrossRef] [PubMed]

140. Fukuoka, S.; Saka, N.; Koga, H.; Ono, K.; Shimizu, T.; Ebana, K.; Hayashi, N.; Takahashi, A.; Hirochika, H.; Okuno, K.; Yano, M. Loss of function of a proline-containing protein confers durable disease resistance in rice. *Science* **2009**, *325*, 998–1001. [CrossRef] [PubMed]

141. Hayashi, N.; Inoue, H.; Kato, T.; Funao, T.; Shirota, M.; Shimizu, T.; Kanamori, H.; Yamane, H.; Hayano-Saito, Y.; Matsumoto, T. Durable panicle blast-resistance gene Pb1 encodes an atypical CC-NBS-LRR protein and was generated by acquiring a promoter through local genome duplication. *Plant J.* **2010**, *64*, 498–510. [CrossRef] [PubMed]

142. Ren, Z.H.; Gao, J.P.; Li, L.G.; Cai, X.L.; Huang, W.; Chao, D.Y.; Zhu, M.Z.; Wang, Z.Y.; Luan, S.; Lin, H.X. A rice quantitative trait locus for salt tolerance encodes a sodium transporter. *Nat. Genet.* **2005**, *37*, 1141–1146. [CrossRef] [PubMed]

143. Fujino, K.; Sekiguchi, H.; Matsuda, Y.; Sugimoto, K.; Ono, K.; Yano, M. Molecular identification of a major quantitative trait locus, qLTG3-1, controlling low-temperature germinability in rice. *Proc. Natl. Acad. Sci. USA* **2008**, *105*, 12623–12628. [CrossRef] [PubMed]

144. Xu, K.; Xu, X.; Fukao, T.; Canlas, P.; Maghirang-Rodriguez, R.; Heuer, S.; Ismail, A.M.; Bailey-Serres, J.; Ronald, P.C.; Mackill, D.J. *Sub1A* is an ethylene-response-factor-like gene that confers submergence tolerance to rice. *Nature* **2006**, *442*, 705–708. [CrossRef] [PubMed]

145. Hattori, U.Y.; Nagai, K.; Furukawa, S.; Song, X.J.; Kawano, R.; Sakakibara, H.; Wu, J.; Matsumoto, T.; Yoshimura, A.; Kitano, H.; Matsuoka, M. The ethylene response factors SNOKEL1 and SNOKEL2 allow rice to adapt to deep water. *Nature* **2009**, *460*, 1026–1030. [CrossRef] [PubMed]

146. Ueno, D.; Koyama, E.; Kono, I.; Ando, T.; Yano, M.; Ma, J.F. Identification of a novel major quantitative trait locus controlling distribution of Cd between roots and shoots in rice. *Plant Cell Physiol.* **2009**, *50*, 2223–2233. [CrossRef] [PubMed]

147. Li, C.; Zhou, A.; Sang, T. Rice domestication by reducing shattering. *Science* **2006**, *311*, 1936–1939. [CrossRef] [PubMed]

148. Konishi, S.; Izawa, T.; Lin, S.Y.; Ebana, K.; Fukuta, Y.; Sasaki, T.; Yano, M. An SNP caused loss of seed shattering during rice domestication. *Science* **2006**, *312*, 1392–1396. [CrossRef] [PubMed]

149. Tan, L.; Li, X.; Liu, F.; Sun, X.; Li, C.; Zhu, Z.; Fu, Y.; Cai, H.; Wang, X.; Xie, D.; Sun, C. Control of a key transition from prostrate to erect growth in rice domestication. *Nat. Genet.* **2008**, *40*, 1360–1364. [CrossRef] [PubMed]

150. Jin, J.; Huang, W.; Gao, J.P.; Yang, J.; Shi, M.; Zhu, M.Z.; Luo, D.; Lin, H.X. Genetic control of rice plant architecture under domestication. *Nat. Genet.* **2008**, *40*, 1365–1369. [CrossRef] [PubMed]

151. Romero, L.E.; Lozano, I.; Garavito, A.; Carabali, S.J.; Triana, M.; Villareal, N.; Reyes, L.; Duque, M.C.; Martinez, C.P. Major QTLs Control Resistance to Rice Hoja Blanca Virus and Its Vector *Tagosodes orizicolus*. *G3 (Bethesda)* **2014**, *4*, 133–142. [CrossRef] [PubMed]

152. Uga, Y.; Yamamoto, E.; Kanno, N.; Kawai, S.; Mizubayashi, T.; Fukuoka, S. A major QTL controlling deep rooting on rice chromosome 4. *Sci. Rep.* **2013**, *3*, 3040. [CrossRef] [PubMed]

153. Ni, J.J.; Wu, P.; Senadhira, D.; Huang, N. Mapping QTLs for phosphorus deficiency tolerance in rice (*Oryza sativa* L.). *Theor. Appl. Genet.* **1998**, *97*, 1361–1369. [CrossRef]

154. Wissuwa, M.; Wegner, J.; Ae, N.; Yano, M. Substitution mapping of *Pup1*: A major QTL increasing phosphorus uptake of rice from a phosphorus-deficient soil. *Theor. Appl. Genet.* **2002**, *105*, 890–897. [PubMed]

155. Shimizu, A.; Yanagihara, S.; Kawasaki, S.; Ikehashi, H. Phosphorus deficiency-induced root elongation and its QTL in rice (*Oryza sativa* L.). *Theor. Appl. Genet.* **2004**, *109*, 1361–1368. [CrossRef] [PubMed]

156. Lang, N.T.; Buu, B.C. Mapping QTLs for phosphorus deficiency tolerance in rice (*Oryza sativa* L.). *Omonrice* **2006**, *14*, 1–9.

157. Shimizu, A.; Kato, K.; Komatsu, A.; Motomura, K.; Ikehashi, H. Genetic analysis of root elongation induced by phosphorus deficiency in rice (*Oryza sativa* L.): Fine QTL mapping and multivariate analysis of related traits. *Theor. Appl. Genet.* **2008**, *117*, 987–996. [CrossRef] [PubMed]

158. Chao, X.; Jie, R.; Xiu-qin, Z.; Zai-song, D.; Jing, Z.; Chao, W.; Jun-wei, Z.; Joseph, C.A.; Qiang, Z.; et al. Genetic Dissection of Low Phosphorus Tolerance Related Traits Using Selected Introgression Lines in Rice. *Rice Sci.* **2015**, *22*, 264–274. [CrossRef]

159. Wang, K.; Cui, K.; Liu, G.; Xie, W.; Yu, H.; Pan, J.; Huang, J.; Nie, L.; Shah, F. Identification of quantitative trait loci for phosphorus use efficiency traits in rice using a high density SNP map. *BMC Genet.* **2014**, *15*, 155. [CrossRef] [PubMed]

160. Feng, M.; Xianwu, Z.; Guohua, M.; He, P.; Zhu, L.; Zhang, F. Identification of quantitative trait loci affecting tolerance to low phosphorus in rice (*Oryza sativa* L.). *Chin. Sci. Bull.* **2000**, *45*, 519–525.

161. Fang, P.; Wu, P. QTL x N-level interaction for plant height in rice (*Oriza sativa* L.). *Plant Soil* **2001**, *236*, 237–242. [CrossRef]

162. Ishimaru, K.; Kobayashi, N.; Ono, K.; Yano, M.; Ohsugi, R. Are contents of Rubisco, soluble protein and nitrogen in flag leaves of rice controlled by the same genetics? *J. Exp. Bot.* **2001**, *52*, 1827–1833. [CrossRef] [PubMed]

163. Ogawa, S.; Valencia, M.O.; Lorieux, M.; Arbelaez, J.D.; McCouch, M.; Ishitani, M.; Selvaraj, M.G. Identification of QTLs associated with agronomic performance under nitrogen-deficient conditions using chromosome segment substitution lines of a wild rice relative, *Oryza rufipogon*. *Acta Physiol. Plant.* **2016**, *38*, 103. [CrossRef]

164. Obara, M.; Kajiura, M.; Fukuta, Y.; Yano, M.; Hayashi, M.; Yamaya, T.; Sato, T. Mapping of QTLs associated with cytosolic glutamine synthetase and NADH- glutamate synthase in rice (*Oryza sativa* L.). *J. Exp. Bot.* **2001**, *52*, 1209–1217. [PubMed]

165. Senthilvel, S.; Govindaraj, P.; Arumugachamy, S.; Latha, R.; Malarvizhi, P.; Gopalan, A.; Maheswaran, M. Mapping genetic loci associated with nitrogen use efficiency in rice (*Oryza sativa* L.). In Proceedings of the 4th International Crop Science Congress, Brisbane, Australia, 26 September–1 October 2004.

166. Tong, H.H.; Mei, H.W.; Yu, X.Q.; Xu, X.Y.; Li, M.S.; Zhang, S.Q.; Luo, L.J. Identification of Related QTLs at Late Developmental Stage in Rice (*Oryza sativa* L.) Under Two Nitrogen Levels. *Acta Genet. Sin.* **2006**, *33*, 458–467. [CrossRef]

167. Wang, Y.; Sun, Y.J.; Chen, D.Y.; Yu, S.B. Analysis of Quantitative Trait Loci in Response to Nitrogen and Phosphorus Deficiency in Rice Using Chromosomal Segment Substitution Lines. *Acta Agron. Sin.* **2009**, *35*, 580–587. [CrossRef]

168. Laza, M.R.; Kondo, M.; Ideta, O.; Barleen, E.; Imbe, T. Identification of quantitative trait loci for d13C and productivity in irrigated lowland rice. *Crop Sci.* **2006**, *46*, 763–773. [CrossRef]

169. MacMillan, K.; Emrich, K.; Piepho, H.P.; Mullins, C.E.; Price, A.H. Assessing the importance of genotype × environment interaction for root traits in rice using a mapping population II: Conventional QTL analysis. *Theor. Appl. Genet.* **2006**, *113*, 953–964. [CrossRef] [PubMed]

170. Cho, Y.I.; Jiang, W.Z.; Chin, J.H.; Piao, Z.; Cho, Y.G.; McCouch, S.; Koh, H.J. Identification of QTLs associated with physiological nitrogen use efficiency in rice. *Mol. Cell* **2007**, *23*, 72–79.

171. Senthilvel, S.; Vinod, K.K.; Malarvizhi, P.; Maheswaran, M. QTL and QTL × environment effects on agronomic and nitrogen acquisition traits in rice. *J. Integr. Plant Biol.* **2008**, *50*, 1108–1117. [CrossRef] [PubMed]

172. Piao, Z.; Li, M.; Li, P.; Zhang, C.; Wang, H.; Luo, Z.; Lee, J.; Yang, R. Bayesian dissection for genetic architecture of traits associated with nitrogen utilization efficiency in rice. *Afr. J. Biotechnol.* **2009**, *8*, 6834–6839.

173. Srividya, A.; Vemireddy, L.R.; Hariprasad, A.S.; Jayaprada, M.; Sakile, S.; Puram, V.R.R.; Anuradha, G.; Siddiq, E.A. Identification and mapping of landrace derived QTL associated with yield and its components in rice under different nitrogen levels and environments. *Int. J. Plant Breed. Genet.* **2010**, *4*, 210–227. [CrossRef]

174. Tong, H.H.; Chen, L.; Li, W.P.; Mei, H.; Xing, Y.; Yu, X.; Xu, X.; Zhang, S.; Luo, L. Identification and characterization of quantitative trait loci for grain yield and its components under different nitrogen fertilization levels in rice (*Oryza sativa* L.). *Mol. Breed.* **2011**, *28*, 495–509. [CrossRef]

175. Yue, F.; Rong-rong, Z.; Ze-chuan, L.; Li-yong, C.; Xing-hua, W.; Shi-hua, C. Quantitative trait locus analysis for rice yield traits under two nitrogen levels. *Rice Sci.* **2015**, *22*, 108–115. [CrossRef]

176. Wei, D.; Cui, K.; Ye, G.; Pan, J.; Xiang, J.; Huang, J.; Nie, L. QTL mapping for nitrogen-use efficiency and nitrogen-deficiency tolerance traits in rice. *Plant Soil* **2012**, *359*, 281–295. [CrossRef]

177. Wu, P.; Ni, J.J.; Luo, A.C. QTLs underlying Rice Tolerance to Low-Potassium Stress in Rice Seedlings. *Crop Sci.* **1998**, *38*, 1458–1462. [CrossRef]

178. Senaratne, R.; Ratnasinghe, D.S. Nitrogen fixation and beneficial effects of some grain legumes and green-manure crops on rice. *Biol. Fer. Soils* **1995**, *19*, 49–54. [CrossRef]

179. Lin, X.Q.; Zhou, W.J.; Zhu, D.F.; Zhang, Y. Effect of water management on photosynthetic rate and water use efficiency of leaves in paddy rice. *Chin. J. Rice Sci.* **2004**, *18*, 333–338, (in Chinese with English abstract).

180. Peng, S.; Cassman, K.G. Upper thresholds of nitrogen uptake rates and associated nitrogen fertilizer efficiencies in irrigated rice. *Agron. J.* **1998**, *90*, 178–185. [CrossRef]

181. Yamaya, T.; Obara, M.; Nakajima, H.; Sasaki, S.; Hayakawa, T.; Sato, T. Genetic manipulation and quantitative trait loci mapping for nitrogen recycling in rice. *J. Exp. Bot.* **2002**, *53*, 917–925. [CrossRef] [PubMed]

182. Lian, X.; Xing, Y.; Yan, H.; Xu, C.; Li, X.; Zhang, Q. QTLs for low nitrogen tolerance at seedling stage identified using a recombinant inbred line population derived from an elite rice hybrid. *Theor. Appl. Genet.* **2005**, *112*, 85–96. [CrossRef] [PubMed]

183. De, M.; Velk, P.L.G. The role of Azolla cover in improving the nitrogen use efficiency of lowland rice. *Plant Soil* **2004**, *263*, 311–321.

184. Ladha, J.K.; Kirk, G.J.D.; Bennett, J.; Peng, S.; Reddy, C.K.; Reddy, P.M.; Singh, U. Opportunities for increased nitrogen use efficiency from improved lowland rice germplasm. *Field Crops Res.* **1998**, *56*, 41–71. [CrossRef]

185. Obara, M.; Sato, T.; Sasaki, S.; Kashiba, K.; Nagano, A.; Nakamura, I.; Ebitani, T.; Yano, M.; Yamaya, T. Identification and characterization of a QTL on chromosome 2 for cytosolic glutamine synthetase content and panicle number in rice. *Theor. Appl. Genet.* **2004**, *110*, 1–11. [CrossRef] [PubMed]

186. Shan, Y.H.; Wang, Y.; Pan, X.B. Mapping of QTLs for nitrogen use efficiency and related traits in rice (*Oryza sativa* L.). *Agric. Sci. China* **2005**, *4*, 721–727.

187. Dong, G.C.; Wang, Y.L.; Zhang, Y.F.; Chen, P.; Yang, L.; Huang, J.; Zuo, B. Characteristics of yield and yield components in conventional *indica* rice cultivars with different nitrogen use efficiencies for grain output. *Acta Agron. Sin.* **2006**, *32*, 1511–1518.

188. Dong, G.C.; Wang, Y.; Yu, X.F. Differences of nitrogen uptake and utilization of conventional rice varieties with different growth duration. *Sci. Agric. Sin.* **2011**, *44*, 4570–4582.

189. Sanchez, P.A.; Salinas, J.G. Low-input technology for managing oxisols and ultisols in tropical America. *Adv. Agron.* **1981**, *34*, 279–406.

190. Dobermann, A.; Fairhurst, T. *Rice: Nutrient Disorders & Nutrient Management*; Potash & Phosphate Institute, Potash & Phosphate Institute of Canada, and International Rice Research Institute: Singapore; Los Baños, Philippines, 2000.

191. Su, J.Y.; Xiao, Y.M.; Li, M.; Liu, Q.; Li, B.; Tong, Y.; Jia, J.; Li, Z. Mapping QTLs for phosphorus-deficiency tolerance at wheat seedling stage. *Plant Soil* **2006**, *281*, 25–36. [CrossRef]

192. Liu, Y.; Li, Z.C.; Mi, G.H.; Zhang, H.L.; Mu, P.; Wang, X. Screen and identification for tolerance to low-phosphorus stress of rice germplasm (*Oryza sativa* L.). *Acta Agron. Sin.* **2005**, *31*, 238–242. (in Chinese with English abstract)

193. Ping, M.U.; Huang, C.; Li, J.X.; Liu, L.F.; Li, Z.C. Yield Trait Variation and QTL Mapping in a DH Population of Rice Under Phosphorus Deficiency. *Acta Agron. Sin.* **2008**, *34*, 1137–1142.

194. Liu, L.J.; Chang, E.H.; Fan, M.M.; Wang, Z.Q.; Yang, J.C. Effects of Potassium and Calcium on Root Exudates and Grain Quality During Grain Filling. *Acta Agron. Sin.* **2011**, *37*, 661–669.

195. Torkashv, M.; Vahed, S. The efficiency of potassium fertilization methods on the growth of rice (*Oryza sativa* L.) under salinity stress. *Afr. J. Biotechnol.* **2011**, *10*, 15946–15952.

196. Ali, J.; Franje, N.J.; Revilleza, J.E.; Acero, B. *Breeding for Low-Input Responsive Green Super Rice (GSR) Varieties for Rainfed Lowlands of Asia and Africa. University Library*; University of the Philippines at Los Baños: Los Baños, Philippines, 2016.

197. Yorobe, J.M.; Ali, J.; Pede, V.; Rejesus, R.M.; Velarde, O.P.; Wang, W. Yield and income effects of rice varieties with tolerance of multiple abiotic stresses: The case of green super rice (GSR) and flooding in the Philippines. *Agric. Econ.* **2016**, *47*, 1–11. [CrossRef]

198. Wu, L.; Yuan, S.; Huang, L.; Sun, F.; Zhu, G.; Li, G.; Fahad, S.; Peng, S.; Wang, F. Physiological Mechanisms Underlying the High-Grain Yield and High-Nitrogen Use Efficiency of Elite Rice Varieties under a Low Rate of Nitrogen Application in China. *Front. Plant Sci.* **2016**, *7*, 1024. [CrossRef] [PubMed]

199. Wang, F.; Peng, S. Yield potential and nitrogen use efficiency of China's super rice. *J. Integr. Agric.* **2017**, *16*, 1000–1008. [CrossRef]

200. Ali, J.; Xu, J.L.; Gao, Y.; Fontanilla, M.; Li, Z.K. Breeding for yield potential and enhanced productivity across different rice ecologies through green super rice (GSR) breeding strategy. In *International Dialogue on Perception and Prospects of Designer Rice*; Muralidharan, K., Siddiq, E.A., Eds.; Society for the Advancement of Rice Research, Directorate of Rice Research: Hyderabad, India, 2013; pp. 60–68.

201. Mortvedt, J.J.; Murphy, L.S.; Follett, R.H. *Fertilizer Technology and Application*; Meister Publishing Co.: Willoughby, OH, USA, 2001.

202. Yadav, R.L. Assessing on-farm efficiency and economics of fertilizer N, P and K in rice-wheat systems of India. *Field Crops Res.* **2003**, *18*, 39–51. [CrossRef]

203. Cassman, K.G.; Gines, G.C.; Dizon, M.A.; Samson, M.I.; Alceantara, J.M. Nitrogen use efficiency in tropical lowland rice systems: Contributions from indigenous and applied nitrogen. *Fields Crops Res.* **1996**, *47*, 1–12. [CrossRef]

204. Dobermann, A.R. *Nitrogen Use Efficiency-State of the Art*; Agronomy-Faculty Publications: Lincoln, NE, USA, 2005; p. 316.

205. Wen-xia, X.; Guang-huo, W.; Qi-chun, Z.; Guo, H.C. Effects of nitrogen fertilization strategies on nitrogen use efficiency in physiology, recovery, and agronomy and redistribution of dry matter accumulation and nitrogen accumulation in two typical rice cultivars in Zhejiang. *China J. Zhejiang Univ. Sci. B* **2007**, *8*, 208–216.

206. Yoshida, S. *Fundamentals of Rice Crop Science*; IRRI: Los Baños, Laguna, Philippines, 1981; 269p.

207. Amanullah; Muhammad, A.; Almas, L.K.; Amanullaj, J.; Zahir, S.; Rahman, H.; Khalil, S.K. Agronomic Efficiency and Profitability of P-Fertilizers Applied at Different Planting Densities of Maize in Northwest. *Pak. J. Plant Nutr.* **2012**, *35*, 331–341. [CrossRef]

208. Rao, T.N. Improving nutrient use efficiency: The role of beneficial management practices. In *Better Crops-India*; IPNI–India Program 133: Gurgaon, India, 2007; Volume 1, pp. 6–7.

209. Singh, D.P. Vermiculture biotechnology and biocomposting. In *Environmental Microbiology and Biotechnology*; Singh, D.P., Dwivedi, S.K., Eds.; New Age International (P) Limited Publishers: Darya Ganj, New Delhi, 2004; pp. 97–112.

7

Molecular Mapping of QTLs for Heat Tolerance in Chickpea

Pronob J. Paul [1,2], Srinivasan Samineni [1], Mahendar Thudi [1], Sobhan B. Sajja [1], Abhishek Rathore [1], Roma R. Das [1], Aamir W. Khan [1], Sushil K. Chaturvedi [3], Gera Roopa Lavanya [2], Rajeev. K. Varshney [1] and Pooran M. Gaur [1,4,*]

[1] International Crops Research Institute for the Semi-Arid Tropics (ICRISAT), Patancheru Hyderabad 502324, India; pronobjpaul@gmail.com (P.J.P.); s.srinivasan@cgiar.org (S.S.); t.mahendar@cgiar.org (M.T.); S.Sobhan@cgiar.org (S.B.S.); a.rathore@cgiar.org (A.R.); r.das@cgiar.org (R.R.D.); A.khan@cgiar.org (A.W.K.); r.k.varshney@cgiar.org (R.K.V.)

[2] Department of Genetics and Plant Breeding, Sam Higginbottom University of Agriculture, Technology and Sciences (SHUATS), Allahabad 211007, India; lavanya.roopa@gmail.com

[3] ICAR-Indian Institute of Pulses Research (ICAR-IIPR), Kanpur 208024, India; sushilk.chaturvedi@gmail.com

[4] The UWA Institute of Agriculture, University of Western Australia, Perth, WA 6009, Australia

* Correspondence: p.gaur@cgiar.org

Abstract: Chickpea (*Cicer arietinum* L.), a cool-season legume, is increasingly affected by heat-stress at reproductive stage due to changes in global climatic conditions and cropping systems. Identifying quantitative trait loci (QTLs) for heat tolerance may facilitate breeding for heat tolerant varieties. The present study was aimed at identifying QTLs associated with heat tolerance in chickpea using 292 F_{8-9} recombinant inbred lines (RILs) developed from the cross ICC 4567 (heat sensitive) × ICC 15614 (heat tolerant). Phenotyping of RILs was undertaken for two heat-stress (late sown) and one non-stress (normal sown) environments. A genetic map spanning 529.11 cM and comprising 271 genotyping by sequencing (GBS) based single nucleotide polymorphism (SNP) markers was constructed. Composite interval mapping (CIM) analysis revealed two consistent genomic regions harbouring four QTLs each on CaLG05 and CaLG06. Four major QTLs for number of filled pods per plot (FPod), total number of seeds per plot (TS), grain yield per plot (GY) and % pod setting (%PodSet), located in the CaLG05 genomic region, were found to have cumulative phenotypic variation of above 50%. Nineteen pairs of epistatic QTLs showed significant epistatic effect, and non-significant QTL × environment interaction effect, except for harvest index (HI) and biomass (BM). A total of 25 putative candidate genes for heat-stress were identified in the two major genomic regions. This is the first report on QTLs for heat-stress response in chickpea. The markers linked to the above mentioned four major QTLs can facilitate marker-assisted breeding for heat tolerance in chickpea.

Keywords: abiotic stress; *Cicer arietinum*; candidate genes; genetics; heat-stress; molecular breeding

1. Introduction

In recent years, the adverse impact of climate change on agriculture is well recognized all over the globe. The ever-increasing day and night temperature is going to affect the production of crops, especially those grown in the winter [1]. In this context, heat-stress due to rise in temperatures remains a challenge in developing crop varieties that are adaptive to changing climatic conditions.

Chickpea is a nutrient-rich grain legume crop cultivated in arid and semi-arid regions. The chickpea grain is an excellent source of proteins along with a wide range of essential amino acids and vitamins. In the fight against hidden hunger all over the globe, the role of legumes such as chickpea is indispensable. Grown in over 60 countries and traded in over 190 countries, chickpea is the second most consumed pulse crop in the world after common bean [2]. Due to global warming, several noticeable changes occurred in the cropping system and intensity in the recent past. These are delaying the cultivation of chickpea to relatively hot conditions [1]. Generally, the crop faces heat-stress during reproductive phase under late sown condition in the tropical and semi-arid regions [3]. Reports state that the exposure to temperature, 35 °C and above, even for a few days, during reproductive phase has a negative impact on optimum yield in chickpea [4,5]. Unlike drought and other abiotic stresses, until recently, the importance of breeding for heat-stress conditions in chickpea has not been realized [1].

Grain yield under heat-stress is considered to be one of the important criteria for assessing heat tolerance in chickpea [3–5]. However, chickpea yield is known to be highly influenced by environments [6]. Due to genotype by environment (G × E) interaction, breeding for heat tolerance through conventional breeding approaches based on yield parameter sometimes limits selection for heat-stress tolerance in chickpea.

In recent years, progress has been made in genomics-enabled trait dissection in several crop plants, including chickpea. Several studies have been carried out earlier to identify the quantitative trait loci (QTLs) for tolerance to various biotic stresses [7,8], and abiotic stresses like drought tolerance [9], and salinity tolerance [10–12] in chickpea. Moreover, genomic regions associated with heat tolerance have been reported in several crops, including wheat, rice, maize, barley, potato, tomato, cowpea, azuki bean, brassica [13]. Pod setting (seed set) and grain yield have been used as proxy traits to detect QTLs for heat tolerance in different crops [14–18]. Similarly, in chickpea, the number of filled pods, total number of seeds, biomass, and harvest index were found to be significantly associated with heat tolerance [3,19]. However, to date, QTLs for heat tolerance have not been reported in chickpea.

In this study, genotyping by sequencing (GBS)-based single nucleotide polymorphism markers were used to identify key genomic regions responsible for heat tolerance. In addition, putative candidate genes for heat tolerance in these genomic regions were identified using the available chickpea genome sequence information [20].

2. Results

2.1. Response of Parents and Recombinant Inbred Lines (RILs) under Heat-Stress and Non-Stress Environments

The descriptive analysis of parents and RILs are presented in Table 1. Predicted means for all the traits in parents differed significantly in both heat-stress environments, except biomass per plot (BM). In the non-stress environment, predicted means for grain yield per plot (GY), BM, harvest index (HI) and %PodSet were non-significant between parents, while filled pods per plot (FPod) and total number of seeds per plot (TS) were significant. The range of variation in all the traits was high in stress environments (Table 1). The combined analysis of variance (ANOVA) for both the stress environments revealed that significant variation existed in RILs for all the traits measured, except BM, whereas under non-stress environment relatively low genetic variability was observed. Transgressive segregants in both directions were observed for several traits in the RIL population (Figure 1a,b).

The potential use of a trait in a breeding program relies on the heritability of that trait. Under both the heat-stress conditions, the heritability of all the traits was high (72.0–90.7%), except BM in summer 2014 (49.8%). Whereas, under non-stress environment the heritability of the traits was moderate (47.6–66.0%) (Table 1).

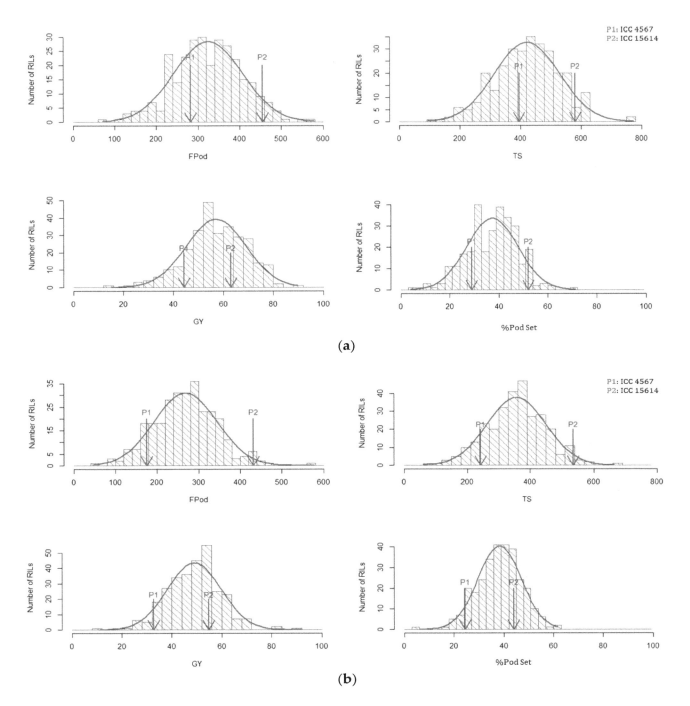

Figure 1. (a) Frequency distribution of Number of Filled Pods per Plot (FPod), Total Number of Seeds per Plot (TS), Grain Yield per Plot (GY, g), and Percent Pod Setting (%PodSet) in RIL population (ICC 4567 × ICC 15614). P1 is heat sensitive parent ICC 4567 and P2 is heat tolerant parent ICC 15614. The left portion of the P1 on the X-axis indicates the negative transgressive segregants, conversely, the right portion of the P2 on the X-axis indicates the positive transgressive segregants in heat-stress environment, 2013; **(b)** Frequency distribution of Number of Filled Pods per Plot (FPod), Total Number of Seeds per Plot (TS), Grain Yield per Plot (GY, g), and Percent Pod Setting (%PodSet) in RIL population (ICC 4567 × ICC 15614). P1 is heat sensitive parent ICC 4567 and P2 is heat tolerant parent ICC 15614. The left portion of the P1 on the X-axis indicates the negative transgressive segregants, conversely, the right portion of the P2 on the X-axis indicates the positive transgressive segregants in heat-stress environment, 2014.

Table 1. Summary statistics and heritability (H^2) values for the measured traits of 292 RILs in non-stress and heat-stress environments.

Trait	Visual Score	Filled Pods Plot^{-1}	Total No. of Seeds Plot^{-1}	Grain Yield Plot^{-1} (g)	Biomass Plot^{-1} (g)	Harvest Index	Percent PodSet (%)
Non-stress Environment, 2013							
ICC 4567 (heat sensitive)	-	406.8	429.2	76.0	144.8	52.1	67.7
ICC 15614 (heat tolerant)	-	538.7	553.0	70.2	132.3	53.9	75.6
Contrast analysis between parents	-	−131.9 *	−123.9 *	5.8 ns	12.5 ns	−1.9 ns	−7.9 ns
Mean of RILs	-	459.0	486.3	73.5	139.7	53.0	68.8
Range of RILs	-	360.8–580.1	378.3–604.7	57.6–93.3	118.1–165.2	45.5–59.2	48.1–84.2
Heritability (%)	-	62.1	60.5	57.6	47.6	63.4	66.0
Heat-stress environment, 2013							
ICC 4567 (heat sensitive)	2	281.3	395.1	44.3	147.6	34.2	28.8
ICC 15614 (heat tolerant)	5	455.6	580.7	62.9	125.9	50.6	52.0
Contrast analysis between parents	−0.5 *	−174.3 *	−185.6 *	−18.6 *	21.7 ns	−16.3 *	−23.1 *
Mean of RILs	3.0	323.9	421.3	57.1	114.6	50.6	37.3
Range of RILs	(1–5)	70.5–578.3	91.9–772.4	14.9–89.8	32.9–185.6	34.5–69.1	3.7–71.3
Heritability (%)	79.8	86.9	86.3	82.2	83.2	72.0	90.7
Heat-stress environment, 2014							
ICC 4567 (heat sensitive)	2	175.3	242.0	32.6	123.2	23.9	24.4
ICC 15614 (heat tolerant)	5	431.2	534.9	54.8	111.6	52.0	43.9
Contrast analysis between parents	−0.6 *	−255.9 *	−292.9 *	−22.1 *	11.7 ns	−28.2 *	−19.6 *
Mean of RILs	3.0	268.0	355.7	49.0	119.7	40.9	38.4
Range of RILs	(1–5)	46.9–576.8	61.8–665.8	11.0–91.6	65.4–142.4	12.8–63.4	5.8–61.6
Heritability (%)	86.5	86.8	86.6	80.9	49.8	91.3	84.7
Pooled environments (Heat-stress environments, 2013 and 2014)							
ICC 4567 (heat sensitive)	2	201.6	278.1	37.5	134.8	28.6	26.1
ICC 15614 (heat tolerant)	5	453.6	570.3	59.6	116.4	51.2	48.7
Contrast analysis between parents	−0.6 *	−252 *	−292.2 *	−22 *	18.4 ns	−22.6 *	−22.6 *
Mean of RILs	3.0	296.0	388.5	53.0	117.2	45.8	37.9
Range of RILs	(1–5)	42.2–516	54.9–672.5	9.01–82.3	37.14–157.5	24.13–58.8	2.61–63.9
Heritability (%)	72.2	81.6	82.3	73.1	19.2	NA	81.6

* significant at $p = 0.05$, ns = Not significant, NA = Not available.

2.2. Relationship between Yield and Yield Determining Traits

Heat tolerance is a complex trait and can be estimated indirectly through yield and yield contributing traits under heat-stress. All the traits- visual score (VS), FPod, TS, BM and %PodSet were positively correlated with yield ($r = 0.51$ **–0.90 **) under both the heat-stress environments and pooled over analysis except HI ($r = 0.32$ **) under heat-stress environment of 2013 (Table 2). In addition, VS had positive association with FPod ($r = 0.68$ **–0.80 **) and TS ($r = 0.67$ **–0.79 **). Likewise, %PodSet was found to have a strong positive correlation with FPod ($r = 0.59$ **–0.77 **) and TS ($r = 0.60$ **–0.78 **) under both the heat-stress environments as well as in pooled analysis (Table 2). In contrast, under non-stress environment, the correlation with yield was low for %PodSet ($r = 0.17$ **) and HI ($r = 0.33$ **), and high for other traits ($r = 0.63$ **–0.91 **) (Table 2). Regression analysis between the traits and yield revealed that all the traits exhibited medium to high variation for yield (25% to 81%) in both stress environments as well as pooled over years (Figure S3a–c). In non-stress environment, %PodSet had low contribution (3%) whereas BM was found to have high variation for yield (82%) (Figure S3d). A significant correlation between the yield and yield contributing traits under heat-stress environment indicated that these traits can be used in direct or indirect selection for improving heat tolerance in chickpea.

Table 2. Correlation among the different traits evaluated in RIL population in two heat-stress environments, non-stress environment and pooled over years.

Environments	Traits	VS	FPod	TS	BM	HI	%PodSet	GY
HSE-2013	VS	1						
HSE-2014	VS	1						
Pooled years	VS	1						
HSE-2013	FPod	0.68 **	1					
HSE-2014	FPod	0.78 **	1					
Pooled years	FPod	0.80 **	1					
HSE-2013	TS	0.67 **	0.97 **	1				
HSE-2014	TS	0.78 **	0.96 **	1				
Pooled years	TS	0.79 **	0.97 **	1				
HSE-2013	BM	0.69 **	0.70 **	0.68 **	1			
HSE-2014	BM	0.15 **	0.40 **	0.38 **	1			
Pooled years	BM	0.61 **	0.67 **	0.65 **	1			
HSE-2013	HI	−0.04 ns	0.22 **	0.25 **	−0.35 **	1		
HSE-2014	HI	0.83 **	0.84 **	0.84 **	0.08 ns	1		
Pooled years	HI	0.62 **	0.70 **	0.72 **	0.24 **	1		
HSE-2013	%PodSet	0.63 **	0.72 **	0.73 **	0.62 **	0.00	1	
HSE-2014	%PodSet	0.61 **	0.59 **	0.60 **	0.05 **	0.62 **	1	
Pooled years	%PodSet	0.71 **	0.77 **	0.78 **	0.50 **	0.59 **	1	
HSE-2013	GY	0.66 **	0.88 **	0.89 **	0.74 **	0.32 **	0.63 **	1
HSE-2014	GY	0.73 **	0.90 **	0.89 **	0.57 **	0.84 **	0.50 **	1
Pooled years	GY	0.79 **	0.89 **	0.88 **	0.78 **	0.76 **	0.69 **	1
	Traits	FPod	TS	BM	HI	%PodSet	GY	
NSE-2013	FPod	1						
NSE-2013	TS	0.94 **	1					
NSE-2013	BM	0.60 **	0.63 **	1				
NSE-2013	HI	0.15 **	0.22 **	−0.07 ns	1			
NSE-2013	%PodSet	0.23 **	0.27 **	0.17 **	0.05 ns	1		
NSE-2013	GY	0.63 **	0.69 **	0.91 **	0.33 **	0.17 **	1	

** Significant at $p < 0.01$, respectively. ns: Non-significant. HSE-2013: Heat-stress environment—2013; HSE-2014: Heat-stress environment-2014; NSE-2013: Non-stress environment-2013; Pooled years: Pooled over HSE-2013 and HSE-2014; VS, Visual Score; FPod, Number of Filled Pods per Plot; TS, Total Number of Seeds Per Plot; BM, Biomass; HI, Harvest Index; %PodSet, Percentage Pod Setting; GY, Grain Yield per Plot.

2.3. Sequencing Data and SNP Discovery

The parents of the mapping population (ICC 4567 × ICC 15614) were sequenced at higher depth (5× coverage), and a total of 19.63 million reads containing 1.70 Gb for ICC 4567, and 15.79 million reads containing 1.37 Gb for ICC 15614, were generated. In addition, 3333.41 million reads containing 289.70 Gb were generated from 292 RILs. The number of reads generated varied from 6.86 million (RIL099) to 20.66 million (RIL112) with an average of 11.42 million per line. The single nucleotide polymorphisms (SNPs), identified using the software SOAP, were analyzed to remove heterozygous SNPs in the parents, and a set of 396 SNPs were identified across 292 RILs. The sequence details of all SNPs have been provided in Table S1a,b.

2.4. Genetic Linkage Map and Marker Distribution

The 396 polymorphic SNPs obtained from GBS were used for genetic map construction. The genetic linkage map covered 529.11 cM of the chickpea genome with an average interval of 1.95 cM between markers (Table S2 and Figure S1). The highest number of markers was in CaLG04 (57), while the lowest number of markers was in CaLG08 (10) (Figure S1). CaLG08 showed the highest marker density with 1.78 markers per cM on average. The lowest marker density was observed for CaLG02, which had 0.29 markers per cM on average. Overall, the map had on average 0.51 markers per cM (Table S2).

2.5. QTL Analysis

2.5.1. Genomic Region on CaLG05

A promising genomic region harbouring major QTLs for four traits—FPod, TS, GY, and %PodSet flanked by markers Ca5_44667768 and Ca5_46955940—was identified on CaLG05 (Table 3). The four QTLs—*qfpod02_5*, *qts02_5*, *qgy02_5*, and *q%podset06_5*—were found in both the stress environments spanning 6.9 cM (corresponding to ~2.28 Mb on physical map) (Figure 2a). The phenotypic variation for GY-QTL (*qgy02_5*) was 16.04% (LOD 11.69) and 16.56% (LOD 12.00) in heat-stress environments I (2013) and II (2014), respectively. QTLs for FPod—*qfpod02_5* in this genomic region demonstrated phenotypic variation of 11.57% (LOD 8.37) and 12.03% (LOD 7.79), respectively, in the consecutive stress environments (Table 3). Similarly, QTLs for the TS *qts02_5* in heat-stress environments I (2013) and II (2014) explained phenotypic variation of 12.0% (LOD 8.54) and 10.0% (LOD 7.30). The QTL for %PodSet (*q% podset06_5*), which has been considered as an important selection criterion for heat tolerance in chickpea, had a phenotypic variation of 11.51% (LOD 8.04) and 13.30% (LOD 9.20) in the heat-stress environments of 2013 and 2014, respectively (Table 3).

All the major QTLs present in the genomic region of CaLG05 were found to exist in the pooled analysis for the two stress environments (Table 3). In CaLG05, two major QTLs for VS and HI were found explaining 15.1% (LOD 11.1) and 18.5% (LOD 13.0) of phenotypic variation, under the heat-stress environment (2014), respectively (Table S3). In contrast, during the stress environment in 2013, one major QTL for VS was found close to the genomic region on CaLG05 with a phenotypic variation of 13.88% (LOD 12.05) (Table S3). Through single marker analysis (SMA), Ca5_44667768 was co-segregated with the four major QTLs in this genomic region.

Table 3. Identification of QTLs associated with heat tolerance in ICC 4567 × ICC 15614 derived RIL population.

LG	Marker Interval	Trait	QTL Name	Heat-Stress Environment, 2013				Heat-Stress Environment, 2014				Pooled Environments			
				Position (cM)	%PVE	LOD	Add	Position (cM)	%PVE	LOD	Add	Position (cM)	%PVE	LOD	Add
CaLG05	Ca5_44667768-Ca5_46955940	FPod	qfpod02_5	4.41	11.57	8.37	27.93	5.41	12.03	7.79	27.31	5.41	12.03	9.41	28.83
		TS	qts02_5	5.41	12.00	8.54	36.14	5.41	10.00	7.30	31.27	5.41	10.00	9.07	35.27
		GY	qgy02_5	4.41	16.04	11.69	4.72	4.41	16.56	12.00	4.61	4.41	16.56	13.17	4.64
		%PodSet	q%podset06_5	6.41	11.51	8.04	3.47	6.41	13.30	9.20	3.40	6.41	13.30	9.48	3.47
CaLG06	Ca6_7846335-Ca6_14353624	VS	qvs05_6	62.41	11.07	9.79	0.05	61.51	9.04	7.26	0.06	61.51	9.04	9.54	0.06
		FPod	qfpod03_6	62.41	6.56	5.10	20.88	63.40	5.92	4.10	19.01	62.41	5.92	5.22	19.91
		GY	qgy03_6	62.41	4.43	3.68	2.48	62.41	3.92	3.21	2.24	62.41	3.92	3.58	2.24
		%PodSet	q%podset08_6	63.41	8.44	6.22	3.00	65.41	6.96	4.61	2.46	64.41	6.96	5.97	2.77

VS, Visual Score; FPod, Number of Filled Pods per Plot; TS, Total Number of Seeds per Plot; %PodSet, Percentage Pod Setting; GY, Grain Yield per Plot; %PVE, Percentage of Phenotypic Variance Explained; Add, additive effect, where a positive value indicates that ICC 15614 allele was favorable, and a negative value ICC 4567 allele was favorable; LOD, likelihood of Odds Ratio; LG, Linkage Group.

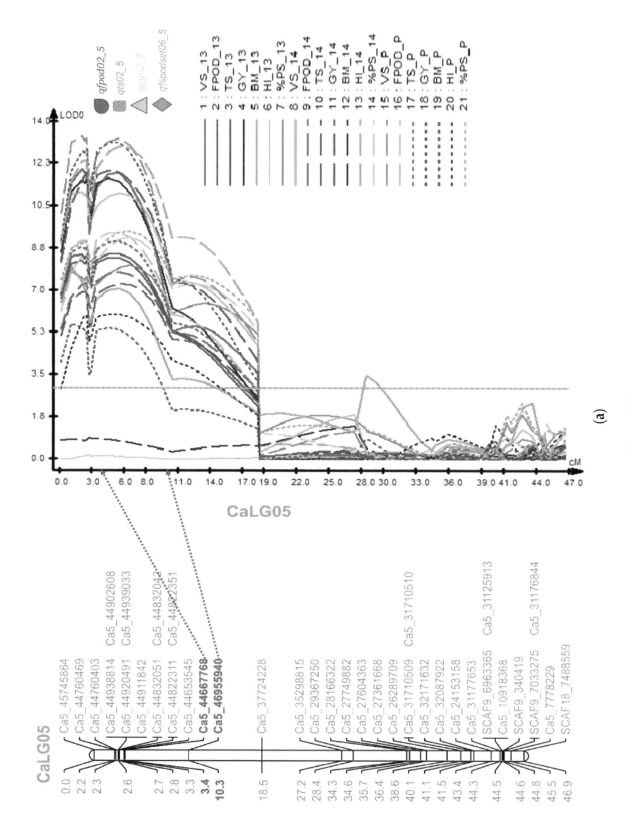

Figure 2. *Cont.*

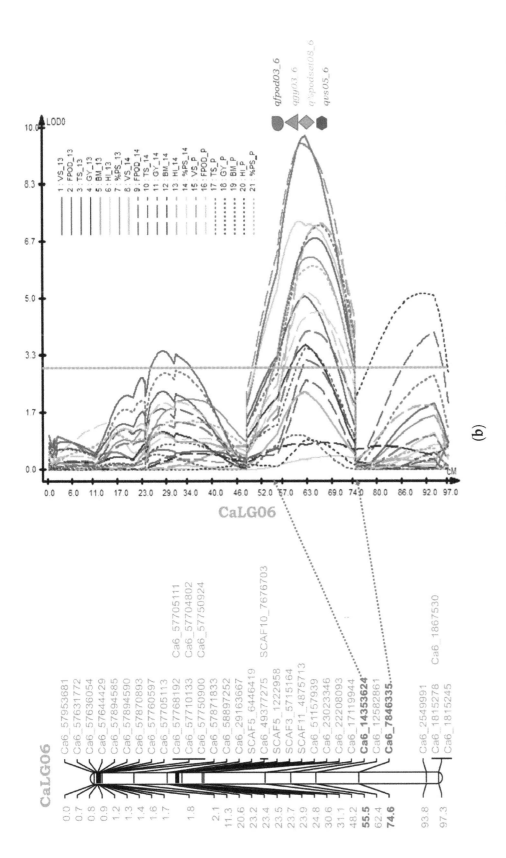

Figure 2. (a) Likelihood of odds ratio (LOD) curves obtained by composite interval mapping for quantitative trait loci (QTL) mapped over two heat-stress environments, 2013, 2014 and their pooled years together. Four major QTLs-*qfpod02_5, qts02_5, qgy02_5, q% podset06_5* of the four traits-Number of Filled Pods per Plot (FPod), Total Number of Seeds per Plot (TS), Grain Yield per Plot (GY) and Percent Pod Setting (%PodSet) in the genomic region on CaLG05 flanked by markers Ca5_44667768 and Ca5_46955940. The vertical lines indicate the threshold LOD value (2.5) determining significant QTL; (b) Likelihood of odds ratio (LOD) curves obtained by composite interval mapping for quantitative trait loci (QTL) mapped over two heat-stress environments, 2013, 2014 and their pooled years together. Four QTLs, *qfpod03_6, qgy03_6, q% podset08_6, qvs05_6* for the traits Number of Filled Pods per Plot (FPod), Grain Yield per Plot (GY), Percent Pod Setting (%PodSet) and visual score on podding behaviour (VS) in the genomic region on CaLG06 with the marker interval Ca6_1435362-Ca6_7846335, in the RIL mapping population of ICC 4567 × ICC 15614. The vertical lines indicating the threshold LOD value (2.5) determining significant QTL.

2.5.2. Genomic Region on CaLG06

A second genomic region, harbouring QTLs for four important traits in this study, was identified having the marker interval Ca6_14353624—Ca6_7846335 (Table 3 and Figure 2b). The QTLs for FPod (*qfpod03_6*), GY (*qgy03_6*), %PodSet (*q% podset08_6*), and VS (*qvs05_6*) spanned a genetic length of 19.14 cM (~6.50 Mb on physical map) in CaLG06. The range of phenotypic variation shown by various traits in this genomic region was from 3.92 to 11.07% (Table 3).

2.5.3. QTLs Identified on Other LGs

In the present work, a total of 13 QTLs were identified consistently across two heat-stress environments showing both major and minor effects for various traits measured. Apart from the QTLs identified in CaLG05 and CaLG06, a QTL for GY (*qgy01_1*) was found in the same position (40.0 cM) demonstrating 7.33% and 10% of phenotypic variation in the first and second year, respectively, on CaLG01 (Table S4).

On CaLG02, QTL for FPod (*qfpod01_2*) occurred at the same position (65.81 cM) in consecutive years with a phenotypic variation of 4.9% (LOD 3.38) and 5.8% (LOD 4.0). Similarly, QTL for TS (*qts01_2*) was found explaining 5.6% and 8.1% phenotypic variation under heat-stress environments (2013 and 2014), respectively. A major QTL (*q%podset03_4*) with phenotypic variation 12.5% (LOD 4.72) for %PodSet in 2013 was also observed in 2014 with 7.8% phenotypic variation and LOD value of 3.6 with same marker interval (Ca4_13699195-Ca4_7818876) on CaLG04 (Table S4).

2.5.4. Mapping of Epistatic QTLs (E-QTLs)

Epistatic interaction analysis revealed that 19 QTL pairs were involved in the epistatic interactions covering seven LGs (Table 4). A significant effect was observed for all the epistatic interactions. However, no significant interaction between epistasis and environment was observed, except for the trait biomass (BM).

Two epistatic QTL pairs for VS were found to have loci distributed on four different LGs accounting for 3.43% phenotypic variation. In the case of FPod, two QTLs were found to be interacting in the same LG, CaLG02. Another QTL pair was found for FPod to interact with each other in two different LGs (Table 4). These two epistatic QTL pairs for FPod together explained a phenotypic variation of 2.94%.

The highest number of epistatic QTL pairs (nine pairs) were detected for TS in this population and have contributed up to 12.38%. The epistatic interaction for TS was found in all the linkage groups, except CaLG03 and CaLG07. One QTL interaction pair was detected for GY interacting from CaLG01 to the locus on CaLG02 with a phenotypic variation 0.83% (Table 4 and Figure S2). Similarly, in the case of %PodSet, four epistatic QTL pairs were found to interact with each other in three linkage groups CaLG01, CaLG03, and CaLG04 showing a phenotypic variation of 5.79%.

In addition, an interaction between non-QTL, and additive and additive × environment-QTL was found in the case of BM, which showed 1.22% phenotypic variation. Concurrently, five loci (loci located at 10.1 cM and 26.4 cM in CaLG01, 2.2 cM and 75.6 cM in CaLG04, and at 44.5 cM in CaLG05) were observed to have interaction simultaneously with several other loci affecting the expression of the particular trait. Two loci (*eqts2_1/eqpodset2_1* in CaLG01 and *neqfpod4_5/neqts9_5* in CaLG05) controlling two or three different traits were also interacted with other loci (Table 4).

Table 4. Epistatic effect, and epistatic × environment interaction QTL found in RIL population (ICC 4567 × ICC 15614) in two heat-stress environments, 2013 and 2014.

SL. No.	Trait	QTL_i	LG	Marker Interval (QTL i)	Position (QTL_i)	QTL_j	LG	Marker Interval (QTL j)	Position (QTL_j)	AA	h^2 (%) (AA)	h^2 (%) (AAE)
1	VS	eqts1_1	1	Ca1_1732919-Ca1_4429044	48.5	eqos4_7	7	Ca7_3634430-Ca7_6584610	4.6	-0.02 ***	1.02	0.12
2	VS	neqts2_4	4	Ca4_48498166-Ca4_48498181	2.6	neqts3_5	5	Ca5_29367250-Ca5_28166322	30.4	0.03 ***	2.41	0.17
3	FPod	eqfpod1_2	2	Ca2_24709295-Ca2_30876552	30.7	eqfpod2_2	2	Ca2_34481663-Ca2_35860429	64.8	-8.85 ***	0.73	0.01
4	FPod	neqfpod3_4	4	Ca4_48497765-Ca4_48458381	2.2	neqfpod4_5/neqts9_5	5	SCAF9_6963365-Ca5_31125913	44.5	13.10 ***	2.21	0.01
5	TS	eqts1_1	1	Ca1_11321839-Ca1_11411540	10.8	eqts11_6	6	Ca6_5115739-Ca6_23023346	27.8	13.15 ***	0.42	0.01
6	TS	eqts2_1/eqpodset2_1	1	Ca1_39746426-Ca1_34727065	26.4	eqts14_8	8	Ca8_14753681-Ca8_14587797	5.6	9.79 ***	0.46	0.02
7	TS	eqts4_2	2	Ca2_34481663-Ca2_35860429	65.8	eqts12_6	6	Ca6_12582861-Ca6_7846335	62.4	-9.79 ***	0.38	0.05
8	TS	eqts4_2	2	Ca2_34481663-Ca2_35860429	65.8	eqts14_8	8	Ca8_14753681-Ca8_14587797	5.6	16.97 ***	0.96	0.01
9	TS	eqts7_5	5	Ca5_45745864-Ca5_44760469	2	eqts13_6	6	Ca6_2549991-Ca6_1815278	93.8	-8.86 ***	0.6	0.00
10	TS	eqts2_1/eqpodset2_1	1	Ca1_39746426-Ca1_34727065	26.4	neqts10_6	6	Ca6_58897252-Ca6_29163667	14.4	17.68 ***	2.22	0.03
11	TS	neqts3_2	2	Ca2_32483185-Ca2_32979328	47.7	neqts6_4	4	Ca4_47243660-Ca4_44753224	22.3	13.47 ***	2.12	0.01
12	TS	neqts5_4	4	Ca4_48458381-Ca4_48475589	2.2	neqts8_5	5	Ca5_27604363-Ca5_27361668	35.7	10.76 ***	2.52	0.03
13	TS	neqts5_4	4	Ca4_48458381-Ca4_48475589	2.2	neqts9_5/neqfpod4_5	5	SCAF9_6963365-Ca5_31125913	44.5	12.02 ***	2.7	0.00
14	GY	eqgy1_1	1	Ca1_1732919-Ca1_4429044	45.5	eqgy2_2	2	Ca2_34481663-Ca2_35860429	63.8	1.41 ***	0.83	0.01
15	BM	aaeqbm1_1	1	Ca1_11685790-Ca1_11372972	9.1	neqbm2_3	3	Ca3_24194574-Ca3_22539683	52.9	-2.09 ***	1.22	0.21
16	%PodSet	eqpodset1_1	1	Ca1_11685790-Ca1_11372972	10.1	eqpodset6_4	4	Ca4_13699195-Ca4_7818876	75.6	-1.33 ***	0.83	0.01
17	%PodSet	eqpodset2_1/eqts2_1	1	Ca1_39746426-Ca1_34727065	26.4	eqpodset6_4	4	Ca4_13699195-Ca4_7818876	75.6	1.89 ***	0.99	0.03
18	%PodSet	eqpodset1_1	1	Ca1_11685790-Ca1_11372972	10.1	neqpodset4_4	4	Ca4_48478303-Ca4_48475461	2.5	-1.38 ***	2.13	0.02
19	%PodSet	neqpodset3_3	3	Ca3_9400875-SCAF14_6484051	63.2	neqpodset5_4	4	Ca4_4826918-Ca4_4724365	11	-1.44 ***	1.84	0.00

VS, Visual Score; FPod, Number of Filled Pods per Plot; TS, Total Number of Seeds per Plot; GY, Grain Yield per Plot; BM, Biomass; %PodSet, Percentage Pod Setting. QTL_i and QTL_j, the two QTL/non-QTL involved in epistatic interaction; AA, additive × additive effect interactions; AAE, epistatic × environment effect interactions, h^2 (AA): the contribution rate of additive x additive effect interactions; h^2 (AAE): the contribution rate of epistatic × environment effect interactions. *** Significant at the 0.001 probability level. The underlined QTLs denotes those with an additive effect. *eqpodset2_1/eqts2_1 or eqts2_1/eqpodset2_1* and *neqts9_5/neqfpod4_5 or neqfpod4_5/neqts9_5/neqts9_5* indicates co-localized loci.

3. Discussion

3.1. Phenotypic Evaluation of RILs and Parents in Field Condition

Sowing during the month of February proved to be an ideal condition to expose chickpea crop to heat-stress and selecting heat tolerance lines in earlier studies under field conditions at ICRISAT, Patancheru, India [19,21]. A recent study on chickpea reported 34 °C as the threshold temperature for pod setting and also observed that at 35 °C, pod set was reduced by 50% in chickpea genotypes [19]. The average maximum temperatures (37.5 °C and 36.7 °C in summer 2013 and summer 2014, respectively) in both the heat-stress environments found were ideal for phenotyping RIL population. An average maximum temperature of 29.4 °C was recorded in non-stress environment, which was considered as control for this study. This temperature was ideal for sowing in the non-stress environment for the timely sown crop [22].

The frequency distribution of measured traits showed the characteristics of continuous variation (Figure 1a,b). Paliwal et al. (2012) [23] in RILs of wheat and Buu et al. (2014) [24] in BC_2F_2 population in rice, reported several transgressive segregants for heat tolerance. Similarly, in this present study, transgressive segregants in both directions were observed, indicating that both parents have contributed alleles for heat tolerance in the RILs (Figure 1a,b). A significant variation found among the RILs for all the traits indicate the presence of genetic diversity in the selected parents for the selected traits under heat-stress condition. Parents differed significantly for all the traits in both the heat-stress environments, except biomass (BM).

High heritability (H^2) values were observed for all the traits measured under both the heat-stress environments, except for biomass in summer 2014, which indicates that there is a high probability of achieving the same kind of results if the trial is repeated under similar growing conditions.

Yield under high temperatures is the key objective for heat tolerance breeding in chickpea. Traits such as FPod, %PodSet and TS contributing to increased yield under high-temperature stress can be treated as a proxy for heat tolerance. The presence of significant correlations between yield and other traits in heat-stress environments indicated that these traits can be used as selection criteria for heat tolerance.

FPod and TS had a strong correlation with yield (88 to 90%) under both the stress environments. Such high correlation of these traits toward yield was reported earlier in chickpea under abiotic stress [10,11]. In addition, VS and %PodSet was also found to have good correlation (50 to 79%) with yield. However, BM and HI showed large difference in correlation with yield in both the heat-stress conditions. Positive and strong association of the four traits-FPod, TS, VS and %PodSet with grain yield revealed the importance of these traits in determining yield under heat-stress environment. Hence, detecting QTLs of these traits under stress would be helpful in heat tolerance programme.

3.2. QTL Mapping for Heat Tolerance

The genomic region in CaLG05 harbours QTLs for FPod, TS, GY, and %PodSet, which were reportedly associated with heat tolerance in chickpea [3,19]. Interestingly, the positions of the QTLs (*qts02_5, qgy02_5, q% podset06_5*) for TS, GY, and %PodSet were identified in the same position over the years, which strongly confirm the QTLs in these positions.

The presence of four major co-localized QTLs (*qfpod02_5, qts02_5, qgy02_5, and q% podset06_5*) suggests tight linkage or the phenomenon of pleiotropy and the phenotypic correlations between these traits were highly significant in both the stress environments. Moreover, the tolerant parent ICC 15614 is contributing the desirable alleles for all the QTLs found in the two genomic regions in CaLG05 and CaLG06.

Identification of QTLs at the same positions in both the heat-stress environments indicate their possible practical utility in breeding for heat-stress tolerance in subsequent studies [25]. Several co-localized QTLs for various traits were found which could possibly due to pleiotropy or tightly linked QTLs. Fine mapping of the target genomic region will further help in resolving the issues of pleiotropy and tight linkage. The incorporation of a higher number of markers into the existing genetic map can further narrow down the genomic regions identified.

QTLs for traits such as FPod, TS, and GY were not expressed under non-stress condition, confirming the fact that these QTLs were only expressed under high-temperature condition. Two major QTLs for HI were identified in CaLG01 and CaLG04 explaining the phenotypic variation of 12.03% (LOD 8.8) and 12.53% (LOD 7.9), respectively. In addition, three minor QTLs including one for HI and two for %PodSet were found in different LGs. The fewer number of detected QTLs and their unique positions in the non-stress environment is a strong evidence that there is no correspondence between QTLs found in non-stress with the QTLs found in heat-stress environment. This phenomenon proves the fact that those QTLs identified in heat-stress condition were independent and exclusive for heat tolerance.

3.3. Epistatic QTLs for Heat Tolerance

Epistatic interaction is one of the key factors controlling the expression of a complex trait. The epistatic interaction analysis of QTLs provides a more comprehensive knowledge of the QTLs and their genetic behaviour underlying the trait [26,27].

In the current study, 19 pairs of digenic epistatic QTLs were found to be associated with the six traits: VS, FPod, TS, GY, BM, and %PodSet. Maximum number epistatic QTLs loci were observed for TS (nine), followed by %PodSet (four). In this study, some loci such as *eqts2_1/eqpodset2_1*, *eqts2_1/eqpodset2_1*, *eqpodset2_1/eqts2_1*, *neqts9_5/neqfpod4_5*, *neqfpod4_5/neqts9_5* were simultaneously controlling more than one trait indicating the pleiotropy nature of the traits.

Four categories of epistatic interaction were found in this study such as, additive × additive, additive × non-QTL, non-QTL × non-QTL, and additive × (additive-environment) × non-QTL interaction. FPod and VS showed two epistatic interactions each. Out of two epistatic interactions, one additive × additive epistatic interaction was found for both FPod and VS.

For GY, one additive × additive QTL epistatic interaction was found. For TS, five additive × additive QTL epistatic interactions, three non-QTL × non-QTL interaction and one additive × non-QTL interactions were observed. Similarly, two additive × additive QTL interactions, one non-QTL × non-QTL interaction and one additive × non-QTL interaction were observed for %PodSet. All the epistatic interactions were found to be significant.

The additive effects were found in both directions for all the traits. Nine interactions had negative additive effects, meaning that recombinant allele combinations could increase the particular trait value. Similarly, ten epistatic QTL interactions having positive additive effects, indicating parental allele combinations, would help to improve the trait [28].

Presence of epistatic interactions for a given trait will make the selection difficult. Interestingly, all major QTLs had no epistatic interaction and this will increase the heritability of the trait and make the selection easy.

3.4. Putative Candidate Genes for Heat Tolerance

Recent progress in functional genomics facilitates the elucidation of the important role of candidate genes for expression of tolerance against abiotic stress in plants [29–31]. In the present study, mining of the candidate genes for heat tolerance revealed 236 genes in 2.28 Mb (44.6–46.9 Mb) region in CaLG05 and 550 genes in 6.50 Mb (7.85–14.35 Mb) in CaLG06 (Tables S5 and S6). Based on functional categorization, many genes were found to be associated with biological processes (168 genes in CaLG05 and 365 genes in CaLG06) in the two genomic regions.

Gene ontology classification revealed a total of 25 putative candidate genes (11 in CaLG05 and 14 in CaLG06) known to function, directly or indirectly, as heat-stress response genes in several plant species (Table S9a,b). Of the 25 candidate genes, five genes encode protein like farnesylated protein 6 (AtFP6), ethylene-responsive transcription factor ERF114, ethylene-responsive transcription factor CRF4, F-box protein SKP2B, and ethylene-responsive transcription factor RAP2-11. These genes were identified to have key roles in heat acclimation and growth of plants under severe heat-stress condition. Many transcription factors, enzyme, and stress responsive element binding factors responsible for heat tolerance in various plant species were reported earlier [32]. Furthermore, various heat shock proteins (HSPs), ethylene forming enzymes (EFEs), and ethylene-responsive element factors (ERFs) were found to be candidate genes for heat tolerance in soybean and cowpea, two of the plant species closest to chickpea [32].

The role of various heat shock proteins and heat-stress transcription factors has been widely accepted and reported in different crops [33]. The role of HSP90 transcription factors under heat-stress conditions was also reported in chickpea [34]. Five putative genes were identified in the two examined genomic regions, encoding for either heat shock proteins or heat shock transcription factors contributing for thermo-tolerance.

Oxidative stress can occur in parallel with heat-stress through the formation of reactive oxygen species (ROS) [35]. Three putative candidate genes were also observed in this study to have a role in defying oxidative stress and recovering plants from heat-stress damage. These genes encode different types of proteins like protein tansparent testa glabra 1, peroxidase 52, and zinc finger protein CONSTANS-LIKE 5. In addition, certain signalling molecules like ethylene, abscisic acid (ABA), and salicylic acid are among a few to have a significant role in the development of heat tolerance [36]. In this study, a few genes—MYB44, AKH3, and RAN1—were found to involve with these signalling molecules through upregulation process to mitigate the heat-stress. Being a preliminary study, evaluation of these putative candidate gene-functions in chickpea through fine mapping and gene expression study is necessary to use them for further study.

4. Materials and Methods

4.1. Plant Material and Treatment Condition

A mapping population of 292 RILs developed from a cross between a heat sensitive parent ICC 4567 and a heat tolerant parent ICC 15614 was used for the study. Field experiments were carried out at ICRISAT, Patancheru, India (17°30′ N; 78°16′ E; altitude 549 m) on a vertisol soil. The F8-9 RIL population was evaluated under two heat-stress environments (in summer, February–May 2013 and February–May 2014) and in one non-heat-stress environment (in winter, November–February 2013).

In all the environments, the field was solarized using polythene mulch during the preceding summer to sanitize the field, especially to avoid incidence of root diseases. Sowing was done on the ridges using ridge and furrow method with inter- and intra-row spacing of 60 × 10 cm. Each plot consisted of a 2 m long row. Need-based insecticide sprays were provided to control pod borer (*Helicoverpa armigera*) and the experimental plots were kept weed-free through manual weeding. Before sowing, seeds were treated with the mixture of fungicides 0.5% Benlate® (E.I. DuPont India Ltd., Gurgaon, India) + Thiram® (Sudhama Chemicals Pvt., Ltd., Gujarat, India).

The experimental design was laid out in a 15 × 20 alpha lattice design with three replications. The sowing for the non-stress environment was done on the residual moisture in the last week of November 2013 and provided with essential irrigation. The planting was done in the first week of February for stress environments to expose the reproductive phase of RILs to heat-stress (>35 °C). The stress experiments were provided with irrigation to avoid the confounding effect of moisture stress during the heat screening.

In chickpea, a temperature higher than 35 °C during reproductive phase adversely affects growth, development, and yield [1,19]. The parents used for developing RIL population for this study showed significant variations at this temperature (35 °C and above) in an earlier study [19] (Devasirvatham et al., 2013). The mean daily day/night temperatures during the reproductive phase of RILs in heat-stress environment 2013 and heat-stress environment 2014 were 37.5/22.5 °C and 36.7/22.9 °C, respectively (Figure 3). Whereas under normal season (non-stress environment), the mean daily temperatures were 29.6/15.5 °C.

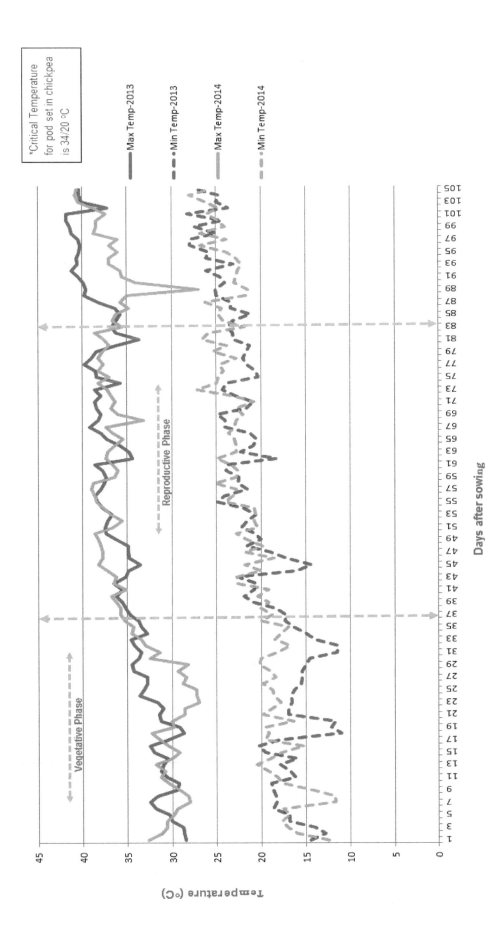

Figure 3. Daily maximum and minimum temperatures (°C) during the late sown crop growing period (stress season) in 2013 and 2014 (34/19 °C is the threshold temperature for the maximum and minimum temperatures for chickpea yield, respectively. The maximum day temperatures were 39.8 °C and 39.0 °C, and maximum night temperatures were 24.9 °C and 27.2 °C in heat-stress environments 2013, and 2014, respectively. Crop growing period was 2nd week of February to 3rd week of May).

4.2. Variables Measured

Number of filled pods per plot (FPod), total number of seeds per plot (TS), grain yield per plot (GY, g), harvest index (HI, %), biomass (BM, g) and percent pod setting (%PodSet), were reportedly found to be associated with heat tolerance in chickpea [3,19]. These six traits along with visual score on podding behaviour (VS) were recorded in the RIL population. The data for FPod, TS, GY, BM, and HI were recorded from a half-meter (0.5 m) long continuous patch out of the 2-m plot. VS at maturity and %PodSet were recorded from the entire plot. For visual scoring, score-1 was considered most sensitive (least number of pod-bearing ability), whereas, score-5 was taken as the most tolerant (maximum number of pod-bearing ability) under heat-stress. In the non-stress environment, all RILs were assumed to behave more or less the same. Hence, no visual score data were recorded in this environment.

4.3. DNA Extraction, Genotyping, and SNP Calling

DNA from 292 RILs, along with the parents, was isolated from 15-day old seedlings following the high-throughput mini-DNA extraction method [37]. Genotyping was done using GBS approach [38]. The GBS libraries from the parental lines and RILs were prepared using ApeKI endonuclease (recognition site: G/CWCG) and were sequenced using the Illumina HiSeq 2000 platform (Illumina Inc, San Diego, CA, USA). The detailed procedure of genotyping approach was described by Jaganathan et al. (2015) [25].

For SNP calling the raw reads obtained were first de-bimultiplexed using sample barcodes, and adapter sequences were removed using a custom Perl script (Figure S5). The reads having more than 50% of low-quality base pairs (Phred < 5%) were discarded and filtered data were used for calling SNPs after due quality check (Q score > 20). The high-quality data from each sample were aligned to the draft genome sequence (CaGAv1.0) of chickpea [20] using SOAP [39]. After SNP calling, the polymorphic loci were determined by following the criteria defined in [25].

4.4. Linkage Map Construction, QTL Detection and Mining of Candidate Genes

By adopting a stringent selection criterion including the missing percentage, minor allele frequency, and percent heterozygosity, the final number of SNPs included in the analysis were 396. The selected panel of robust SNPs were used for construction of genetic maps.

A linkage map was constructed with the 396 SNPs using JoinMap 4.1 [40]. Composite interval mapping in QTL Cartographer-V 2.5 [41] was employed to identify the QTLs responsible for heat tolerance with a forward and backward stepwise regression (threshold p-value < 0.05). A window size of 10 cM, along with a walking speed of 1.0 cM, and 1000 permutations for $p < 0.05$ were chosen for the QTL analysis. QTL × QTL and QTL × E interactions were estimated using the QTL Network version 2.0 (http://ibi.zju.edu.cn/software/qtlnetwork/) which is based on a mixed linear model.

First-dimensional genome scan (with the option to map epistasis) and second-dimensional genome scan (to detect epistatic interactions with or without single-locus effect) were applied. A significance level of 0.05 with 1000 permutations, 1.0 cM walk speed, 10.0 cM testing window and filtration window size were employed for the epistatic QTL analysis. QTL was named with prefix "q" for main-effect QTL, "eq" for epistatic QTL and "neq" for non-QTL epistasis followed by the abbreviated trait name and the identity of the linkage group involved.

The identified markers along with the flanking sequences were mapped on the chickpea reference genome CaGAv1.0 [20]. The genes present within the physical locations of these markers were extracted from the genome features file and were searched against TrEMBL and Swiss-Prot databases. Further functional annotation was done using UniProtKB. The Gene Ontology annotations were categorized into three categories: biological processes (BP), molecular function (MF) and cellular components (CC).

4.5. Statistical Analyses

Analysis of Variance, Predicted Means (BLUP), Heritability, and Correlations

The analysis of variance (ANOVA) for the RIL population was performed using GenStat (17th Edition), for individual environments using mixed model analysis. For each trait and environment, the analysis was performed considering entry and block (nested within replication) as random effects and replication as fixed effect.

To pool the data across environments, and to make the error variances homogeneous, individual variances were estimated and modelled for the error distribution using residual maximum likelihood (ReML) procedure. Z value and F value were calculated for random effects and fixed effects, respectively. For single and multi-environment, QTL mapping was performed using predicted means (BLUP-Best Linear Unbiased Prediction) [42].

Broad-sense heritability was estimated by following Falconer et al., 1996 [43] as

$$H^2 = Vg/(Vg + Ve/nr);$$

and pooled broad-sense heritability was estimated by following Hill et al., 2012 [44] as

$$H^2 = Vg/\{(Vg) + (Vge/ne + Ve/(ne \times nr))\}$$

Whereas, H^2 is broad-sense heritability, Vg is genotypic variance, Vge is $G \times E$ interaction variance, Ve is residual variance, ne is number of environments, and nr is number of replications. Pearson correlation analysis and linear regressions were fitted using Microsoft Excel 2016 (Microsoft Corp., 1985, Redmond, WA, USA).

5. Conclusions

The present study identified two potential genomic regions harbouring major QTLs for several heat responsive traits that are directly related to heat tolerance in chickpea. The two regions consistently appeared at the same map position across two years. Epistatic effects were not observed for major QTLs and no QTL $\times$ E interaction in the CaLG05 region. The results laid a foundation in understanding heat tolerance and increases the confidence of breeders to proceed with early generation selection for heat tolerance through marker-assisted breeding. In addition, the candidate genes identified in the two genomic regions further help to understand the mechanism of heat tolerance.

Supplementary Materials
Figure S1: Intra-specific genetic map of chickpea RIL population (ICC 4567 × ICC 15614) with 271 GBS-based SNPs covering 529.11 cM. Genetic distances (cM) were shown on the left side and the markers were shown on the right side of the bars. The map was constructed using JoinMap 4.1 and Kosambi function, Figure S2: The epistatic QTLs on linkage groups detected by QTLNetwork v 2.0 in the RIL population (ICC 4567 × ICC 15614). Lines joining two QTLs represents the epistatic interaction between them, Figure S3: Relationship of visual score on podding behaviour (VS), Number of Filled Pods per Plot (FPod), Total Number of Seeds per Plot (TS), Biomass (BM), Harvest Index (HI) and Percent Pod Setting (%PodSet) with Grain Yield per Plot (GY) (a) during heat-stress environment of 2013 (b) during heat-stress environment of 2014 (c) of pooled environments (heat-stressed environments, 2013 and 2014) (d) during non-stress environment of 2013 (Due to non-availability of VS data, no relationship of VS with GY is presented in non-stress environment, 2013). X-axis represents yield components traits e.g., VS, FPod, TS, BM HI and %PodSet; Y-axis represents GY; (No. of RILs-292), Figure S4: Likelihood of odds ratio (LOD) curves obtained by composite interval mapping for quantitative trait loci (QTL) mapped for the traits-visual score on podding behaviour (VS), Number of Filled Pods per Plot(FPod), Total Number of Seeds per Plot (TS), Grain Yield per Plot (GY), Biomass (BM), Harvest Index (HI), and Percent Pod Setting (%PodSet) in RIL population (ICC 4567 × ICC 15614) (a) in the heat-stress environment-2013 (b) in the heat-stress environment-2014 (c) in the pooled environments (heat-stress environments, 2013, and 2014) (d) in the non-stress environment-2013 (Due to non-availability of VS data, VS was not mapped in non-stress environment, 2013). The vertical lines indicating the threshold LOD value (2.5) determining significant QTL, Figure S5: Pipeline of Bioinformatics analysis: GBS data processing and SNP calling, Table S1: (a) Summary sequence data generated genotyping-on 292 RILs and two parents (ICC 4567 and ICC 15614) using GBS approach; (b) Summary of called SNPs on 292 RILs and two parents (ICC 4567 and ICC 15614) using GBS approach, Table S2: Features

of intra-specific genetic map developed using 271 SNPs and RIL population ICC 4567 × ICC 15614, Table S3: Summary of QTLs identified in two heat-stress environments, pooled environments and non-stressed environment in RIL population (ICC 4567 × ICC 15614), Table S4: Consistent QTLs found across heat-stress environments (2013 and 2014) in RIL population (ICC 4567 × ICC 15614), Table S5: Gene ontology classification for CaLG05, Table S6: Gene ontology classification for CaLG06, Table S7: Gene ontology categorization of 236 genes identified on the genomic region flanked by markers Ca5_44667768-Ca5_46955940 on CaLG05, Table S8: Gene ontology categorization of 550 genes identified on the genomic region flanked by markers Ca6_7846335-Ca6_14353624 on CaLG06, Table S9: (a) List of putative candidate genes found to be associated with heat stress on CaLG05 in chickpea, (b) List of putative candidate genes found to be associated with heat stress on CaLG06 in chickpea.

Author Contributions: P.M.G. conceived the idea and coordinated this project. P.M.G. and S.S. were involved in developing the mapping population. P.M.G., S.S., S.K.C., S.B.S. and G.R.L. provided guidance to P.J.P. in conducting field experiments and phenotyping. A.R., R.R.D. and S.S. helped P.J.P. in statistical data analysis. P.J.P. and M.T. were involved in genotyping of the mapping population, construction of linkage maps and QTL analysis. P.J.P., A.W.K. and M.T. were involved in bioinformatics work. P.J.P., P.M.G., MT, S.B.S. and S.S. contributed to writing of the manuscript, and R.K.V. and G.R.L. provided their inputs. All the authors reviewed and approved the final manuscript.

Abbreviations

%PodSet	Pod Setting Percentage
ANOVA	Analysis of Variance
BLUP	Best Linear Unbiased Prediction
BM	Biomass
CaLG	*Cicer arietinum* Linkage Group
CIM	Composite Interval Mapping
cM	Centimorgan
FPod	Number of Filled Pods Per Plot
GY	Grain Yield Per Plot
HI	Harvest Index
ICRISAT	International Crops Research Institute for the Semi-Arid Tropics
LG	Linkage Group
QTL	Quantitative Trait Loci
ReML	Residual Maximum Likelihood
RIL	Recombinant Inbred Line
TS	Total Number of Seeds Per Plot
VS	Visual Scoring

References

1. Gaur, P.M.; Jukanti, A.K.; Samineni, S.; Chaturvedi, S.K.; Basu, P.S.; Babbar, A.; Jayalakshmi, V.; Nayyar, H.; Devasirvatham, V.; Mallikarjuna, N.; et al. *Climate Change and Heat Stress Tolerance in Chickpea. Climate Change and Plant Abiotic Stress Tolerance*; Wiley-VCH Verlag GmbH & Co. KGaA: Weinheim, Germany, 2014; pp. 837–856.
2. Food and Agriculture Organization (FAO). Food and Agricultural Organization of the United Nation, FAO Statistical Database. 2015. Available online: http://faostat3.fao.org/download/Q/QC/E (accessed on 8 February 2018).
3. Krishnamurthy, L.; Gaur, P.M.; Basu, P.S.; Chaturvedi, S.K.; Tripathi, S.; Vadez, V.; Rathore, A.; Varshney, R.K.; Gowda, C.L.L. Large genetic variation for heat tolerance in the reference collection of chickpea (*Cicer arietinum* L.) germplasm. *Plant Genet. Resour.* **2011**, *9*, 59–69. [CrossRef]
4. Devasirvatham, V.; Gaur, P.M.; Mallikarjuna, N.; Tokachichu, R.N.; Trethowan, R.M.; Tan, D.K.Y. Effect of high temperature on the reproductive development of chickpea genotypes under controlled environments. *Funct. Plant. Biol.* **2012**, *39*, 1009–1018. [CrossRef]

5.   Wang, J.; Gan, Y.T.; Clarke, F.; McDonald, C.L. Response of chickpea yield to high temperature stress during reproductive development. *Crop Sci.* **2006**, *46*, 2171–2178. [CrossRef]

6.   Dehghani, H.; Sabaghpour, S.H.; Ebadi, A. Study of genotype × environment interaction for chickpea yieldin Iran. *Agron. J.* **2010**, *102*, 1–8. [CrossRef]

7.   Gaur, P.M.; Thudi, M.; Samineni, S.; Varshney, R.K. Advances in chickpea genomics. In *Legumes in the Omic Era*; Springer: New York, NY, USA, 2014; pp. 73–94.

8.   Sabbavarapu, M.M.; Sharma, M.; Chamarthi, S.K.; Swapna, N.; Rathore, A.; Thudi, M.; Gaur, P.M.; Pande, S.; Singh, S.; Kaur, L.; et al. Molecular mapping of QTLs for resistance to Fusarium wilt (race 1) and Ascochyta blight in chickpea (*Cicer arietinum* L.). *Euphytica* **2013**, *193*, 121–133. [CrossRef]

9.   Varshney, R.K.; Thudi, M.; Nayak, S.N.; Gaur, P.M.; Kashiwagi, J.; Krishnamurthy, L.; Jaganathan, D.; Koppolu, J.; Bohra, A.; Tripathi, S.; et al. Genetic dissection of drought tolerance in chickpea (*Cicer arietinum* L.). *Theor. Appl. Genet.* **2014**, *127*, 445–462. [CrossRef] [PubMed]

10. Pushpavalli, R.; Krishnamurthy, L.; Thudi, M.; Gaur, P.M.; Rao, M.V.; Siddique, K.H.; Colmer, T.D.; Turner, N.C.; Varshney, R.K.; Vadez, V. Two key genomic regions harbour QTLs for salinity tolerance in ICCV 2× JG 11 derived chickpea (*Cicer arietinum* L.) recombinant inbred lines. *BMC Plant Biol.* **2015**, *15*, 124. [CrossRef] [PubMed]

11. Vadez, V.; Krishnamurthy, L.; Thudi, M.; Anuradha, C.; Colmer, T.D.; Turner, N.C.; Siddique, K.H.; Gaur, P.M.; Varshney, R.K. Assessment of ICCV 2 × JG 62 chickpea progenies shows sensitivity of reproduction to salt stress and reveals QTL for seed yield and yield components. *Mol. Breed.* **2012**, *30*, 9–21. [CrossRef]

12. Samineni, S. Physiology, Genetics and QTL Mapping of Salt Tolerance in Chickpea (*Cicer arietinum* L.). Ph.D. Thesis, The University of Western Australia, Perth, Australia, 2011.

13. Jha, U.C.; Bohra, A.; Singh, N.P. Heat stress in crop plants: Its nature, impacts and integrated breeding strategies to improve heat tolerance. *Plant Breed.* **2014**, *133*, 679–701. [CrossRef]

14. Jagadish, S.V.K.; Craufurd, P.Q.; Wheeler, T.R. Phenotyping parents of mapping populations of rice for heat tolerance during anthesis. *Crop Sci.* **2008**, *48*, 1140–1146. [CrossRef]

15. Ye, C.; Argayoso, M.A.; Redoña, E.D.; Sierra, S.N.; Laza, M.A.; Dilla, C.J.; Mo, Y.; Thomson, M.J.; Chin, J.; Delaviña, C.B.; et al. Mapping QTL for heat tolerance at flowering stage in rice using SNP markers. *Plant Breed.* **2012**, *131*, 33–41. [CrossRef]

16. Xiao, Y.; Pan, Y.; Luo, L.; Zhang, G.; Deng, H.; Dai, L.; Liu, X.; Tang, W.; Chen, L.; Wang, G.L. Quantitative trait loci associated with seed set under high temperature stress at the flowering stage in rice (*Oryza sativa* L.). *Euphytica* **2011**, *178*, 331–338. [CrossRef]

17. Pinto, R.S.; Reynolds, M.P.; Mathews, K.L.; McIntyre, C.L.; Olivares-Villegas, J.J.; Chapman, S.C. Heat and drought adaptive QTL in a wheat population designed to minimize confounding agronomic effects. *Theor. Appl. Genet.* **2010**, *121*, 1001–1021. [CrossRef] [PubMed]

18. Zhang, G.L.; Chen, L.Y.; Xiao, G.Y.; Xiao, Y.H.; Chen, X.B.; Zhang, S.T. Bulked segregant analysis to detect QTL related to heat tolerance in rice (*Oryza sativa* L.) using SSR markers. *Agric. Sci. China* **2009**, *8*, 482–487. [CrossRef]

19. Devasirvatham, V.; Gaur, P.M.; Mallikarjuna, N.; Raju, T.N.; Trethowan, R.M.; Tan, D.K.Y. Reproductive biology of chickpea response to heat stress in the field is associated with the performance in controlled environments. *Field Crop Res.* **2013**, *142*, 9–19. [CrossRef]

20. Varshney, R.K.; Song, C.; Saxena, R.K.; Azam, S.; Yu, S.; Sharpe, A.G.; Cannon, S.; Baek, J.; Rosen, B.D.; Tar'an, B.; et al. Draft genome sequence of chickpea (*Cicer arietinum*) provides a resource for trait improvement. *Nat. Biotechnol.* **2013**, *31*, 240–246. [CrossRef] [PubMed]

21. Gaur, P.M.; Srinivasan, S.; Gowda, C.L.L.; Rao, B.V. Rapid generation advancement in chickpea. *J. SAT Agric. Res.* **2007**, *3*, 3.

22. Berger, J.D.; Milroy, S.P.; Turner, N.C.; Siddique, K.H.M.; Imtiaz, M.; Malhotra, R. Chickpea evolution has selected for contrasting phenological mechanisms among different habitats. *Euphytica* **2011**, *180*, 1–15. [CrossRef]

23. Paliwal, R.; Röder, M.S.; Kumar, U.; Srivastava, J.P.; Joshi, A.K. QTL mapping of terminal heat tolerance in hexaploid wheat (*T. aestivum* L.). *Theor. Appl. Genet.* **2012**, *125*, 561–575. [CrossRef] [PubMed]

24. Buu, B.C.; Ha, P.T.T.; Tam, B.P.; Nhien, T.T.; Van Hieu, N.; Phuoc, N.T.; Giang, L.H.; Lang, N.T. Quantitative trait loci associated with heat tolerance in rice (*Oryza sativa* L.). *Plant Breed. Biotechnol.* **2014**, *2*, 14–24. [CrossRef]

25. Jaganathan, D.; Thudi, M.; Kale, S.; Azam, S.; Roorkiwal, M.; Gaur, P.M.; Kishor, P.K.; Nguyen, H.; Sutton, T.; Varshney, R.K. Genotyping-by-sequencing based intra-specific genetic map refines a "QTL-hotspot" region for drought tolerance in chickpea. *Mol. Genet. Genom.* **2015**, *290*, 559–571. [CrossRef] [PubMed]

26. Bocianowski, J. Epistasis interaction of QTL effects as a genetic parameter influencing estimation of the genetic additive effect. *Genet. Mol. Biol.* **2013**, *36*, 093–100. [CrossRef] [PubMed]

27. Gowda, S.J.M.; Radhika, P.; Mhase, L.B.; Jamadagni, B.M.; Gupta, V.S.; Kadoo, N.Y. Mapping of QTLs governing agronomic and yield traits in chickpea. *J. Appl. Genet.* **2011**, *52*, 9–21. [CrossRef] [PubMed]

28. Qi, L.; Mao, L.; Sun, C.; Pu, Y.; Fu, T.; Ma, C.; Shen, J.; Tu, J.; Yi, B.; Wen, J. Interpreting the genetic basis of silique traits in *Brassica napus* using a joint QTL network. *Plant Breed.* **2014**, *133*, 52–60. [CrossRef]

29. Urano, K.; Kurihara, Y.; Seki, M.; Shinozaki, K. 'Omics' analyses of regulatory networks in plant abiotic stress responses. *Curr. Opin. Plant Biol.* **2010**, *13*, 132–138. [CrossRef] [PubMed]

30. Sreenivasulu, N.; Sopory, S.K.; Kishor, P.K. Deciphering the regulatory mechanisms of abiotic stress tolerance in plants by genomic approaches. *Gene* **2007**, *388*, 1–13. [CrossRef] [PubMed]

31. Vij, S.; Tyagi, A.K. Emerging trends in the functional genomics of the abiotic stress response in crop plants. *Plant Biotechnol. J.* **2007**, *5*, 361–380. [CrossRef] [PubMed]

32. Pottorff, M.; Roberts, P.A.; Close, T.J.; Lonardi, S.; Wanamaker, S.; Ehlers, J.D. Identification of candidate genes and molecular markers for heat-induced brown discoloration of seed coats in (*Vigna unguiculata* (L.) Walp). *BMC Genom.* **2014**, *15*, 328. [CrossRef] [PubMed]

33. Maestri, E.; Klueva, N.; Perrotta, C.; Gulli, M.; Nguyen, H.T.; Marmiroli, N. Molecular genetics of heat tolerance and heat shock proteins in cereals. *Plant Mol. Biol.* **2002**, *48*, 667–681. [CrossRef] [PubMed]

34. Agarwal, G.; Garg, V.; Kudapa, H.; Doddamani, D.; Pazhamala, L.T.; Khan, A.W.; Thudi, M.; Lee, S.H.; Varshney, R.K. Genome-wide dissection of AP2/ERF and HSP90 gene families in five legumes and expression profiles in chickpea and pigeonpea. *Plant Biotechnol. J.* **2016**, *14*, 1563–1577. [CrossRef] [PubMed]

35. Wahid, A.; Gelani, S.; Ashraf, M.; Foolad, M.R. Heat tolerance in plants: An overview. *Environ. Exp. Bot.* **2007**, *61*, 199–223. [CrossRef]

36. Larkindale, J.; Huang, B. Effects of abscisic acid, salicylic acid, ethylene and hydrogen peroxide in thermotolerance and recovery for creeping bentgrass. *Plant Growth Regul.* **2005**, *47*, 17–28. [CrossRef]

37. Cuc, L.M.; Mace, E.S.; Crouch, J.H.; Quang, V.D.; Long, T.D.; Varshney, R.K. Isolation and characterization of novel microsatellite markers and their application for diversity assessment in cultivated groundnut (*Arachis hypogaea*). *BMC Plant Biol.* **2008**, *8*, 55. [CrossRef] [PubMed]

38. Elshire, R.J.; Glaubitz, J.C.; Sun, Q.; Poland, J.A.; Kawamoto, K.; Buckler, E.S.; Mitchell, S.E. A robust, simple genotyping-by-sequencing (GBS) approach for high diversity species. *PLoS ONE* **2011**, *6*, e19379. [CrossRef] [PubMed]

39. Li, R.; Yu, C.; Li, Y.; Lam, T.W.; Yiu, S.M.; Kristiansen, K.; Wang, J. SOAP2: An improved ultrafast tool for short read alignment. *Bioinformatics* **2009**, *25*, 1966–1967. [CrossRef] [PubMed]

40. Van Ooijen, J.J. *JoinMap®4.1, Software for the Calculation of Genetic Linkage Maps in Experimental Populations*; Kyazma BV: Wageningen, The Netherlands, 2006.

41. Wang, S.; Basten, C.J.; Zeng, Z.B. *Windows QTL Cartographer 2.5*; Department of Statistics, North Carolina State University: Raleigh, NC, USA, 2012.

42. Searle, S. *Linear Models*; John Wiley & Sons, Inc.: New York, NY, USA, 1971.

43. Falconer, D.S.; Mackay, T.F.; Frankham, R. *Introduction to Quantitative Genetics. Trends in Genetics*; Longman Frankel: Harlow, UK, 1996; Volume 12, p. 280.

44. Hill, J.; Becker, H.C.; Tigerstedt, P.M. *Quantitative and Ecological Aspects of Plant Breeding*; Springer Science & Business Media: Berlin, Germany, 2012.

8

Identification, Classification and Functional Analysis of *AP2/ERF* Family Genes in the Desert Moss *Bryum argenteum*

Xiaoshuang Li [1,†], Bei Gao [2,†], Daoyuan Zhang [1,*], Yuqing Liang [1,3], Xiaojie Liu [1,3], Jinyi Zhao [4], Jianhua Zhang [5] and Andrew J. Wood [6]

[1] Key Laboratory of Biogeography and Bioresource in Arid Land, Xinjiang Institute of Ecology and Geography, Chinese Academy of Sciences, Urumqi 830011, China; lixs@ms.xjb.ac.cn (X.L.); liangyuqing14@mails.ucas.ac.cn (Y.L.); liuxiaojie215@mails.ucas.ac.cn (X.L.)

[2] School of Life Sciences and State Key Laboratory of Agrobiotechnology, The Chinese University of Hong Kong, Hong Kong, China; gaobei@link.cuhk.edu.hk

[3] University of Chinese Academy of Sciences, Beijing 100049, China

[4] School of Life Science, University of Liverpool, Liverpool L169 3BX, UK; j.zhao46@student.liverpool.ac.cn

[5] Department of Biology, Hong Kong Baptist University, Hong Kong, China; jzhang@hkbu.edu.hk

[6] Department of Plant Biology, Southern Illinois University, Carbondale, IL 62901-6899, USA; wood@plant.siu.edu

* Correspondence: zhangdy@ms.xjb.ac.cn

† These authors contributed equally to this paper.

Abstract: *Bryum argenteum* is a desert moss which shows tolerance to the desert environment and is emerging as a good plant material for identification of stress-related genes. *AP2/ERF* transcription factor family plays important roles in plant responses to biotic and abiotic stresses. *AP2/ERF* genes have been identified and extensively studied in many plants, while they are rarely studied in moss. In the present study, we identified 83 *AP2/ERF* genes based on the comprehensive dehydrationrehydration transcriptomic atlas of *B. argenteum*. BaAP2/ERF genes can be classified into five families, including 11 AP2s, 43 DREBs, 26 ERFs, 1 RAV, and 2 Soloists. RNA-seq data showed that 83 *BaAP2/ERFs* exhibited elevated transcript abundances during dehydration–rehydration process. We used RT-qPCR to validate the expression profiles of 12 representative *BaAP2/ERFs* and confirmed the expression trends using RNA-seq data. Eight out of 12 BaAP2/ERFs demonstrated transactivation activities. Seven BaAP2/ERFs enhanced salt and osmotic stress tolerances of yeast. This is the first study to provide detailed information on the identification, classification, and functional analysis of the *AP2/ERFs* in *B. argenteum*. This study will lay the foundation for the further functional analysis of these genes in plants, as well as provide greater insights into the molecular mechanisms of abiotic stress tolerance of *B. argenteum*.

Keywords: *AP2/ERF* genes; *Bryum argenteum*; transcriptome; gene expression; stress tolerance

1. Introduction

Bryum argenteum is an important component of the desert biological soil crusts in the Gurbantunggut and Tengger Deserts of northwestern China [1,2]. *B. argenteum* has gained increasing attention as a model organism due to its comprehensive tolerances to the desert environment, such as frequent desiccation–rehydration events and high UV radiation [3,4]. Wood et al. (2007) reported that *B. argenteum* is among the most desiccation tolerant (DT) moss species and is classified as category "A" [5]. Studies on *B. argenteum* have focused on the ecological aspects of vegetative desiccation tolerance, including morphological, structural, and physiological responses to adapt to the desert environment [3,4,6,7]. *B. argenteum* is emerging as a model moss for studying the molecular mechanisms of DT and as a source of stress-related genes [8].

APETALA2/Ethylene Responsive Factor (AP2/ERF) is one of the largest transcription factor (TF) families of plants, and the family members have been demonstrated to play important roles in plant metabolism, development, and stresses response [9]. *AP2/ERF* genes have been identified and studied extensively in the context of plant stress tolerance in many plants [10–12]. The *AP2/ERF* gene family has been rarely studied in moss species, however, the largest TFs families found in the plant transcription factor databases (TFDB) are *AP2/ERF* genes annotated in the mosses *Physcomitrella patens* and *Sphagnum fallax* [13,14]. Moreover, *AP2/ERFs* were demonstrated to be regulated in response to multiple stresses, such as salinity and UV in *P. patens* [15], and *PpDBF1* gene was reported to confer drought, salt, and cold tolerances in transgenic tobacco [16]. Additionally, AP2/ERFs also demonstrated to be the most abundant TFs in the DT moss *Syntrichia caninervis* [17]. The majority of *DREB* (Dehydration-Responsive Element-Binding Protein) genes in *S. caninervis* responded to dehydration and/or rehydration treatments [18,19], indicating that AP2/ERF transcriptional factors also play a central regulatory role during stress responses in moss species.

AP2/ERF classification employs a well-established method based on Arabidopsis and rice [20] and this method has been widely used to classify the *AP2/ERF* family genes in many plant species [11,21]. Two classic classification methods have been proposed for the plant AP2/ERF superfamily based upon the number of AP2 domains and sequence similarities [22]. Sakuma et al. classified the AP2/ERF superfamily into five families: AP2, RAV (Related to ABI3/VP1), DREB ERF and Soloists [22]. Furthermore, *DREBs* are further classified into A1–A6 groups and *ERFs* are divided into the groups B1–B6 [22]. Nakano et al. (2006) classified AP2/ERF proteins into three major families: AP2, ERF (include both DREBs and ERFs) and RAV [20]. The ERF family is then further sub-divided into twelve groups in Arabidopsis and fifteen groups in rice according to the structure and similarity of the AP2 domain [20].

High-throughput sequencing has been an effective tool to identify stress-related genes, and transcriptome-based identification and selection of *AP2/ERF* superfamily genes have been widely used in many non-model plants, such as *Hevea brasiliensis* [23], tea [24], as well as the desiccation tolerant moss *S. caninervis* [19]. Previously, we generated a de novo transcriptome for *B. argenteum* and established a desiccation–rehydration transcriptomic atlas which covered five different hydration stages [8,25]. Using this combined dataset, we established a fully comprehensive desiccation–rehydration transcriptome dataset containing 76,206 high-quality *B. argenteum* transcripts. We found that AP2/ERF was the second largest TF family in *B. argentum*. Moreover, *AP2/ERF* genes were also the most abundant differentially expressed TFs during the desiccation and rehydration process (Figure S1), indicating that the *AP2/ERF* family genes play key roles in *B. argentum* response to moss specific desiccation–rehydration process.

Hence, in this study, we aimed to identify and classify the *AP2/ERF* gene family in *B. argentum* based on a comprehensive transcriptome dataset. Additionally, we investigated the gene expression patterns of *BaAP2/ERF* genes under desiccation–rehydration treatment based on RNA-seq data as well as real-time quantitative PCR (RT-qPCR) assay, and their transactivation activities were also analyzed. We further evaluated the stress tolerance ability of eight representative *BaAP2/ERF* genes in yeast system. This is the first report on identification, classification, characterization, and evaluation the stress tolerance functions of the *AP2/ERF* gene family in desiccation tolerant moss *B. argentums*. This study will provide candidate genes for molecular breeding to improve crop stress tolerance, and could be helpful for understanding the molecular mechanisms of the stress responses in *B. argentum*.

2. Results

2.1. Identification of the AP2/ERF Family Genes in B. argenteum

Based on Hidden Markov model (HMM) profiles and BLAST search, 83 AP2/ERF predicted proteins were identified from the *B. argenteum* transcriptome. The unigenes ranged from 559 to 8076 bp in length, and the corresponding deduced polypeptide sequences ranged from 145 to 1983 aa. The sequences of 83 *AP2/ERFs* were submitted to Genbank under accession numbers MK170284-MK170366. Among these 83 unigenes, 73 genes have an intact open reading frame (ORF) (87%), ranging from 163 to 1983 aa. AP2 domain analysis demonstrated that all 83 predicted proteins had full-length AP2 domains (ca. 60 aa) (100%). Seven out of 83 genes had two AP2 domains, one gene had both an AP2 and a B3 domain, while the other 75 genes had a single AP2 domain. Based on the number of AP2 domains, BaAP2/ERFs can be preliminarily classified as follows: 7 genes having two AP2 domains were classified as members of the AP2 family, 1 gene with both an AP2 and B3 domain was classified as a member of RAV family, and the 75 remaining genes with only one AP2 domain were considered as ERF family members (including both DREBs and ERFs).

2.2. Classification of the AP2/ERF Genes in B. argenteum

Family classification was further confirmed by constructing the phylogenetic tree using the AP2 domain of AP2/ERF in *B. argenteum*, the model plant Arabidopsis and the model moss *P. patens*. We first generated the gene tree of AP2/ERFs with *B. argenteum* and Arabidopsis. The tree showed that *BaAP2/ERF* genes were divided into five families according to Arabidopsis' classification method, including AP2, DREB, ERF, RAV and Soloist (Figure 1). Almost half of the *B. argenteum BaAP2/ERF* genes grouped into the DREB family (with 43 members). The second largest group was the ERF family (27 genes), followed by the AP2 family (10 genes), RAV (1 gene) and Soloists (2 genes). However, the number of the AP2 family genes classified by gene tree was inconsistent with the classification result based on the AP2 domain counting. There were two inconsistencies: one was the TR14027|c0_g1_i1 gene which had two AP2 domains, therefore should be classified as AP2 family gene, while it was grouped together with ERF family in the gene tree, and the other was five genes had single AP2 domain but were clustered together with AP2 family in Arabidopsis. To further confirm their classification, we generated another two gene trees using only *BaAP2/ERF* genes (Figure S2), and with *AP2/ERF* genes both in *B. argenteum* and the model moss *P. patens* (Figure S3). We conclude that there were 11 AP2s, 43 DREBs, 26 ERFs, 1 RAV and 2 Soloists in *B. argenteum* based on the available transcriptome data.

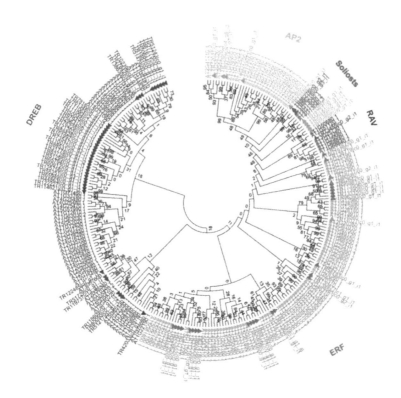

Figure 1. Phylogenetic analysis of AP2/ERF superfamily genes in *B. argenteum* and Arabidopsis. The gene tree was constructed using the neighbor-joining method using 83 BaAP2/ERFs and 176 AtAP2/ERFs, the evolutionary distances were computed using the Poisson correction method with pairwise deletion. Bootstrap values from 1000 replicates were used to assess the robustness of the tree. Different subfamilies were marked with various colors, the BaAP2/ERFs were labeled with rectangles to distinguish from AtAP2/ERF.

DREB and ERF are the two major families in the plant AP2/ERF superfamily, and which are demonstrated to play important roles in abiotic and biotic stress response. DREB and ERF can be subdivided into twelve subfamilies, namely A1–A6 and B1–B6 [22], or ten groups named Groups I–X [20], and each group/subfamily have different functions [11]. To detail classify the *DREB* and *ERF* family genes, we constructed a phylogenetic tree using 69 BaERFs (including 43 DREBs and 26 ERFs) and 31 ERFs in Arabidopsis representative of each subfamily/group. We identified 10 out of 12 DREB/ERF subfamilies in *B. argenteum* based upon the classification method of Sakamu et al. [22], A2 (3 genes), A3 (1 gene), A4 (1 gene), A5 (10 genes), A6 (3 genes), B1 (8 genes), B2 (4 genes), B3 (8 genes), B4 (4 genes), and B6 (2 genes), while no members of the A1 and B5 subfamilies were found in BaAP2/ERFs (Figure 2). Accordingly, based on Nakano et al.'s classification method [20], *BaAP2/ERF* superfamily genes contained 9 out of 10 groups (lacking Group VI), including Groups I–X, and the members for each group were Group I (3 members), Group II (10 members), Group III (one member), Group IV (4 members), Group V (2 members), Group VII (4 members), Group VIII (8 members), Group IX (8 members) and Group X (4 members) (Figure 2). Moreover, we found that 26 *ERFs* can be clearly classified into specific subfamilies/groups, while more than half of *DREBs* (25/43 genes) cannot be classified to any exist subfamilies, which was clustered into one separate clade and was named as *Bryum* group (Ba-clade) (Figure 2). To know whether these Ba-clade genes are specific to Bryum or moss species, we first performed BLASTp search in NCBI database using the full-length sequence of 25 Ba-clade DREBs/ERF and found that all 25 Ba-clade DREBs have homologies in other plants, with the highest sequence identities to model moss *P. patens*. We then performed BLASTp search in ONE KP database with mosses, liverworts, and hornworts transcriptome data. The result also show that all Ba-clade genes have homologies in mosses, liverworts, and hornworts. Moreover, half of Ba-clade DREBs shared highest sequence identities to *Funaria* (Table S1).

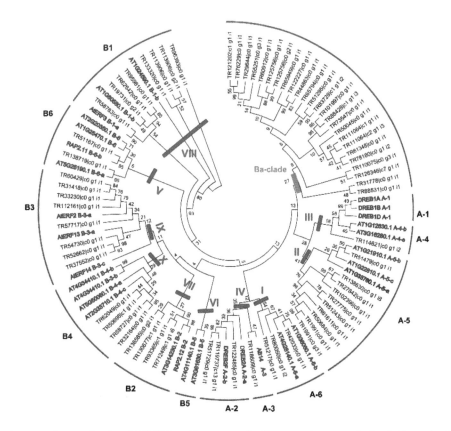

Figure 2. Phylogenetic analysis of ERF family genes in *B. argenteum* and Arabidopsis. The gene tree was constructed using 69 BaERFs and 31 AtERFs representative of each subfamily or group of ERF family genes in Arabidopsis. The evolutionary distances were computed using the neighbor-joining method and Poisson model with pairwise deletion. Bootstrap values from 1000 replicates were used to assess the robustness of the tree. To distinguish ERFs from *B. argenteum* and Arabidopsis, AtERFs and BaERFs were marked in blue and dark, respectively. Previously reported subfamily names (A1–A6 and B1–B6) and group names (Group I to Xb–L) were employed [20,22]. The Bryum-unique clade (Ba-unique) was labeled in green, and other groups were labeled in red.

2.3. Conserved Amino Acids and Motifs Analysis of BaERF Genes

To analyze the amino acids conservation of the AP2 domains, 69 BaERF deduced polypeptide sequences were aligned with 31 AtERFs representative of each gene subfamily in Arabidopsis. Multiple sequence alignment showed that BaERF sequences shared significant amino acid similarity with AtERFs except Ba-clade DREBs (Figure 3). Ba-clade DREBs did not belong to either subfamily/group based on existing classification. Ba-clade DREBs had more diverse amino acids composition in the AP2 domain, especially in the region between two β-sheets and the α-helix (marked with pink boxes). For example, a consensus sequence "TAE" in the C-terminal of α-helix was very conserved among AtERFs and BaERFs, however, in Ba-clade DREBs, TPE and TEE/Q/I patterns also existed (Figure 3). The motif composition analyses also supported this phenomenon. Eight motifs were detected in 69 BaERFs in total; among them, motifs 1–3 represented the typical AP2 domains in Arabidopsis as well as most of the well-classified BaERFs, of which motif 1 contained the β3 sheet and α-helix of AP2 domain, motif 2 corresponded to β1 and β2 sheet, and motif 3 was located in the very C-terminal of the AP2 domain (Figure 4). Motif 4 was similar to motif 1 which corresponded to the β3 sheet and α-helix; in the same way, motif 5 was similar to motif 2 which represent β1 and β2 sheet, and motif 8 was similar to motif 3, but their amino acids compositions differed from the classic AP2 domains in Arabidopsis. Interestingly, motif 4, 5 and 8 constituted the unique Ba-clade DREB AP2 domain. Additionally, we identified motif 7 as an A-5 DREB specific amino acids pattern.

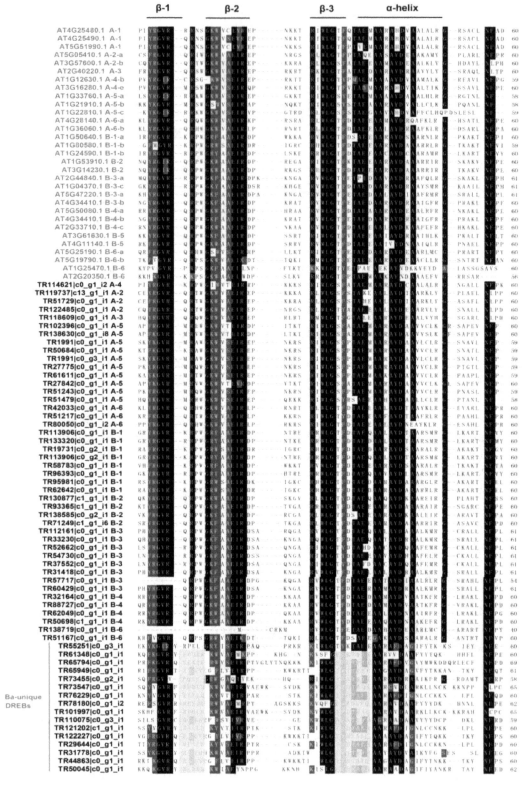

Figure 3. Sequence alignments of AP2 domains of representative ERF proteins in *B. argenteum* and Arabidopsis. Thirty-one AtERF genes representative of each ERF subfamily/group were aligned with 69 BaERFs. The subfamily for each ERF is depicted on the right. The locus names of AtERFs and BaERFs are marked in blue and black, respectively. The identical and conserved amino acid residues are indicated with black and light gray shading, respectively. The black bars represent three β sheets and α helix regions. The Ba-unique DREBs are grouped in pink bar, and two regions representing diverse amino acids compositions are marked with pink boxes.

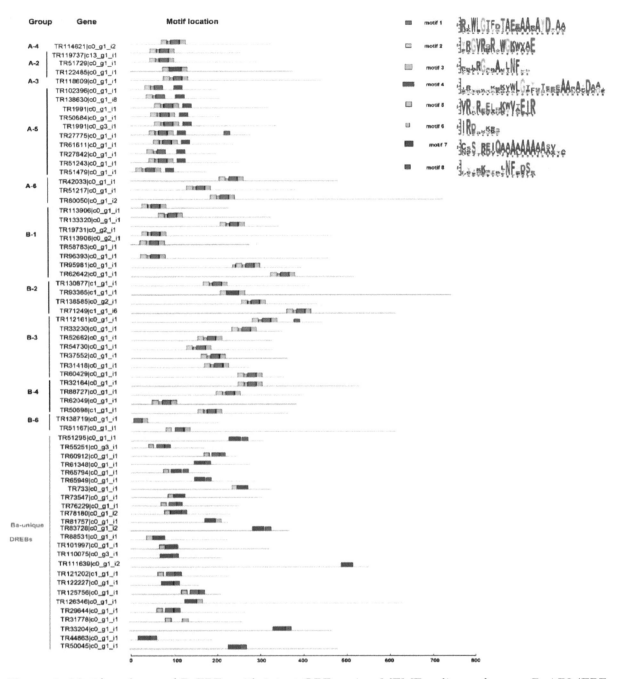

Figure 4. Motif analyses of BaERFs with intact ORFs using MEME online software. BaAP2/ERF proteins with complete ORFs were used for motif prediction. Parameters are as follows: any number of repetitions per sequence, motif width ranges of 6–50 amino acids, and 8 as the maximum number of motifs. Each of the sequence has an E-value less than 10. Motif composition and deduced amino acid sequence of each motif are presented.

2.4. Gene Expression Analysis of all the BaAP2/ERFs during Moss Specific Dehydration–Rehydration Process Using RNA-seq Data

To evaluate the potential function of 83 *BaAP2/ERFs* genes under dehydration–rehydration stress treatment, we investigated the gene expression pattern based on the RNA-seq datasets (H0, D2, D24, R2, and R48). The results show that all 83 *BaAP2/ERFs* belonged to AP2, DREB, ERF, RAV and Soloists families exhibited elevated transcript amounts during dehydration–rehydration process (Figure 5). The majority of *BaAP2/ERF* transcripts were more abundant in both dehydration (D2 and D24) and early-rehydration (R2) stages. For example, 6 out of 11 *AP2* family transcripts and 13 out of 17

DREB transcripts were more abundant in dehydration–rehydration stages. Some transcripts also showed different expression patterns of accumulation. TR88531 | c0_g1_i1 and TR76229 | c0_g1_i1 (Bryum-unique *DREB* genes) transcripts were more abundant in D2 and D24, while TR110575 | c0_g3_i1 (Bryum-unique *DREB* gene) and TR130877 | c0_g1_i1 (*ERF* gene) were more abundant in R2 and R48.

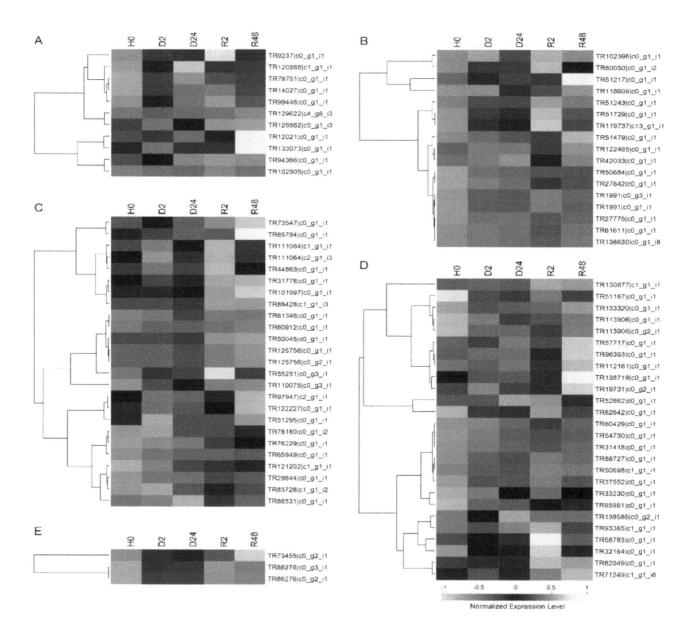

Figure 5. Heat map of the relative expression levels of all identified BaAP2/EFR genes during dehydration–rehydration process of *B. argenteum*. Color scores were normalized by the log2 transformed counts of RPKM values. Yellow represents high expression, while blue represents low expression. Expression differences in the transcripts were clustered by the hierarchical complete linkage clustering method using an uncentered correlation similarity matrix. Prior to the clustering analysis, expression data in unit of FPKM were pretreated using the standardization tools in Cluster 3.0. The heat maps were drawn using the Java Treeview package. Expression profiles (in log2 based values) of the: (**A**) *AP2*; (**B**) *DREB*; (**C**) unclassified *DREB* group (Ba-unique clade); (**D**) *ERF*; and (**E**) *RAV* and *Soloist* genes in *B. argenteum* response to dehydration–rehydration treatment.

2.5. Diverse Gene Expression Patterns of Twelve BaAP2/ERFs during Dehydration–Rehydration Process Using RT-qPCR

The expression profiles of 12 transcripts, representative of different family members of *BaAP2/ERF* (one *AP2*, four *DREBs*, two Ba-unique *DREBs*, three *ERFs* and two *Soloists*) that demonstrated a diverse pattern of induced gene expression were validated by RT-qPCR (Figure 6). RT-qPCR results confirmed the expression trends observed with RNA-seq data. RT-qPCR demonstrated that 9 out of 12 genes increased, reached a peak and then decreased, while the other three genes changed slightly and rapidly increased to the maximum fold at rehydration (R48) stage. Most genes reached an expression peak at rehydration stage (R2 or R48), except TR27842 I c0_g1_i1 and TR42033 I c0_g1_i1, which peaked at D2 and D24, respectively (Figure 6). The expression pattern can be divided into four types: (1) transcripts that accumulate in response to desiccation stress (e.g., TR27842 I c0_g1_i1 which was strongly induced by desiccation treatment (almost 20-fold compared to H0) and rapidly reduced after rehydration); (2) transcripts which modestly accumulate in response to both desiccation and rehydration (e.g.,TR42033 I c0_g1_i1 gene, the gene expression level of which changed within two-fold during desiccation and rehydration process); (3) transcripts which accumulate in response to desiccation and remain elevated upon rehydration (e.g., TR29644 I c0_g1_i1, which was 10-fold increased after desiccation treatment, and then reached a peak (more than 40-fold) at R2 stage); and (4) transcripts which accumulate in response to rehydration (e.g., TR138719 I c0_g1_i1, which was highly induced by rehydration (R48)).

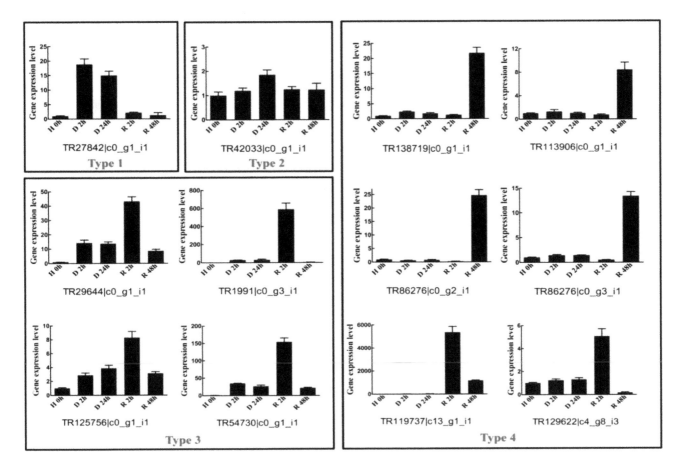

Figure 6. RT-qPCR validation of gene expression patterns of 12 representative *BaAP2/ERF* genes during *B. argenteum* dehydration–rehydration process. RT-qPCR quantitative gene expression data are shown as the mean $\pm$ SE. The relative gene expression levels were calculated relative to 0 h and using the $2^{-\Delta\Delta Ct}$ method.

2.6. Transactivation Activity Analyses of Twelve BaAP2/ERFs

We further investigated the transactivation activity of the above 12 *BaAP2/ERFs* using a yeast-based transcriptional activity assay. The results show that 8 out of 12 BaAP2/ERFs proteins can grow well on SD-Trp, SD-Trp-His medium and exhibit α-galactosidase activity on SD-Trp-His medium containing x-α-gal (Figure 7). This indicates that these AP2/ERF proteins (one AP2 (TR129622 | c4_g8_i3), four DREBs (TR119737 | c13_g1_i1, TR27842 | c0_g1_i1, TR1991 | c0_g3_i1, and TR42033 | c0_g1_i1), one Ba-unique DREB (TR29644 | c0_g1_i1) and two ERFs (TR54730 | c0_g1_i1 and TR138719 | c0_g1_i1)) demonstrate transactivation activities. Four proteins (one Ba-unique DREB (TR125756 | c0_g1_i1), one ERF (TR113906 | c0_g1_i1) and two Soloists (TR86276 | c0_g2_i1 and TR86276 | c0_g3_i1)) grew similarly to the negative control indicating that these proteins might not function as transcriptional activators in this yeast heterologous system.

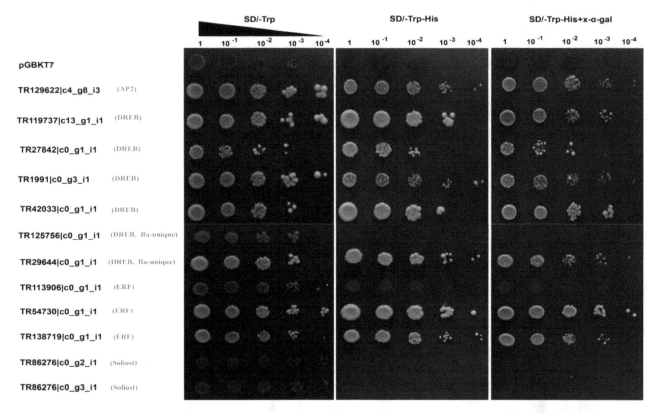

Figure 7. Transactivation activities of 12 BaAP2/ERF proteins in yeast. Yeast cells Y2H expressing the fusion proteins were cultured and adjusted to an OD600 of 2.0, then series diluted and dropped with 2 μL on nutritional selective medium SD/−Trp, SD/−Trp−His and SD/−Trp−His+x-α-gal. Yeast cells expressing the empty vector pGBKT7 was used as negative control. Photos were taken after incubating at 30 °C for 2–4 days.

2.7. Stress Tolerance Ability Evaluation in Transgenic Yeast

To investigate the ability of BaAP2/ERF proteins to enhance abiotic stress tolerance in heterologous expression system, eight representative *BaAP2/ERFs*, driven by a galactose-inducible promoter (pYES2), were introduced into *S. cerevisiae* (INVSc1). After 45 h of exposure to 5 M NaCl and 3 M Sorbitol, the growth patterns of all BaAP2/ERF transformed *S. cerevisiae* (pYES2-*BaAP2/ERF*) were similar to the empty vector (pYES2) under non-stress conditions (Figure 8). Seven out of eight *BaAP2/ERF* (except TR86276 | c0_g3_i1) transformed *S. cerevisiae* survived better than the empty vector under salt and osmotic stresses, especially under salt stress, indicating that these seven BaAP2/ERF proteins (TR119737 | c13_g1_i1, TR27842 | c0_g1_i1, TR1991 | c0_g3_i1, TR29644 | c0_g1_i1,

TR54730 | c0_g1_i1, TR138719 | c0_g1_i1, and TR86276 | c0_g2_i1) were functional in yeast cells and improved the yeast tolerance to salt and osmotic stresses.

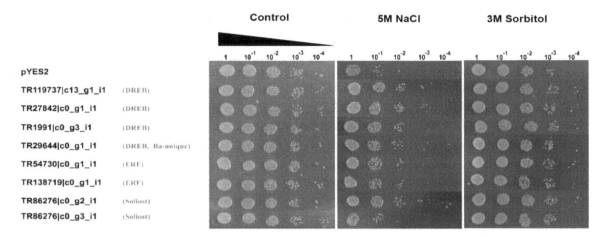

Figure 8. Growth of *S. cerevisiae* yeast cells transformed with the pYES2-BaAP2/ERFs under salt and osmotic stress conditions. To test the salt and osmotic tolerances, the same quantity of yeast culture sample was re-suspended in 5 M NaCl and 3 M Sorbitol, respectively, at 30 °C for 45 h. For non-stress control, an equivalent number of yeast cells was re-suspended in 200 μL of sterile water and incubated at 30 °C for 45 h. Serial dilutions of 1:10 transformed yeast cells were grown on SC-ura medium for two days.

3. Discussion

AP2/ERF genes play central roles during plant stress responses and have been widely identified in both dicotyledonous and monocotyledonous plants using genomic or transcriptomic data [20,22], however, little is known about the functions of *AP2/ERF* genes in moss species. Several recent studies have demonstrated that AP2/ERF transcription factors play an important role in the stress responses of bryophytes [15,16,19]. *B. argenteum* is extremely tolerant to desiccation stress and is a promising model for the identification of stress related genes [8], however, no *AP2/ERF* gene in *B. argenteum* has been reported until now.

The AP2/ERF family of TFs in Arabidopsis comprises five subfamilies of TFs, classified based on sequence similarity, number of AP2 domains, and the presence of other characteristic domains [26]. It is reported that different gene families have different functions. The *AP2* gene family is associated with plant flower development, while *ERF* and *DREB* family genes accumulate in response to biotic and abiotic stress, respectively [9]. Genes annotated to a specific group within the same gene family are also reported to have different functions. A-1 type *DREB* transcripts accumulate in response to cold stress, while A-2 type *DREB* transcripts accumulate in response to osmotic and heat stresses [27]. Hence, a precise and detailed classification of *AP2/ERF* genes within a genome is an important tool for predicting gene expression and function.

Classification of the AP2/ERF superfamily is based on the number of AP2 domains and by constructing a phylogenetic tree comparing the moss AP2 domains to Arabidopsis or rice. The resulting gene tree construction has been a classic and reliable method of annotation which is widely employed in many non-model plants [24,28–31]. However, classification by gene tree should be used cautiously as the results can be inconsistent with an AP2 domain counting based classification. *Soloist* genes with a single AP2 domain always grouped together with *AP2* family genes which have two domains (as reported in *Hevea brasiliensis* and *Vitis vinifera* [23,28]). In this study, we found that *BaSoloists* clustered with *AtSoloists* and were mixed together with *AP2* family genes. In addition, five genes which contained a single AP2 domain also clustered with *AP2* family genes in Arabidopsis. To confirm the classification as AP2 family members, we constructed two more phylogenetic trees: one using only *AP2/ERF* genes in *B. argenteum* and another one using *AP2/ERF* genes from the model moss

P. patens. Finally, these five genes were classified into *AP2* family given their greater homology with the *AP2* family genes.

The AP2 domain of *AP2/ERF* genes is conserved in plants, however amino acid variations within the AP2 domain have been documented and can refine classification of the gene [23,29]. For example, the motifs "HLG" and "WLG" in the β3 sheet can distinguish a *Soloist* gene from an *ERF* gene. Li et al. (2017) demonstrated the "EVR" motif pattern was only present in the A-1 group of *DREB* genes and "ERK" was specific to the B-6 subfamily of *ERF* genes in the β2 sheet of AP2 domain [19]. Based on this A-1 DREB-specific amino acid, in the present study, we finally confirmed the TR11462 | c0_g1_i2 gene was A-4 type of DREB rather than A-1 type. Specific motif elements are also helpful for robust gene classification. In the DREB family, ERF-associated amphiphilic repression (EAR) motif was specifically present in the A-5a group genes, which contained (L/F) DLN (L/F) xP residues and may be essential for repression function [32–34].

Moss species have unique genes which are challenging to annotate compared to other organisms [19]. In *S. caninervis*, the majority of *ScERF* genes can be classified while few *ScERF* genes are not clustered with any Arabidopsis group, and clustered as a unique clade [19]. Similarly, in this study, half of *DREBs* cannot be classified relative to other plant genes and clustered as a Ba-unique clade. The amino acids compositions of AP2 domains were also supported that Ba-unique clade genes have diverse amino acids composition and showed more diverse motif patterns compared with other *DREB* genes. Furthermore, BLASTp search in One KP and NCBI database showed that, although Ba-unique *DREBs* have homolog genes in angiosperm, they shared very low amino acid identities. Some Ba-unique *DREBs* had very high sequence identities with other moss genes, while these moss genes were rarely characterized and no functional analysis were reported until now. Our results extend the idea that *AP2/ERF* genes in moss species can be different from angiosperm genes, and moss-unique genes may have novel and/or altered functions. It is necessary to explore their functions in future work.

Gene expression pattern was considered to be directly connected with the gene function [35], and, in this study, the expression of all 83 *BaAP2/ERFs* genes were induced during dehydration–rehydration process. Moreover, the 12 representative *BaAP2/ERF* genes in different families exhibited differential expression in response to dehydration and rehydration treatment. Within the same family, the gene expression patterns were different suggesting a functional diversity of *AP2/ERF* genes in response to dehydration and rehydration stress in *B. argenteum*. AP2/ERF proteins are important transcriptional factors which can activate many down-stream genes, thus improving the overall stress tolerance of plants [10]. In the present study, 8 out of 12 BaAP2/ERFs proteins demonstrated transactivation activity in the yeast system. Based upon patterns of gene expression and transactivation activity analysis, we selected eight representative *BaAP2/ERF* genes for further functional test in yeast and the result showed that seven of them improved the yeast tolerance to salt and osmotic stresses. Our results demonstrated that *BaAP2/ERFs* genes play crucial roles in *B. argenteum* response to stresses.

4. Materials and Methods

4.1. Identification of the AP2/ERF Protein Family in B. argenteum

B. argenteum transcripts (76,206) were obtained from a hydration–dehydration–rehydration transcriptome [25] (data were deposited at NCBI-SRA with accession SRP077772, https://www.ncbi.nlm.nih.gov/sra/?term=SRP077772) and served as the source for the *AP2/ERF* gene identification and presented in this study. Two methods were used together to identify the putative *AP2/ERF* genes from *B. argenteum*. Firstly, 176 Arabidopsis AP2 predicted amino acid sequences and 171 *P. patens* AP2 predicted amino acid sequences were downloaded from the plant transcription factor database (PlantTFDB v3.0) (http://planttfdb.cbi.edu.cn/) [14], and used as queries to search against the *B. argenteum* transcriptome database using tBLASTn program (E value of 1×10^{-3}. Second, the HMM profiles PF00847 (AP2 domain) and PF02362 (B3 domain) were downloaded from Pfam database v27.0 (http://pfam.sanger.ac.uk/) [36], and the profiles were queried using hmm search command included in the HMMER (v3.0) software (E value cutoff at 1×10^{-3}). All candidate *BaAP2/ERF* genes identified through these two methods were confirmed with Conserved Domain Database (CDD http://www.ncbi.nlm.nih.gov/cdd/) [37] and SMART (http://smart.embl-heidelberg.de/) [38] searches to ensure the presence of an AP2 domain. An AP2 domain, length of approximately 60 amino acids was considered to be a full-length AP2 domain [23]. All the predicted peptide sequences were filtered with a minimum length of 80 amino acids. Sequences which shared >98% matches were considered redundant.

4.2. Sequence Analysis and Classification of BaAP2/ERF Genes Using Phylogenetic Tree

ORFs were predicted with the ORF Finder at NCBI (http://www.ncbi.nlm.nih.gov/gorf/gorf. html). Protein sequence motif detection was performed with MEME program (http://meme-suite.org/index.html) [39] using the parameters: zero or one repetition per sequence, motif width ranges of 6–40 amino acids, and 8 as the maximum number of motifs. Multiple sequence alignment was performed with ClustalW [40], phylogenetic trees were constructed by the neighbor-joining method (with 1000 bootstrap replicates) using MEGA 6.06 the evolutionary distances were computed using the Poisson correction method with pairwise deletion. Sequence similarity was analyzed using BLASTp search with NCBI and ONE KP (https://db.cngb.org/onekp/) database. All the *BaAP2/ERF* sequences were submitted to the GenBank database using *Bank*It (http://www.ncbi.nlm.nih.gov/BankIt/).

4.3. Gene Expression Analysis of BaAP2/ERF Genes Using RNA-seq Data

Expression differences in the transcripts under dehydration–rehydration condition were clustered by the hierarchical complete linkage clustering method using an uncentered correlation similarity matrix. Prior to the clustering analysis, expression data in unit of Fragments Per Kilobase of transcript per Million fragments mapped (FPKM) were pretreated using the standardization tools in Cluster 3.0: (a) log transform data; (b) center genes (mean); and (c) normalize genes [41]. The heat maps were drawn by using the Java Treeview package [42,43].

4.4. Gene Expression Pattern Analysis of BaAP2/ERF Genes Using RT-qPCR Assay

B. argenteum gametophytes were cultured in solid Knop medium at 25 °C with 16 h/8 h photoperiod in a climate chamber as described previously [8]. For desiccation–rehydration treatment, the well-hydrated gametophytes in Knop solid medium were transferred to 90 cm open Petri dish and air-dried for 2 h (D2) and 24 h (D24), and the desiccated gametophytes (D24) samples were subsequently rehydrated with deionized water for 2 h (R2) and 48 h (R48). All treatments were performed at 25 °C with RH ≈ 25–27%, and the well-hydrated gametophores in Knop medium without any treatment was served as the control (H0). Three biological replicates were collected for each of the time point of different treatments.

Total RNAs of *B. argenteum* gametophytes were extracted using MiniBEST plant RNA kit (Takara, Japan). Gel electrophoresis and a NanoDrop 2000 spectrophotometer (Thermo Fisher Scientific, Waltham, MA, USA) were used for RNA quality test and quantitative analysis. High quality RNA samples were used for subsequent reverse transcription. First strand cDNA was synthesized using PrimeScript™ RT reagent kit (Takara, Shiga Prefecture, Japan).

Twelve *BaAP2/ERF* genes representative different groups/subfamilies were selected to verify the gene expression pattern obtained from transcriptome data under desiccation and rehydration condition. RT-qPCR primers were designed with Primer Premier 5.0 and the primer specificities were tested by running BLAST search against the local *B. argenteum* transcriptional data. Each primer pair was further assessed using melting-curve analysis after RT-qPCR. All primer information for RT-qPCR is shown in Table S2. RT-qPCR experiments were carried out using CFX96 Real-Time PCR Detection System (Bio-Rad, Hercules, CA, USA) with SYBR *Premix Ex Taq*™ kit (Takara, Shiga Prefecture, Japan). The PCR reaction mixture consisted of 2 μL cDNA sample (1:5 diluted), 0.4 μL each of the forward and reverse primers (10 μM), 10 μL master mix and 7.2 μL PCR-grade water in a final volume of 20 μL. Three biological replicates and three technical replicates of each biological replicate were used for all samples. The RT-qPCR program was as follows: initial denaturation step of 30 s at 95 °C and 40 cycles of PCR (94 °C for 5 s and 60–62 °C for 30 s). The gene relative expression levels were calculated relative to the H0 samples using the $2^{-\Delta\Delta Ct}$ method. The *ACT* gene was used to normalize the RT-qPCR data [8]. Figures were generated using Sigmaplot 12.0.

4.5. Gene Cloning, Vector Construction and Transcriptional Activation Analysis in Yeast Cells

To further evaluation of transactivation activity of 12 *BaAP2/ERF* genes, we cloned these 12 genes into the pMD18-T clone vector. After sequence analysis, the PCR products of these genes were cloned separately into the pGBKT7 vector using the in-fusion PCR cloning system (Clontech, Mountain View, CA, USA). Positive plasmids containing different *BaAP2/ERF* genes were transformed into the Y2H yeast strain (Clontech, Mountain View, CA, USA). All primers used for cloning and vector construction are listed in Tables S3 and S4. The cell concentration of yeast positive transformants were adjusted to an OD600 of 2.0, the yeast cells were then diluted serially ($1, 10^{-1}, 10^{-2}, 10^{-3}$, and 10^{-4}) and dropped with 2 μL on synthetic dropout (SD) medium without tryptophan (SD/−Trp), without tryptophan and histidine (SD/−Trp−His), and with SD/−Trp−His plates containing x-α-gal with the final concentration of 40 mg/L (SD/−Trp−His+x-α-gal). Yeast cells expressing the empty vector pGBKT7 was used as negative control. The plates were incubated at 30 °C for 2–4 days before photographing. Adobe Illustrator CS5 was used for image processing.

4.6. Stress Tolerance Studies in Yeast

Eight representative *BaAP2/ERF* genes including four *DREBs* (TR119737|c13_g1_i1, TR27842|c0_g1_i1, TR1991|c0_g3_i1 and TR29644|c0_g1_i1), two *ERFs* (TR54730|c0_g1_i1 and TR138719|c0_g1_i1) and two *Soloists* (TR86276|c0_g2_i1 and TR86276|c0_g3_i1) were selected to study the stress tolerance ability under salt and osmotic stress conditions in yeast. The ORF of TR1991|c0_g3_i1, TR29644|c0_g1_i1, TR54730|c0_g1_i1, TR138719|c0_g1_i1, TR86276|c0_g2_i1 and TR86276|c0_g3_i1 were amplified from pGBKT7-*BaAP2/ERF* plasmids of transcriptional activation assay, using primers shown in Table S5, and inserted into the yeast expression vector pYES2 using the in-fusion PCR cloning system. The ORF of two *DREB* genes TR119737|c13_g1_i1 and TR27842|c0_g1_i1 were obtained from pGBKT7-*BaAP2/ERF* plasmids using *Not*I and *Eco*RI restriction enzymes digestion and inserted into the *Not*I and *Eco*RI sites of the yeast expression vector pYES2, which contains a URA3 selection marker driven by the GAL1 promoter. Subsequently, eight pYES2-*BaAP2/ERF* plasmids and the empty pYES2 control plasmids were introduced into yeast strain INVSc1 (Invitrogen, Carlsbad, CA, USA) using a lithium acetate procedure, according to the pYES2 vector kit instructions (Invitrogen, Carlsbad, CA, USA). The transformants were screened by growth on a uracil-deficient synthetic complete (SC-ura) medium with 2% (*w/v*) glucose at 30 °C for 2 days.

For the stress assay, yeast cells harboring both pYES2-BaAP2/ERFs and the empty pYES2 vector (control) were incubated in SC-ura liquid medium containing 2% glucose at 30 °C for approximately 20 h with shaking (180 rpm). After incubation, the optical densities of the yeast cell were determined at OD600. The culture samples were adjusted to contain an equal OD600 of 0.4 as a starting concentration in 10 mL of induction SC-ura medium (supplemented with 2% *w/v* galactose). After incubation for approximately 24 h, the yeast cell densities were recalculated and adjusted to contain an equal number of cells (OD600 = 2) in 200 μL solutions with 5 M NaCl or 3M Sorbitol for the salt or osmotic stress, and the same quantity of yeast cells was re-suspended in 200 μL of sterile water was served as the control. After incubating at 30 °C for 45 h, the cells were 10-fold serially diluted with sterile water, and 2 μL aliquots of each dilution were spread on SC-ura medium containing 2% (*w/v*) glucose and growth performance was compared after growing at 30 °C for 2 days [18,44].

5. Conclusions

This is the first report on identification, classification, characterization, and functional evaluation of the *AP2/ERF* gene family in the desiccation tolerant moss *B. argentums*. Eighty-three AP2/ERF predicted proteins were identified from the *B. argenteum* transcriptome and classified within the *AP2/ERF* gene family. The gene expression pattern was analyzed in response to a well characterized dehydration–rehydration based upon RT-qPCR and RNA-seq data. We verified the transactivation activities of 12 representative *BaAP2/ERF* genes. Furthermore, eight *BaAP2/ERF* genes were tested for stress tolerance functions in yeast. We conclude that TR29644 | c0_g1_i1 (*DREB*-Ba-unique), TR119737 | c13_g1_i1 (*DREB*), TR54730 | c0_g1_i1 (*ERF*), TR27842 | c0_g1_i1 (*DREB*) and TR86276 | c0_g2_i1 (*Soloist*) genes strongly respond to environmental stress and that these genes are correlated with enhanced salt- and osmotic-stress tolerance in transgenic yeast. These genes are promising candidate genes for further functional analysis and demonstrate great potential in plant molecular breeding.

Supplementary Materials
Figure S1: The 20 most abundant predicted transcription factor families in the *B. argenteum* transcripotome datasets; Figure S2: Phylogenetic analysis of *AP2/ERF* family genes in *B. argenteum*; Figure S3: Phylogenetic analysis of *AP2/ERF* family genes in *B. argenteum* and *P. patens*; Table S1: Sequence alignment results of 25 Ba-clade DREBs using IKP BLAST; Table S2: Primer information of 12 *BaAP2/ERF* genes for RT-qPCR analysis; Table S3: Primer information of 12 *BaAP2/ERF* genes for gene cloning; Table S4: Primer information of 12 *BaAP2/ERF* genes for fusing to pGBKT7 vector; Table S5: Primer information of *BaAP2/ERF* genes for infusing to pYES2 vector.

Author Contributions: Formal analysis, B.G.; Funding acquisition, D.Z.; Investigation, X.L. (Xiaoshuang Li), Y.L., X.L. (Xiaojie Liu) and J.Z.; Supervision, D.Z.; Validation, A.J.W.; Visualization, B.G.; Writing—original draft, X.L. (Xiaoshuang Li); and Writing—review and editing, J.Z. and A.J.W.

Abbreviations

AP2/ERF	APETALA2/Ethylene responsive factor
DREB	dehydration-responsive element-binding protein
ERF	ethylene-responsive factor
RAV	related to ABI3/VP1
RT-qPCR	reverse transcription quantitative real-time polymerase chain reaction
TF	transcription factor
DT	desiccation tolerance

References

1. Zhang, Y.M.; Chen, J.; Wang, L.; Wang, X.Q.; Gu, Z.H. The spatial distribution patterns of biological soil crusts in the Gurbantunggut Desert, Northern Xinjiang, China. *J. Arid Environ.* **2007**, *68*, 599–610. [CrossRef]

2. Li, X.R.; Zhou, H.Y.; Wang, X.P.; Zhu, Y.G.; O'Conner, P.J. The effects of sand stabilization and revegetation on cryptogam species diversity and soil fertility in the Tengger Desert, Northern China. *Plant Soil* **2003**, *251*, 237–245. [CrossRef]

3. Hui, R.; Li, X.R.; Chen, C.Y.; Zhao, X.; Jia, R.L.; Liu, L.C.; Wei, Y.P. Responses of photosynthetic properties and chloroplast ultrastructure of *Bryum argenteum* from a desert biological soil crust to elevated ultraviolet-B radiation. *Physiol. Plant.* **2013**, *147*, 489–501. [CrossRef] [PubMed]

4. Li, J.H.; Li, X.R.; Chen, C.Y. Degradation and reorganization of thylakoid protein complexes of *Bryum argenteum* in response to dehydration and rehydration. *Bryologist* **2014**, *117*, 110–118. [CrossRef]

5. Wood, A.J. Invited essay: New frontiers in bryology and lichenology—The nature and distribution of vegetative desiccation-tolerance in hornworts, liverworts and mosses. *Bryologist* **2007**, *110*, 163–177. [CrossRef]

6. Li, J.H.; Li, X.R.; Zhang, P. Micro-morphology, ultrastructure and chemical composition changes of *Bryum argenteum* from a desert biological soil crust following one-year desiccation. *Bryologist* **2014**, *117*, 232–240. [CrossRef]

7.   Stark, L.R.; McLetchie, D.N.; Eppley, S.M. Sex ratios and the shy male hypothesis in the moss *Bryum argenteum* (Bryaceae). *Bryologist* **2010**, *113*, 788–797. [CrossRef]

8.   Gao, B.; Zhang, D.Y.; Li, X.S.; Yang, H.L.; Zhang, Y.M.; Wood, A.J. De novo transcriptome characterization and gene expression profiling of the desiccation tolerant moss *Bryum argenteum* following rehydration. *BMC Genom.* **2015**, *16*. [CrossRef] [PubMed]

9.   Licausi, F.; Ohme-Takagi, M.; Perata, P. APETALA2/Ethylene Responsive Factor (AP2/ERF) transcription factors: Mediators of stress responses and developmental programs. *New Phytol.* **2013**, *199*, 639–649. [CrossRef] [PubMed]

10. Xu, Z.S.; Chen, M.; Li, L.C.; Ma, Y.Z. Functions and Application of the AP2/ERF Transcription Factor Family in Crop Improvement. *J. Integr. Plant Biol.* **2011**, *53*, 570–585. [CrossRef] [PubMed]

11. Mizoi, J.; Shinozaki, K.; Yamaguchi-Shinozaki, K. AP2/ERF family transcription factors in plant abiotic stress responses. *Biochim. Biophys. Acta Gene Regul. Mech.* **2012**, *1819*, 86–96. [CrossRef] [PubMed]

12. Bhatta, M.; Morgounov, A.; Belamkar, V.; Baenziger, P. Genome-wide association study reveals novel genomic regions for grain yield and yield-related traits in drought-stressed *Synthetic hexaploid* Wheat. *Int. J. Mol. Sci.* **2018**, *19*, 3011. [CrossRef] [PubMed]

13. Jin, J.P.; Tian, F.; Yang, D.C.; Meng, Y.Q.; Kong, L.; Luo, J.C.; Gao, G. PlantTFDB 4.0: Toward a central hub for transcription factors and regulatory interactions in plants. *Nucleic Acids Res.* **2017**, *45*, D1040–D1045. [CrossRef] [PubMed]

14. Jin, J.; Zhang, H.; Kong, L.; Gao, G.; Luo, J. PlantTFDB 3.0: A portal for the functional and evolutionary study of plant transcription factors. *Nucleic Acids Res.* **2014**, *42*, D1182–D1187. [CrossRef] [PubMed]

15. Hiss, M.; Laule, O.; Meskauskiene, R.M.; Arif, M.A.; Decker, E.L.; Erxleben, A.; Frank, W.; Hanke, S.T.; Lang, D.; Martin, A.; et al. Large-scale gene expression profiling data for the model moss *Physcomitrella patens* aid understanding of developmental progression, culture and stress conditions. *Plant J.* **2014**, *79*, 530–539. [CrossRef] [PubMed]

16. Liu, N.; Zhong, N.Q.; Wang, G.L.; Li, L.J.; Liu, X.L.; He, Y.K.; Xia, G.X. Cloning and functional characterization of PpDBF1 gene encoding a DRE-binding transcription factor from *Physcomitrella patens*. *Planta* **2007**, *226*, 827–838. [CrossRef] [PubMed]

17. Gao, B.; Zhang, D.; Li, X.; Yang, H.; Wood, A.J. De novo assembly and characterization of the transcriptome in the desiccation-tolerant moss *Syntrichia caninervis*. *BMC Res. Notes* **2014**, *7*, 490. [CrossRef] [PubMed]

18. Li, H.; Zhang, D.; Li, X.; Guan, K.; Yang, H. Novel DREB A-5 subgroup transcription factors from desert moss (*Syntrichia caninervis*) confers multiple abiotic stress tolerance to yeast. *J. Plant Physiol.* **2016**, *194*, 45–53. [CrossRef] [PubMed]

19. Li, X.; Zhang, D.; Gao, B.; Liang, Y.; Yang, H.; Wang, Y.; Wood, A.J. Transcriptome-Wide Identification, Classification, and Characterization of AP2/ERF Family Genes in the Desert Moss *Syntrichia caninervis*. *Front. Plant Sci.* **2017**, *8*, 262. [CrossRef] [PubMed]

20. Nakano, T.; Suzuki, K.; Fujimura, T.; Shinshi, H. Genome-wide analysis of the ERF gene family in Arabidopsis and rice. *Plant Physiol.* **2006**, *140*, 411–432. [CrossRef] [PubMed]

21. Song, X.M.; Li, Y.; Hou, X.L. Genome-wide analysis of the AP2/ERF transcription factor superfamily in Chinese cabbage (*Brassica rapa* ssp pekinensis). *BMC Genom.* **2013**, *14*. [CrossRef] [PubMed]

22. Sakuma, Y.; Liu, Q.; Dubouzet, J.G.; Abe, H.; Shinozaki, K.; Yamaguchi-Shinozaki, K. DNA-binding specificity of the ERF/AP2 domain of Arabidopsis DREBs, transcription factors involved in dehydration- and cold-inducible gene expression. *Biochem. Biophys. Res. Commun.* **2002**, *290*, 998–1009. [CrossRef] [PubMed]

23. Duan, C.; Argout, X.; Gebelin, V.; Summo, M.; Dufayard, J.F.; Leclercq, J.; Kuswanhadi; Piyatrakul, P.; Pirrello, J.; Rio, M.; et al. Identification of the Hevea brasiliensis AP2/ERF superfamily by RNA sequencing. *BMC Genom.* **2013**, *14*, 30. [CrossRef] [PubMed]

24. Wu, Z.J.; Li, X.H.; Liu, Z.W.; Li, H.; Wang, Y.X.; Zhuang, J. Transcriptome-based discovery of AP2/ERF transcription factors related to temperature stress in tea plant (*Camellia sinensis*). *Funct. Integr. Genom.* **2015**, *15*, 741–752. [CrossRef] [PubMed]

25. Gao, B.; Li, X.; Zhang, D.; Liang, Y.; Yang, H.; Chen, M.; Zhang, Y.; Zhang, J.; Wood, A.J. Desiccation tolerance in bryophytes: The dehydration and rehydration transcriptomes in the desiccation-tolerant bryophyte *Bryum argenteum*. *Sci. Rep.* **2017**, *7*, 7571. [CrossRef] [PubMed]

26. Agarwal, P.K.; Gupta, K.; Lopato, S.; Agarwal, P. Dehydration responsive element binding transcription factors and their applications for the engineering of stress tolerance. *J. Exp. Bot.* **2017**, *68*, 2135–2148. [CrossRef] [PubMed]

27. Lata, C.; Prasad, M. Role of DREBs in regulation of abiotic stress responses in plants. *J. Exp. Bot.* **2011**, *62*, 4731–4748. [CrossRef] [PubMed]

28. Licausi, F.; Giorgi, F.M.; Zenoni, S.; Osti, F.; Pezzotti, M.; Perata, P. Genomic and transcriptomic analysis of the AP2/ERF superfamily in *Vitis vinifera*. *BMC Genom.* **2010**, *11*. [CrossRef] [PubMed]

29. Zhuang, J.; Cai, B.; Peng, R.H.; Zhu, B.; Jin, X.F.; Xue, Y.; Gao, F.; Fu, X.Y.; Tian, Y.S.; Zhao, W.; et al. Genome-wide analysis of the AP2/ERF gene family in *Populus trichocarpa*. *Biochem. Biophys. Res. Commun.* **2008**, *371*, 468–474. [CrossRef] [PubMed]

30. Chen, L.H.; Han, J.P.; Deng, X.M.; Tan, S.L.; Li, L.L.; Li, L.; Zhou, J.F.; Peng, H.; Yang, G.X.; He, G.Y.; et al. Expansion and stress responses of AP2/EREBP superfamily in *Brachypodium Distachyon*. *Sci. Rep.* **2016**, *6*. [CrossRef] [PubMed]

31. Lakhwani, D.; Pandey, A.; Dhar, Y.V.; Bag, S.K.; Trivedi, P.K.; Asif, M.H. Genome-wide analysis of the AP2/ERF family in Musa species reveals divergence and neofunctionalisation during evolution. *Sci. Rep.* **2016**, *6*. [CrossRef] [PubMed]

32. Ohta, M.; Matsui, K.; Hiratsu, K.; Shinshi, H.; Ohme-Takagi, M. Repression domains of class II ERF transcriptional repressors share an essential motif for active repression. *Plant Cell* **2001**, *13*, 1959–1968. [CrossRef] [PubMed]

33. Huang, B.; Liu, J.Y. A cotton dehydration responsive element binding protein functions as a transcriptional repressor of DRE-mediated gene expression. *Biochem. Biophys. Res. Commun.* **2006**, *343*, 1023–1031. [CrossRef] [PubMed]

34. Dong, C.J.; Liu, J.Y. The Arabidopsis EAR-motif-containing protein RAP2.1 functions as an active transcriptional repressor to keep stress responses under tight control. *BMC Plant Biol.* **2010**, *10*. [CrossRef] [PubMed]

35. Hao, Y.J.; Wei, W.; Song, Q.X.; Chen, H.W.; Zhang, Y.Q.; Wang, F.; Zou, H.F.; Lei, G.; Tian, A.G.; Zhang, W.K.; et al. Soybean NAC transcription factors promote abiotic stress tolerance and lateral root formation in transgenic plants. *Plant J.* **2011**, *68*, 302–313. [CrossRef] [PubMed]

36. Punta, M.; Coggill, P.C.; Eberhardt, R.Y.; Mistry, J.; Tate, J.; Boursnell, C.; Pang, N.; Forslund, K.; Ceric, G.; Clements, J.; et al. The Pfam protein families database. *Nucleic Acids Res.* **2012**, *40*, D290–D301. [CrossRef] [PubMed]

37. Marchler-Bauer, A.; Derbyshire, M.K.; Gonzales, N.R.; Lu, S.; Chitsaz, F.; Geer, L.Y.; Geer, R.C.; He, J.; Gwadz, M.; Hurwitz, D.I.; et al. CDD: NCBI's conserved domain database. *Nucleic Acids Res.* **2015**, *43*, D222–D226. [CrossRef] [PubMed]

38. Letunic, I.; Doerks, T.; Bork, P. SMART: Recent updates, new developments and status in 2015. *Nucleic Acids Res.* **2015**, *43*, D257–D260. [CrossRef] [PubMed]

39. Bailey, T.L.; Boden, M.; Buske, F.A.; Frith, M.; Grant, C.E.; Clementi, L.; Ren, J.; Li, W.W.; Noble, W.S. MEME SUITE: Tools for motif discovery and searching. *Nucleic Acids Res.* **2009**, *37*, W202–W208. [CrossRef] [PubMed]

40. Tamura, K.; Peterson, D.; Peterson, N.; Stecher, G.; Nei, M.; Kumar, S. MEGA5: Molecular evolutionary genetics analysis using maximum likelihood, evolutionary distance, and maximum parsimony methods. *Mol. Biol. Evol.* **2011**, *28*, 2731–2739. [CrossRef] [PubMed]

41. Eisen, M.B.; Spellman, P.T.; Brown, P.O.; Botstein, D. Cluster analysis and display of genome-wide expression patterns. *Proc. Natl. Acad. Sci. USA* **1998**, *95*, 14863–14868. [CrossRef] [PubMed]

42. Brock, G.; Pihur, V.; Datta, S.; Datta, S. clValid, an R package for cluster validation. *J. Stat. Softw.* **2011**. [CrossRef]

43. Saldanha, A.J. Java Treeview-extensible visualization of microarray data. *Bioinformatics* **2004**, *20*, 3246–3248. [CrossRef] [PubMed]

44. Li, X.; Zhang, D.; Li, H.; Wang, Y.; Zhang, Y.; Wood, A.J. EsDREB2B, a novel truncated DREB2-type transcription factor in the desert legume *Eremosparton songoricum*, enhances tolerance to multiple abiotic stresses in yeast and transgenic tobacco. *BMC Plant Biol.* **2014**, *14*, 44. [CrossRef] [PubMed]

9

Importance of the Interaction between Heading Date Genes *Hd1* and *Ghd7* for Controlling Yield Traits in Rice

Zhen-Hua Zhang, Yu-Jun Zhu, Shi-Lin Wang, Ye-Yang Fan and Jie-Yun Zhuang *

State Key Laboratory of Rice Biology and Chinese National Center for Rice Improvement, China National Rice Research Institute, Hangzhou 310006, China; zhangzhenhua@caas.cn (Z.-H.Z.); yjzhu2013@163.com (Y.-J.Z.); 15621566500@163.com (S.-L.W.); fanyeyang@caas.cn (Y.-Y.F.)
* Correspondence: zhuangjieyun@caas.cn

Abstract: Appropriate flowering time is crucial for successful grain production, which relies on not only the action of individual heading date genes, but also the gene-by-gene interactions. In this study, influences of interaction between *Hd1* and *Ghd7* on flowering time and yield traits were analyzed using near isogenic lines derived from a cross between *indica* rice cultivars ZS97 and MY46. In the non-functional $ghd7^{ZS97}$ background, the functional $Hd1^{ZS97}$ allele promoted flowering under both the natural short-day (NSD) conditions and natural long-day (NLD) conditions. In the functional $Ghd7^{MY46}$ background, $Hd1^{ZS97}$ remained to promote flowering under NSD conditions, but repressed flowering under NLD conditions. For *Ghd7*, the functional $Ghd7^{MY46}$ allele repressed flowering under both conditions, which was enhanced in the functional $Hd1^{ZS97}$ background under NLD conditions. With delayed flowering, spikelet number and grain weight increased under both conditions, but spikelet fertility and panicle number fluctuated. Rice lines carrying non-functional $hd1^{MY46}$ and functional $Ghd7^{MY46}$ alleles had the highest grain yield under both conditions. These results indicate that longer growth duration for a larger use of available temperature and light does not always result in higher grain production. An optimum heading date gene combination needs to be carefully selected for maximizing grain yield in rice.

Keywords: flowering time; gene-by-gene interaction; *Hd1*; *Ghd7*; rice; yield trait

1. Introduction

Flowering time is a pivotal factor in the adaption of cereals to various ecogeographic environments and agricultural practices, which is controlled by an intricate genetic network. Florigens are at the core of the network, which are encoded by *Hd3a* and *RFT1* in rice [1,2]. The expression of *Hd3a* and *RFT1* are regulated by two important pathways mediating by *Hd1* and *Ehd1*, respectively [3]. *Hd1* has dual functions, which enhances florigen genes expressions under short-day (SD) conditions but inhibits florigen genes expressions under long-day (LD) conditions. The function conversion of *Hd1* is related to *PhyB*, *Se5*, *Ghd7* and *Ghd8* [4–8]. Function loss of any of these genes attenuates the conversion and maintains *Hd1* as an activator under any day-length conditions. *Ehd1* activates florigen genes expressions to promote flowering under both the SD and LD conditions [9]. *Ehd1* likely acts as a signal integrator, and its expression is regulated by many genes [3]. Recent studies revealed that *Hd1* represses expression of *Ehd1* through interaction with *Ghd7* or *DTH8* [6–8].

Flowering time is closely related to the grain yield for crop, owing to its key role in maintaining an appropriate balance between full use of resources and avoidance of environmental stresses. Many heading date (HD) genes were reported to affect yield traits, and their natural variations have been used in rice breeding, such as *Ghd7* [10], *DTH8/Ghd8* [11,12], *Hd1* [13,14],

OsPRR37/Ghd7.1/DTH7/Hd2 [15–17], *RFT1* [18] and *OsMADS51* [19,20]. Abiotic stresses during flowering, such as high temperature, low temperature, and drought, can pose a serious threat to spikelet fertility and consequently induce yield loss. The relationship between HD gene and abiotic stress has been given attention in recent years. The *Ehd1-Hd3a/RFT1* pathway responses stress signals mediated by *Ghd7* [21], *OsABF* [22] or *OsMADS51* [20]. They integrate low temperature, high temperature, and drought signals, respectively, into HD pathway, which induce or repress floral transition to avoid flowering in the stress environments. Moreover, *Ghd7* and other four HD genes, including *Ghd2* [23], *OsHAL3* [24], *OsWOX13* [25] and *OsJMJ703* [26], were found to be involved in drought or salt tolerance during vegetative phase.

When the pleiotropic effects of individual HD genes on yield traits have become recognized, the role of gene-by-gene interaction remains to be explored. In the present study, influences of *Hd1* and *Ghd7* on HD and yield traits were analyzed using near isogenic lines (NILs) and NIL-F_2 populations derived from a cross between *indica* rice cultivars Zhenshan 97 (ZS97) and Milyang 46 (MY46). Our results showed that *Hd1* and *Ghd7* could independently promote and repress flowering, respectively, whereas the flowering-repressor function of *Hd1* under natural long-day (NLD) conditions required functional *Ghd7*. With delayed flowering, spikelet number and grain weight increased under both natural short-day (NSD) and NLD conditions, but the spikelet fertility and panicle number fluctuated. Rice lines with genotype of *hd1Ghd7* produced the highest grain yield under both conditions.

2. Results

2.1. Effects of Hd1 and Ghd7 on Heading Date

In this study, effects of *Hd1* and *Ghd7* on HD were investigated using three populations derived from the rice cross ZS97/MY46//MY46///MY46. ZS97 carries functional *Hd1* and non-functional *ghd7*, whereas MY46 carries non-function *hd1* and functional *Ghd7* [14,17]. The three populations included two NIL populations, namely R1-NIL and R2-NIL, and one NIL-F_2 population namely R2-F_2 (Figure 1). Each NIL population comprised all the four homozygous genotypic combinations of *Hd1* and *Ghd7*, i.e., $hd1^{MY46}ghd7^{ZS97}$, $Hd1^{ZS97}ghd7^{ZS97}$; $hd1^{MY46}Ghd7^{MY46}$ and $Hd1^{ZS97}Ghd7^{MY46}$. The NIL-F_2 population consisted of all the nine genotypic combinations, i.e., $hd1^{MY46}ghd7^{ZS97}$, $Hd1^{heterozygous}ghd7^{ZS97}$, $Hd1^{ZS97}ghd7^{ZS97}$, $hd1^{MY46}Ghd7^{heterozygous}$, $Hd1^{heterozygous}Ghd7^{heterozygous}$, $Hd1^{ZS97}Ghd7^{heterozygous}$, $hd1^{MY46}Ghd7^{MY46}$, $Hd1^{heterozygous}Ghd7^{MY46}$, and $Hd1^{ZS97}Ghd7^{MY46}$. The R1-NIL population was tested under both the NSD and NLD conditions, and the R2-F_2 and R2-NIL populations were tested in NLD conditions only. All the rice materials matured in seasons that are appropriate for rice growth.

The R1-NIL population consisted of 10, 7, 12, and 20 lines of $hd1^{MY46}ghd7^{ZS97}$, $Hd1^{ZS97}ghd7^{ZS97}$, $hd1^{MY46}Ghd7^{MY46}$, and $Hd1^{ZS97}Ghd7^{MY46}$, respectively. In the genetic background tested by whole-genome resequencing and marker analysis, this population was segregated at *Hd16* but homozygous at all the remaining 11 cloned quantitative trait loci (QTL) for HD, including *OsMADS51*, *DTH2*, *OsMADS50/DTH3*, *Hd6*, *Hd17*, *RFT1*, *Hd3a*, *OsPRR37/Ghd7.1/DTH7/Hd2*, *Hd18*, *DTH8/Ghd8* and *Ehd1*. The effects of *Hd1* and *Ghd7* on HD were tested under NSD conditions in Lingshui from Dec. 2016 to Apr. 2017 (16LS) and from Dec. 2017 to Apr. 2018 (17LS), and under NLD conditions in Hangzhou from May to Sep. in 2017 (17HZ).

Highly significant effects ($p < 0.0001$) of *Hd1* and *Ghd7* on HD were detected in all the three trials (Table 1). In the two trials under NSD conditions (16LS and 17LS), the functional $Hd1^{ZS97}$ and $Ghd7^{MY46}$ alleles promoted and delayed flowering, respectively, no matter whether its counterpart was functional or non-functional (Figure 2a,b). In 16LS and 17LS, the proportion of phenotypic variance explained (R^2) were estimated to be 80.74% and 75.69% for *Hd1*, and 5.79% and 6.50% for *Ghd7*, respectively. The interaction between *Hd1* and *Ghd7* was non-significant in the 17LS trial and significant in the 16LS trial with a small R^2 of 1.30%. Overall, *Hd1* and *Ghd7* largely act additively in regulating HD under NSD conditions.

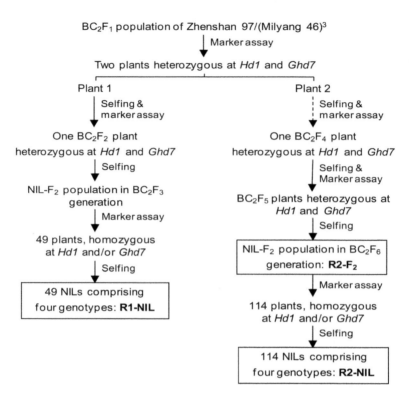

Figure 1. Development of the rice populations used in this study.

Table 1. The effects of *Hd1* and *Ghd7* on heading date and six yield traits.

Population	Trial	Trait	*Hd1*			*Ghd7*			*Hd1* × *Ghd7*		
			P	*A*	R^2%	*P*	*A*	R^2%	*P*	I-effect	R^2%
R1-NIL	16LS	HD	<0.0001	10.09	80.74	<0.0001	0.51	5.79	<0.0001	−1.30	1.30
	17LS	HD	<0.0001	7.95	75.69	<0.0001	0.77	6.50	0.2586		
		NP	0.5940			0.0014	−0.44	7.51	0.0103		
		NSP	<0.0001	7.57	35.99	<0.0001	4.35	20.36	0.0438		
		NGP	<0.0001	7.60	41.05	<0.0001	3.62	17.38	0.1490		
		SF	0.0015	0.94	10.18	0.7543			0.1061		
		TGW	<0.0001	0.99	51.36	<0.0001	0.32	11.31	0.0479		
		GY	<0.0001	3.26	45.02	0.0122			0.0575		
	17HZ	HD	<0.0001	−3.30	3.03	<0.0001	6.08	56.54	<0.0001	3.06	16.43
		NP	<0.0001	0.72	4.93	<0.0001	−1.08	20.84	0.3167		
		NSP	0.0442			<0.0001	5.45	13.87	0.2970		
		NGP	0.1371			0.0677			0.3892		
		SF	0.7773			0.0198			0.8471		
		TGW	0.9515			<0.0001	1.03	38.44	0.0002	0.46	3.49
		GY	0.4677			0.8271			0.5233		
R2-NIL	18HZ	HD	<0.0001	−2.34	6.60	<0.0001	6.15	62.57	<0.0001	4.20	28.27
		NP	0.8453			0.0304			<0.0001	−0.45	9.97
		NSP	0.0253			<0.0001	9.50	43.27	<0.0001	4.07	7.81
		NGP	0.8697			<0.0001	4.05	16.34	0.0610		
		SF	<0.0001	0.99	3.10	<0.0001	−3.25	46.28	<0.0001	−1.80	14.12
		TGW	0.5604			<0.0001	0.33	24.38	<0.0001	0.29	18.28
		GY	0.0097	0.66	2.68	0.6200			<0.0001	−1.14	7.33

16LS, the trial conducted under natural short-day (NSD) conditions in Lingshui from Dec. 2016 to Apr. 2017; 17LS, the trial conducted under NSD conditions in Lingshui from Dec. 2017 to Apr. 2018; 17HZ, the trial conducted under the natural long-day (NLD) conditions in Hangzhou from May to Sep. in 2017; 18HZ, the trial conducted under the NLD conditions in Hangzhou from Apr. to Aug. in 2018. HD, heading date; NP, number of panicles per plant; NSP, number of spikelets per panicle; NGP, number of grains per panicle; SF, spikelet fertility (%); TGW, 1000-grain weight (g); GY, grain weight per plant (g). *A*, additive effect of replacing a Zhenshan 97 allele with a Milyang 46 allele. R^2%, proportion of phenotypic variance explained by the QTL effect. I-effect, positive value: parental type < recombinant type; negative value: parental type > recombinant type.

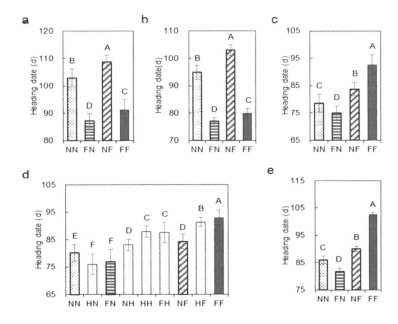

Figure 2. Heading date of rice lines classified based on the genotype of *Hd1* and *Ghd7*. (**a**) R1-NIL population under the NSD conditions in the 16LS trial. (**b**) R1-NIL population under the NSD conditions in the 17LS trial. (**c**) R1-NIL population under the NLD conditions the 17HZ trial. (**d**) R2-F$_2$ population under the NLD conditions in the 17HZ trial. (**e**) R2-NIL population under the NLD conditions in the 18HZ trial. NN, *hd1*^{MY46}*ghd7*^{ZS97}; HN, *Hd1*^{heterozygous}*ghd7*^{ZS97}; FN, *Hd1*^{ZS97}*ghd7*^{ZS97}; NH, *hd1*^{MY46}*Ghd7*^{heterozygous}; HH, *Hd1*^{heterozygous}*Ghd7*^{heterozygous}; FH, *Hd1*^{ZS97}*Ghd7*^{heterozygous}; NF, *hd1*^{MY46}*Ghd7*^{MY46}; HF, *Hd1*^{heterozygous}*Ghd7*^{MY46}; FF, *Hd1*^{ZS97}*Ghd7*^{MY46}. Data are presented in mean ± sd. Bars with different letters are significantly different at $p < 0.01$ based on Duncan's multiple range tests.

In the 17HZ trial under NLD conditions, the effects of *Hd1*, *Ghd7* and their interaction were all highly significant ($p < 0.0001$). The R^2 were estimated to be 3.03% for *Hd1*, 56.54% for *Ghd7*, and 16.43% for the interaction between the two genes (Table 1). Compared with NILs having the *hd1*^{MY46}*ghd7*^{ZS97} genotype, those having the *Hd1*^{ZS97}*ghd7*^{ZS97} genotypes flowered earlier by 3.51 d; compared with NILs having the *hd1*^{MY46}*Ghd7*^{ZS97} genotype, those having the *Hd1*^{ZS97}*Ghd7*^{MY46} genotype flowered later by 8.75 d (Figure 2c; Table 2). These indicated that *Hd1* regulates flowering dependent on *Ghd7* under NLD conditions, and its flowering-repressor activity requires the functional allele of *Ghd7*. For *Ghd7*, it delays flowering regardless of genotype of *Hd1* but its effect is enhanced by *Hd1*. HD was longer by 5.24 d in lines of *hd1*^{MY46}*Ghd7*^{MY46} than *hd1*^{MY46}*ghd7*^{ZS97}, whereas it was longer by 17.49 d in lines of *Hd1*^{ZS97}*Ghd7*^{MY46} than of *Hd1*^{ZS97}*ghd7*^{ZS97} (Table 2).

Table 2. Heading date and six yield traits of the four homozygous genotypes of *Hd1* and *Ghd7*.

Population	Trial	Group	HD	NP	NSP	NGP	SF	TGW	GY
R1-NIL	17LS	FN	87.2 ± 2.6 ^{Dd}	11.8 ± 1.1 ^{ABb}	79.3 ± 4.4 ^{Cc}	69.3 ± 4.7 ^{Cd}	87.4 ± 3.1 ^{Bb}	25.4 ± 1.1 ^{Cd}	20.7 ± 2.8 ^{Cc}
		FF	91.2 ± 3.8 ^{Cc}	11.6 ± 1.1 ^{ABb}	87.9 ± 6.9 ^{Bb}	77.6 ± 5.9 ^{Bc}	88.3 ± 2.6 ^{ABb}	27.0 ± 1.0 ^{Bc}	24.0 ± 2.9 ^{Bb}
		NN	103.0 ± 3.3 ^{Bb}	12.6 ± 0.9 ^{Aa}	92.0 ± 5.8 ^{Bb}	83.3 ± 5.5 ^{Bb}	90.6 ± 2.4 ^{Aa}	28.2 ± 0.9 ^{Ab}	29.4 ± 2.6 ^{Aa}
		NF	108.6 ± 2.5 ^{Aa}	11.0 ± 0.8 ^{Bb}	108.2 ± 7.2 ^{Aa}	96.7 ± 7.0 ^{Aa}	89.3 ± 2.4 ^{ABab}	28.9 ± 1.0 ^{Aa}	29.9 ± 3.0 ^{Aa}
	17HZ	FN	75.0 ± 1.6 ^{Dd}	15.5 ± 1.2 ^{ABa}	107.6 ± 8.6 ^{ABb}	89.0 ± 5.7 ^{Aab}	82.8 ± 3.9 ^{Aa}	22.8 ± 0.8 ^{Cb}	30.4 ± 1.5 ^{Aa}
		FF	78.5 ± 2.8 ^{Cc}	16.1 ± 1.2 ^{Aa}	100.1 ± 6.8 ^{Bc}	83.5 ± 4.0 ^{Ab}	83.6 ± 3.9 ^{Aa}	23.8 ± 0.8 ^{BCb}	30.5 ± 1.8 ^{Aa}
		NN	83.7 ± 1.5 ^{Bb}	14.5 ± 1.2 ^{BCb}	112.6 ± 9.1 ^{Aab}	89.8 ± 9.1 ^{Aab}	79.7 ± 5.7 ^{Aa}	24.9 ± 1.1 ^{ABa}	30.9 ± 3.1 ^{Aa}
		NF	92.5 ± 1.6 ^{Aa}	13.1 ± 1.4 ^{Cc}	115.0 ± 10.7 ^{Aa}	91.3 ± 11.8 ^{Aa}	79.6 ± 8.5 ^{Aa}	25.8 ± 1.2 ^{Aa}	29.9 ± 4.5 ^{Aa}
R2-NIL	18HZ	FN	81.6 ± 1.2 ^{Dd}	12.6 ± 1.0 ^{Aa}	115.4 ± 5.1 ^{Dd}	105.3 ± 4.9 ^{Bb}	91.3 ± 1.6 ^{Aa}	25.0 ± 0.3 ^{Cc}	31.8 ± 2.9 ^{ABb}
		FF	86.0 ± 1.4 ^{Cc}	11.7 ± 0.9 ^{BCbc}	120.4 ± 6.7 ^{Cc}	107.6 ± 5.9 ^{Bb}	89.3 ± 2.1 ^{Bb}	25.6 ± 0.4 ^{Bb}	30.8 ± 2.9 ^{Bbc}
		NN	90.0 ± 1.0 ^{Bb}	12.2 ± 1.2 ^{ABab}	131.5 ± 8.8 ^{Bb}	113.5 ± 7.9 ^{Aa}	86.4 ± 2.1 ^{Cc}	25.7 ± 0.6 ^{Bb}	33.3 ± 2.7 ^{Aa}
		NF	102.5 ± 0.7 ^{Aa}	11.3 ± 1.2 ^{Cc}	142.7 ± 7.5 ^{Aa}	115.7 ± 6.7 ^{Aa}	81.1 ± 2.7 ^{Dd}	26.3 ± 0.5 ^{Aa}	29.8 ± 3.2 ^{Bc}

FN, *Hd1*^{ZS97}*ghd7*^{ZS97}; FF, *Hd1*^{ZS97}*Ghd7*^{MY46}; NN, *hd1*^{MY46}*ghd7*^{ZS97}; NF, *hd1*^{MY46}*Ghd7*^{MY46}. Values are mean ± sd. Uppercase and lowercase letters following the values represent significant differences at $p < 0.01$ and $p < 0.05$, respectively, based on Duncan's multiple range tests.

2.2. Expressions of Genes Involved in the Photoperiod Pathway

The transcript levels of *Hd1*, *Ghd7*, *Ehd1*, *Hd3a* and *RFT1* at 2 h after sunrise were examined in seven-week-old rice lines in the R1-NIL population grown in the 17LS and 17HZ trials (Figure 3). In the 17LS trial under NSD conditions (Figure 3a), expression of *Hd1* and *Ghd7* was not affected by each other. The *Ehd1* expression was also not affected by either *Hd1* or *Ghd7*. For florigen genes, the expression of *Hd3a* was 7.87 times larger in lines of $Hd1^{ZS97}ghd7^{ZS97}$ than $hd1^{MY46}ghd7^{ZS97}$, and 12.46 times larger in lines of $Hd1^{ZS97}Ghd7^{MY46}$ than $hd1^{MY46}Ghd7^{MY46}$. These results indicate that *Hd1* promotes *Hd3a* expression regardless of *Ghd7* function, which was in accordance with that *Hd1* promotes flowering regardless of *Ghd7* function under NSD conditions. In addition, *Hd1* was also found to promote *RFT1* in the *Ghd7* background. At the same time, slightly repression of *Hd3a* by *Ghd7* was detected in the *hd1* background. These were consistent with the small effect of *Ghd7* under NSD conditions.

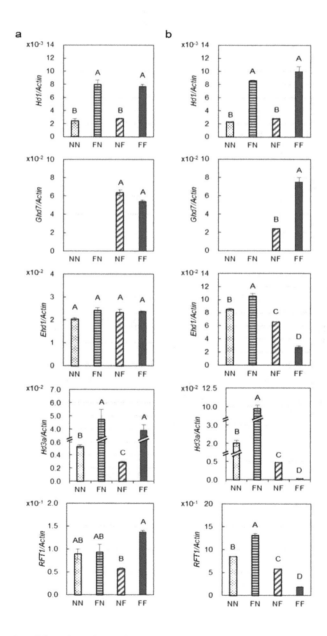

Figure 3. Transcript levels of five heading date genes in the R1-NIL population. (**a**) Under the NSD conditions in Lingshui. (**b**) Under the NLD conditions in Hangzhou. Data are presented in mean ± s. e. m (*n* = 3). Bars with different letters are significantly different at *p* < 0.01 based on Duncan's multiple range tests.

In the 17HZ trial conducted under NLD conditions (Figure 3b), expression of $Hd1$ was not affected by $Ghd7$, but $Hd1$ up-regulated $Ghd7$ expression. The $Ghd7$ expression was 2.12 times larger in lines of $Hd1^{ZS97}Ghd7^{MY46}$ than $hd1^{ZS97}Ghd7^{MY46}$. The expression of $Ehd1$ in lines of $Hd1^{ZS97}ghd7^{ZS97}$ was 1.24 times as large as that in lines of $hd1^{MY46}ghd7^{ZS97}$, but the expression in lines of $Hd1^{ZS97}Ghd7^{MY46}$ was only 0.42 times as large as that in lines of $hd1^{MY46}Ghd7^{MY46}$. These suggest that $Hd1$ significantly represses $Ehd1$ expression in the $Ghd7$ background. For florigen genes, the expressions of $Hd3a$ and $RFT1$ in lines of $Hd1^{ZS97}ghd7^{ZS97}$ were 4.86 and 1.55 times as large as that in lines of $hd1^{MY46}ghd7^{ZS97}$, indicating $Hd1$ promotes expressions of florigen genes in the $ghd7$ background. However, $Hd1$ was converted to severely repress the florigen gene expressions in the $Ghd7$ background. The expressions of $Hd3a$ and $RFT1$ in lines of $Hd1^{ZS97}Ghd7^{MY46}$ were only 0.07 and 0.32 times as large as those in lines of $hd1^{MY46}Ghd7^{MY46}$. In the meantime, significant repression of the $Ehd1$, $Hd3a$ and $RFT1$ expressions by $Ghd7$ were detected in both the $Hd1$ and $hd1$ background, and the effect were larger in the $Hd1$ background. The expressions of the three genes in lines of $hd1^{MY46}Ghd7^{MY46}$ were 0.77, 0.24 and 0.68 times as large as those in lines of $hd1^{MY46}ghd7^{ZS97}$; and the expressions in lines of $Hd1^{ZS97}Ghd7^{MY46}$ were 0.26. 0.004 and 0.14 times as large as those in lines of $Hd1^{ZS97}ghd7^{ZS97}$. These agreed with that flowering-repressor function of $Ghd7$ could be enhanced by $Hd1$.

2.3. Influence of Hd1 and Ghd7 on Yield Traits and Its Relationship with HD

Grain yield per plant (GY), and five yield components traits including number of panicles per plant (NP), number of spikelets per panicle (NSP), number of grains per panicle (NGP), spikelet fertility (SF), 1000-grain weight (TGW), were measured in the R1-NIL population grown in the 17LS and 17HZ trials.

In the 17LS trial under NSD conditions, $Hd1$ showed significant effects ($p < 0.01$) on all the six yield traits except NP; and $Ghd7$ showed significant influences ($p < 0.01$) on all the six yield traits except SF and GY (Table 1). Interaction between the two genes were all non-significant at $p < 0.01$. Relationships between HD and the yield traits were further investigated (Table 2). The lines of $Hd1^{ZS97}ghd7^{ZS97}$ had the shortest HD, followed by $Hd1^{ZS97}Ghd7^{MY46}$, $hd1^{MY46}ghd7^{ZS97}$ and $hd1^{MY46}Ghd7^{MY46}$. Significant differences ($p < 0.05$) were detected for all the five yield determinants among the four genotypic groups. Three of the traits, NSP, NGP, and TGW, were positively correlated with HD, having correlation coefficients (r) of 0.823, 0.828, and 0.614, respectively (Table S1). Values of these three traits increased with delayed heading. On the other hand, NP and SF were not significantly correlated with HD. For GY, the values increased with delayed flowering among the three genotypic groups having the shortest to third shortest HD, and then remained stable when the HD became longer. Consequently, the two genotypic groups having the longest and second longest HD, $hd1^{MY46}Ghd7^{MY46}$ and $hd1^{MY46}ghd7^{ZS97}$, had little difference on GY.

In the 17HZ trial under NLD conditions, $Hd1$ showed significant effects only on NP; and $Ghd7$ showed significant influences on NP, NSP, and TGW ($p < 0.0001$). Significant interaction between the two genes was detected on TGW ($p < 0.001$). The interaction acted for increasing the values of the recombinant types, which was in accordance with the epistasis on HD. The HD and six yield traits were also compared among the four homozygous genotype groups (Table 2). The lines of $Hd1^{ZS97}ghd7^{ZS97}$ had the shortest HD, followed by $hd1^{MY46}ghd7^{ZS97}$, $hd1^{MY46}Ghd7^{MY46}$ and $Hd1^{ZS97}Ghd7^{MY46}$. Significant differences ($p < 0.05$) among the four genotypic groups were detected on four yield determinants, including NP, NGP, NSP, and TGW. Variations of TGW and NSP were positively correlated with HD, having r values of 0.708 and 0.355, respectively (Table S1). The two traits tended to increase with delayed heading. Similar tendency was observed for NGP though it was not significantly correlated with HD. Conversely, NP was negatively correlated with HD ($p < 0.05$), having r value of -0.670. SF also appeared to decrease with delayed heading though no significant difference was observed. Consequently, the largest value of GY in the four genotypic groups was observed for $hd1^{MY46}Ghd7^{MY46}$ which had the second longest HD.

2.4. Validation of the Influences of Hd1 and Ghd7 on HD and Yield Traits under NLD Conditions

The relationship between *Hd1* and *Ghd7* was further analyzed using the R2-F_2 population, which was segregated at *Hd1* and *Ghd7* loci but homozygous at all the remaining 12 cloned flowering QTL mentioned above. The 775 plants of this population were grown in Hangzhou in 2017 under NLD conditions. Significant effects were identified for both genes. The additive effect, dominance effect and R^2 were estimated to be 1.89 d, -0.89 d and 6.4% for *Hd1*, and 6.04 d, 1.91 d and 59.3% for *Ghd7*, respectively. The plants were classified into nine genotypic groups based on the *Hd1* and *Ghd7* alleles, and the HD values were compared (Figure 2d). *Hd1* promoted flowering in the *ghd7* background, but delayed heading when the genotype of *Ghd7* was functional or heterozygous. *Ghd7* delayed flowering regardless of the genotype of *Hd1* but its effect was enhanced by the functional *Hd1* allele.

Plants that were homozygous at *Hd1* and/or *Ghd7* were selected from the R2-F_2 population and selfed. The resultant R2-NIL population, consisting of 29, 26, 29, and 30 lines of $hd1^{MY46}ghd7^{ZS97}$, $Hd1^{ZS97}ghd7^{ZS97}$, $hd1^{MY46}Ghd7^{MY46}$, and $Hd1^{ZS97}Ghd7^{MY46}$, respectively, was tested in Hangzhou in 2018 under NLD conditions. Both the *Hd1* and *Ghd7*, as well as their interaction, had highly significant effects ($p < 0.0001$) on HD (Table 1), which were similar to those observed previously under NLD condition. $Hd1^{ZS97}$ promoted and repressed flowering in the $Ghd7^{MY46}$ and $ghd7^{ZS97}$ backgrounds, respectively, while $Ghd7^{MY46}$ delayed flowering regardless of the *Hd1* function (Figure 2e).

GY and five yield components traits were also measured in the R2-NIL population. *Hd1* showed significant effects on SF ($p < 0.0001$) and GY ($p < 0.01$), and *Ghd7* exhibited highly significant effects on NSP, NGP, SF and TGW ($p < 0.0001$) (Table 1). Highly significant epistatic effects of the two genes were detected on all the traits except NGP ($p < 0.0001$). For NSP and TGW, the interactions acted for increasing the values of the recombinant types, which were consistent with the epistasis on HD. For NP, SF, and GY, the opposite direction was found. The relationships between HD and the yield traits were further analyzed (Table 2). Lines of $Hd1^{ZS97}ghd7^{ZS97}$ had the shortest HD, followed by $hd1^{MY46}ghd7^{ZS97}$, $hd1^{MY46}Ghd7^{MY46}$, and $Hd1^{ZS97}Ghd7^{MY46}$. Significant differences were detected for all the yield traits among the four genotypic groups. NSP, NGP and TGW were positively correlated with HD ($p < 0.05$), having r values of 0.806, 0.507 and 0.672, respectively (Table S1). Values of these traits increased with delayed heading. On the other hand, SF and NP were negatively correlated with HD ($p < 0.05$), having r values of -0.855 and -0.349, respectively. SF decreased with delayed heading; compared with lines having the shortest HD, SF in lines having the third longest, the second longest, and the longest HD decreased by 1.9%, 4.9% and 10.2%, respectively. Similar tendency was observed for NP. Consequently, lines in the $hd1^{MY46}Ghd7^{MY46}$ genotypic group having the second longest HD produced the highest GY.

3. Discussion

The bi-functional action of *Hd1* has been well recognized, promoting flowering under SD conditions and inhibiting flowering under LD conditions [27]. Recent studies revealed that flowering repressing function of *Hd1* is dependent on *Ghd7* [6,7]. In the present study, this relationship between *Hd1* and *Ghd7* was confirmed. Under NSD conditions, *Hd1* always up-regulated expressions of the two florigen genes (Figure 3) and promoted flowering regardless of *Ghd7* genotype (Figure 2). Under NLD conditions, *Hd1* still promoted flowering (Figure 2) by up-regulating florigen genes in the *ghd7* background (Figure 3). In the *Ghd7* background, however, *Hd1* was found to up-regulate *Ghd7*, and down-regulate *Ehd1* and florigen genes, consequently leading to late flowering. For *Ghd7*, its flowering-repressor action was observed under both NSD and NLD conditions regardless of *Hd1* function. Taken together, our results suggest that *Hd1* and *Ghd7* could promote and repress flowering independently, whereas flowering-repressor function of *Hd1* under LD conditions requires the functional *Ghd7*.

Among the four homozygous genotypic combinations of *Hd1* and *Ghd7*, the *Hd1ghd7* group exhibited the shortest HD under NLD conditions. Compared to *Hd1ghd7*, heading was delayed by 3.4–4.3 d and 7.5–8.7 d in the *hd1ghd7* and *hd1Ghd7* groups, respectively. Strikingly, HD in the *Hd1Ghd7* group was delayed by 16.1–20.9 d, owing to the genetic interaction between *Hd1* and *Ghd7* under NLD condition. This is likely the reason the *Hd1Ghd7* genotype was hardly carried by early season *indica* cultivars grown in middle-lower regions of the Yangtze River and South China regions [28] and *japonica* cultivars in northeast China [29], where early flowering is essential to ensure sufficient grown period for late season *indica* cultivars or secure a harvest before cold weather approaches.

It is generally accepted that long growth duration is associated with high-yielding production in rice [29,30], if varieties are harvested before cold weather approaches. A larger number of HD genes were found to have pleiotropic effects on yield traits, and their late-flowering alleles were frequently used to enhance grain yield mainly by increasing spikelet number and partially by increasing grain weight [10–20]. As expected, NSP and TGW gradually increased with delayed flowering under both the NSD and NLD conditions in this study. However, SF and NP tended to decrease under NLD conditions when the HD has become relatively long. As a consequence of trade-off among different yield components, rice lines having the *hd1Ghd7* genotype which had the second longest HD produced the highest grain yield, rather than the lines having the *Hd1Ghd7* genotype which had the longest HD. These results indicate that longer growth duration for a more use of available temperature and light does not always result in higher grain production.

Spikelet sterility is a key determinant of grain yield and frequently used as an indicator for stress tolerance. Two alternative explanations could be given to the decrease of spikelet sterility with delayed flowering. Firstly, alteration of time of flowering causes some loss of seasonal adaptability of rice. Secondly, *Ghd7* and *Hd1* participate in the stress tolerance of rice. *Ghd7* has been found to respond to multiple abiotic stress, such as high temperature, low temperature, and drought. Moreover, overexpression of *Ghd7* increases drought sensitivity, whereas knock-down of *Ghd7* enhances drought tolerance [21]. Our study showed that *Ghd7* expression was dramatically up-regulated in the *Hd1* background. This may be a reason that caused low SF in lines of *Hd1Ghd7*. Moreover, alteration of SF by *Hd1* was also observed in the *ghd7* background (Table 2), suggesting *Hd1* could be involved in stress response independently.

Panicle number is generally recognized as an unstable trait among yield traits. Few genes were reported to have pleiotropic effects on flowering time and panicle number [21,31,32]. *Ghd7* is found to regulate panicle number in a density-dependent manner. It decreases and increases panicle number at normal field condition and low-density conditions, respectively, though it always suppresses flowering time [21]. In the NIL populations used in our study, negative correlation between NP and HD was detected in both trials conducted under NLD conditions at normal planting density (Table 2, Table S1). The lines of *Hd1Ghd7* with the longest HD always produced the least NP (Table 2), indicating that combination of *Hd1* and *Ghd7* could cause decrease of panicle number under NLD conditions.

Although late-flowering alleles of flowering genes generally increase spikelet number, their influences on panicle number and spikelet sterility are not necessarily positive. Thus, an optimum HD genes combination needs to be carefully selected for maximizing grain yield in rice. In the present study, lines carried *hd1* and *Ghd7* alleles from MY46 produced the highest grain yield in both trials conducted in Hangzhou (Table 2) where is in the middle-lower region of the Yangtze River. Among the 14 middle-season *indica* rice cultivars tested by Wei et al [28], MY46 is one the 10 cultivars having the combination of non-functional *hd1* and functional *Ghd7*. These indicate that this combination could have undergone intensive artificial selection and play a significant role in the adaption of middle-season rice.

4. Materials and Methods

4.1. Plant Material

Three rice populations segregating at both the *Hd1* and *Ghd7* loci were used in this study. The developing process was illustrated in Figure 1 and described below. One F_9 plant of ZS97/MY46 was crossed with MY46 for two generations. Two BC_2F_1 plants which were heterozygous at both the $Hd1$ and $Ghd7$ loci were identified and selfed. In one of the two BC_2F_2 populations produced, a plant which was heterozygous for both the genes was identified and selfed. The resultant BC_2F_3 population was assayed with functional or closely linked DNA markers for the two genes. A total of 49 plants which were homozygous at $Hd1$ and/or $Ghd7$ loci were identified and selfed. One NIL population namely R1-NIL, comprising all the four homozygous genotypic combinations of $Hd1$ and $Ghd7$, was constructed.

Another BC_2F_2 population was advanced to the BC_2F_4 generation. A BC_2F_4 plant which was heterozygous for both the genes was identified. In the resultant BC_2F_5 population, plants which were heterozygous for both the genes were selected and selfed. A NIL-F_2 population in the BC_2F_6 generation, namely R2-F_2 population, was constructed. A total of 114 plants which were homozygous at $Hd1$ and/or $Ghd7$ loci were selected and selfed. One NIL population namely R2-NIL, which consisted of all the four homozygous genotypic groups, was constructed.

4.2. Field Experiments and Phenotyping

The rice populations were tested in the experimental stations of the China National Rice Research Institute located at either Hangzhou or Lingshui. During the period of floral transition in the rice materials tested, day length in Hangzhou and Lingshui were corresponding to NLD and NSD conditions, respectively [14]. In all the trials, the planting density was 16.7 cm × 26.7 cm. Field management followed the normal agricultural practice. For NIL sets, the experiments followed a randomized complete block design with two replications. In each replication, one line was grown in a single row of ten plants. HD was recorded for each plant. At maturity, five middle plants in each row were harvested in bulk and measured for six yield traits, including NP, NSP, NGP, SF (%), TGW (g) and GY (g). Of which TGW was evaluated using fully filled grain followed the procedure reported by Zhang et al. [33].

4.3. DNA Marker Genotyping and Quantitative Real-time PCR Analysis

For population development and QTL mapping, total DNA was extracted using 2 cm-long leaf sample following the method of Zheng et al. [34]. PCR amplification was performed according to Chen et al. [35]. The products were visualized on 6% non-denaturing polyacrylamide gels using silver staining or on 2% agarose gels using Gelred staining. Three DNA markers were used, including functional marker Si9337 for $Hd1$, functional marker Se9153 and closely linked marker RM5436 for $Ghd7$ [10,17].

For expression analysis, penultimate leaves of rice lines in the R1-NIL population were harvested at 7:00 am in 17HZ and 9:00 am in 17LS, 2 h after sunrise. Total RNA was extracted using RNeasy Plus Mini Kit (QIAGEN, Hilden, German). First-strand cDNA was synthesized using ReverTra AceR Kit (Toyobo, Osaka, Japan). Quantitative real-time PCR was performed on Applied Biosystems 7500 using SYBR qPCR Mix Kit (Toyobo, Osaka, Japan) according to the manufacturer's instructions. $Actin1$ was used as the endogenous control. The data were analyzed according to the $2^{-\Delta Ct}$ method. Three biological replicates and three technical replicates were used. The primers were selected from previous studies [10,20,36].

4.4. Data Analysis

For the NIL-F_2 population, QTL analysis was performed with single marker analysis in Windows QTL Cartgrapher 2.5 [37]. For the NIL populations, two-way ANOVA was conducted to test the main and epistatic effects. Duncan's multiple range test was used to examine the phenotypic differences among genotypic groups. The analysis was performed using the SAS procedure GLM [38].

Author Contributions: J.-Y.Z. conceived and designed the experiments. Z.-H.Z. and Y.Y.F. performed laboratory experiments. Z.-H.Z., Y.-J.Z. and S.-L.W. performed the field experiments. Z.-H.Z. and J.-Y.Z. analyzed the data and drafted the manuscript. Z.-H.Z. and J.-Y.Z. revised the manuscript. All authors read and approved the final manuscript.

Abbreviations

NSDs	Natural short-day conditions
NLDs	Natural long-day conditions
SD	Short-day conditions
LD	Long-day conditions
NIL	Near isogenic lines
ZS97	Zhenshan 97
MY46	Milyang 46
HD	Heading date
QTL	Quantitative trait locus
R^2	The proportion of phenotypic variance explained
GY	Grain yield per plant
NP	Number of panicles per plant
NSP	Number of spikelets per panicle
NGP	Number of grains per panicle
SF	Spikelet fertility
TGW	1000-grain weight
r	Correlation coefficient

References

1. Tamaki, S.; Matsuo, S.; Wong, H.L.; Yokoi, S.; Shimamoto, K. Hd3a protein is a mobile flowering signal in rice. *Science* **2007**, *316*, 1033–1036. [CrossRef] [PubMed]

2. Komiya, R.; Yokoi, S.; Shimamoto, K. A gene network for long-day flowering activates *RFT1* encoding a mobile flowering signal in rice. *Development* **2009**, *136*, 3443–3450. [CrossRef] [PubMed]

3. Hori, K.; Matsubara, K.; Yano, M. Genetic control of flowering time in rice: Integration of Mendelian genetics and genomics. *Theor. Appl. Genet.* **2016**, *129*, 2241–2252. [CrossRef] [PubMed]

4. Izawa, T.; Oikawa, T.; Sugiyama, N.; Tanisaka, T.; Yano, M.; Shimamoto, K. Phytochrome mediates the external light signal to repress *FT* orthologs in photoperiodic flowering of rice. *Genes Dev.* **2002**, *16*, 2006–2020. [CrossRef] [PubMed]

5. Ishikawa, R.; Aoki, M.; Kurotani, K.; Yokoi, S.; Shinomura, T.; Takano, M.; Shimamoto, K. Phytochrome B regulates *Heading date 1 (Hd1)*-mediated expression of rice florigen *Hd3a* and critical day length in rice. *Mol. Genet. Genomics* **2011**, *285*, 461–470. [CrossRef]

6. Nemoto, Y.; Nonoue, Y.; Yano, M.; Izawa, T. *Hd1*, a CONSTANS ortholog in rice, functions as an *Ehd1* repressor through interaction with monocot-specific CCT-domain protein Ghd7. *Plant J.* **2016**, *86*, 221–233. [CrossRef]

7. Du, A.; Tian, W.; Wei, M.; Yan, W.; He, H.; Zhou, D.; Huang, X.; Li, S.; Ouyang, X. The DTH8-Hd1 module mediates day-length-dependent regulation of rice flowering. *Mol. Plant* **2017**, *10*, 948–961. [CrossRef]

8. Zhang, Z.; Hu, W.; Shen, G.; Liu, H.; Hu, Y.; Zhou, X.; Liu, T.; Xing, Y. Alternative functions of Hd1 in repressing or promoting heading are determined by Ghd7 status under long-day conditions. *Sci. Rep.* **2017**, *7*, 5388. [CrossRef]

9. Doi, K.; Izawa, T.; Fuse, T.; Yamanouchi, U.; Kubo, T.; Shimatani, Z.; Yano, M.; Yoshimura, A. *Ehd1*, a B-type response regulator in rice, confers short-day promotion of flowering and controls *FT-like* gene expression independently of *Hd1*. *Genes Dev.* **2004**, *18*, 926–936. [CrossRef]

10. Xue, W.; Xing, Y.; Weng, X.; Zhao, Y.; Tang, W.; Wang, L.; Zhou, H.; Yu, S.; Xu, C.; Li, X.; et al. Natural variation in *Ghd7* is an important regulator of heading date and yield potential in rice. *Nat. Genet.* **2008**, *40*, 761–767. [CrossRef]

11. Wei, X.; Xu, J.; Guo, H.; Jiang, L.; Chen, S.; Yu, C.; Zhou, Z.; Hu, P.; Zhai, H.; Wan, J. *DTH8* suppresses flowering in rice, influencing plant height and yield potential simultaneously. *Plant Physiol.* **2010**, *153*, 1747–1758. [CrossRef] [PubMed]

12. Yan, W.H.; Wang, P.; Chen, H.X.; Zhou, H.J.; Li, Q.P.; Wang, C.R.; Ding, Z.H.; Zhang, Y.S.; Yu, S.B.; Xing, Y.Z.; et al. A major QTL, *Ghd8*, plays pleiotropic roles in regulating grain productivity, plant height, and heading date in rice. *Mol. Plant* **2011**, *4*, 319–330. [CrossRef] [PubMed]

13. Endo-Higashi, N.; Izawa, T. Flowering time genes *Heading date 1* and *Early heading date 1* together control panicle development in rice. *Plant Cell Physiol.* **2011**, *52*, 1083–1094. [CrossRef] [PubMed]

14. Zhang, Z.-H.; Wang, K.; Guo, L.; Zhu, Y.-J.; Fan, Y.-Y.; Cheng, S.-H.; Zhuang, J.-Y. Pleiotropism of the photoperiod-insensitive allele of *Hd1* on heading date, plant height and yield traits in rice. *PLoS ONE* **2012**, *7*, e52538. [CrossRef] [PubMed]

15. Yan, W.; Liu, H.; Zhou, X.; Li, Q.; Zhang, J.; Lu, L.; Liu, T.; Liu, H.; Zhang, C.; Zhang, Z.; et al. Natural variation in *Ghd7.1* plays an important role in grain yield and adaptation in rice. *Cell Res.* **2013**, *23*, 969–971. [CrossRef] [PubMed]

16. Gao, H.; Jin, M.; Zheng, X.-M.; Chen, J.; Yuan, D.; Xin, Y.; Wang, M.; Huang, D.; Zhang, Z.; Zhou, K.; et al. *Days to heading 7*, a major quantitative locus determining photoperiod sensitivity and regional adaptation in rice. *Proc. Natl. Acad. Sci. USA* **2014**, *111*, 16337–16342. [CrossRef] [PubMed]

17. Zhang, Z.-H.; Cao, L.-Y.; Chen, J.-Y.; Zhang, Y.-X.; Zhuang, J.-Y.; Cheng, S.-H. Effects of *Hd2* in the presence of the photoperiod-insensitive functional allele of *Hd1* in rice. *Biol. Open* **2016**, *5*, 1719–1726. [CrossRef] [PubMed]

18. Zhu, Y.-J.; Fan, Y.-Y.; Wang, K.; Huang, D.-R.; Liu, W.-Z.; Ying, J.-Z.; Zhuang, J.-Y. *Rice Flowering Locus T 1* plays an important role in heading date influencing yield traits in rice. *Sci. Rep.* **2017**, *7*, 4918. [CrossRef]

19. Chen, J.-Y.; Guo, L.; Ma, H.; Chen, Y.-Y.; Zhang, H.-W.; Ying, J.-Z.; Zhuang, J.-Y. Fine mapping of *qHd1*, a minor heading date QTL with pleiotropism for yield traits in rice (*Oryza sativa* L.). *Theor. Appl. Genet.* **2014**, *127*, 2515–2524. [CrossRef]

20. Chen, J.-Y.; Zhang, H.-W.; Zhang, H.-L.; Ying, J.-Z.; Ma, L.-Y.; Zhuang, J.-Y. Natural variation at *qHd1* affects heading date acceleration at high temperatures with pleiotropism for yield traits in rice. *BMC Plant Biol.* **2018**, *18*, 112. [CrossRef]

21. Weng, X.; Wang, L.; Wang, J.; Hu, Y.; Du, H.; Xu, C.; Xing, Y.; Li, X.; Xiao, J.; Zhang, Q. *Grain number, plant height, and heading date7* is a central regulator of growth, development, and stress response. *Plant Physiol.* **2014**, *164*, 735–747. [CrossRef] [PubMed]

22. Zhang, C.; Liu, J.; Zhao, T.; Gomez, A.; Li, C.; Yu, C.; Li, H.; Lin, J.; Yang, Y.; Liu, B.; et al. A drought-inducible transcription factor delays reproductive timing in rice. *Plant Physiol.* **2016**, *171*, 334–343. [CrossRef] [PubMed]

23. Liu, J.; Shen, J.; Xu, Y.; Li, X.; Xiao, J.; Xiong, L. *Ghd2*, a *CONSTANS*-like gene, confers drought sensitivity through regulation of senescence in rice. *J. Exp. Bot.* **2016**, *67*, 5785–5798. [CrossRef] [PubMed]

24. Su, L.; Shan, J.X.; Gao, J.P.; Lin, H.X. OsHAL3, a blue light-responsive protein, interacts with the floral regulator Hd1 to activate flowering in rice. *Mol. Plant* **2016**, *9*, 233–244. [CrossRef] [PubMed]

25. Minh-Thu, P.T.; Kim, J.S.; Chae, S.; Jun, K.M.; Lee, G.S.; Kim, D.E.; Cheong, J.J.; Song, S.I.; Nahm, B.H.; Kim, Y.K. A WUSCHEL homeobox transcription factor, OsWOX13, enhances drought tolerance and triggers early flowering in rice. *Mol. Cells* **2018**, *41*, 781–798. [PubMed]

26. Song, T.; Zhang, Q.; Wang, H.; Han, J.; Xu, Z.; Yan, S.; Zhu, Z. *OsJMJ703*, a rice histone demethylase gene, plays key roles in plant development and responds to drought stress. *Plant Physiol. Biochem.* **2018**, *132*, 183–188. [CrossRef]

27. Yano, M.; Katayose, Y.; Ashikari, M.; Yamanouchi, U.; Monna, L.; Fuse, T.; Baba, T.; Yamamoto, K.; Umehara, Y.; Nagamura, Y.; et al. *Hd1*, a major photoperiod sensitivity quantitative trait locus in rice, is closely related to the Arabidopsis flowering time gene *CONSTANS*. *Plant Cell* **2000**, *12*, 2473–2484. [CrossRef]

28. Wei, X.-J.; Xu, J.-F.; Jiang, L.; Wang, J.-J.; Zhou, Z.-L.; Zhai, H.-Q.; Wan, J.-M. Genetic analysis for the diversity of heading date of cultivated rice in China. *Acta Agron. Sin.* **2012**, *38*, 10–12. [CrossRef]

29. Ye, J.; Niu, X.; Yang, Y.; Wang, S.; Xu, Q.; Yuan, X.; Yu, H.; Wang, Y.; Wang, S.; Feng, Y.; et al. Divergent *Hd1*, *Ghd7*, and *DTH7* alleles control heading date and yield potential of *japonica* rice in northeast China. *Front. Plant Sci.* **2018**, *9*, 35. [CrossRef]

30. Hu, Y.; Li, S.; Xing, Y. Lessons from natural variations: Artificially induced heading date variations for improvement of regional adaptation in rice. *Theor. Appl. Genet.* **2018**. [CrossRef]

31. Wang, Q.; Zhang, W.; Yin, Z.; Wen, C.K. Rice CONSTITUTIVE TRIPLE-RESPONSE2 is involved in the ethylene-receptor signalling and regulation of various aspects of rice growth and development. *J. Exp. Bot.* **2013**, *64*, 4863–4875. [CrossRef] [PubMed]

32. Xu, Q.; Saito, H.; Hirose, I.; Katsura, K.; Yoshitake, Y.; Yokoo, T.; Tsukiyama, T.; Teraishi, M.; Tanisaka, T.; Okumoto, Y. The effects of the photoperiod-insensitive alleles, *se13*, *hd1* and *ghd7*, on yield components in rice. *Mol. Breed.* **2014**, *33*, 813–819. [CrossRef] [PubMed]

33. Zhang, H.-W.; Fan, Y.-Y.; Zhu, Y.-J.; Chen, J.-Y.; Yu, S.-B.; Zhuang, J.-Y. Dissection of the *qTGW1.1* region into two tightly-linked minor QTLs having stable effects for grain weight in rice. *BMC Genet.* **2016**, *17*, 98–107. [CrossRef] [PubMed]

34. Zheng, K.L.; Huang, N.; Bennett, J.; Khush, G.S. *PCR-Based Marker-Assisted Selection in Rice Breeding*; IRRI Discussion Paper Series No. 12; International Rice Research Institute: Los Banos, CA, USA, 1995.

35. Chen, X.; Temnykh, S.; Xu, Y.; Cho, Y.G.; McCouch, S.R. Development of a microsatellite framework map providing genome-wide coverage in rice. *Theor. Appl. Genet.* **1997**, *95*, 553–567. [CrossRef]

36. Dong, Q.; Zhang, Z.-H.; Wang, L.-L.; Zhu, Y.-J.; Fan, Y.-Y.; Mou, T.-M.; Ma, L.-Y.; Zhuang, J.-Y. Dissection and fine-mapping of two QTL for grain size linked in a 460-kb region on chromosome 1 of rice. *Rice* **2018**, *11*, 44. [CrossRef] [PubMed]

37. Wang, S.; Basten, C.J.; Zeng, Z.-B. *Windows QTL Cartographer 2.5*; Department of Statistics, North Carolina State University: Raleigh, NC, USA, 2012.

38. SAS Institute Inc. *SAS/STAT User's Guide*; SAS Institute: Cary, NC, USA, 1999.

10

Genome-Wide Association Studies of 39 Seed Yield-Related Traits in Sesame (*Sesamum indicum* L.)

Rong Zhou [1,†]**, Komivi Dossa** [1,2,†]**, Donghua Li** [1]**, Jingyin Yu** [1]**, Jun You** [1]**, Xin Wei** [1,3,*] **and Xiurong Zhang** [1,*]

[1] Key Laboratory of Biology and Genetic Improvement of Oil Crops, Oil Crops Research Institute of the Chinese Academy of Agricultural Sciences, Ministry of Agriculture, No. 2 Xudong 2nd Road, Wuhan 430062, China; rongzzzzzz@126.com (R.Z.); dossakomivi@gmail.com (K.D.); ldh360681@163.com (D.L.); yujingyin@caas.cn (J.Y.); youjunbio@163.com (J.Y.)

[2] Centre d'Etude Régional Pour l'Amélioration de l'Adaptation à la Sécheresse (CERAAS), Route de Khombole, Thiès, Thiès Escale Thiès BP3320, Senegal

[3] College of Life and Environmental Sciences, Shanghai Normal University, Shanghai 200234, China

* Correspondence: weixin@caas.cn (X.W.); zhangxr@oilcrops.cn (X.Z.)

† These authors contributed equally to this work.

Abstract: Sesame is poised to become a major oilseed crop owing to its high oil quality and adaptation to various ecological areas. However, the seed yield of sesame is very low and the underlying genetic basis is still elusive. Here, we performed genome-wide association studies of 39 seed yield-related traits categorized into five major trait groups, in three different environments, using 705 diverse lines. Extensive variation was observed for the traits with capsule size, capsule number and seed size-related traits, found to be highly correlated with seed yield indexes. In total, 646 loci were significantly associated with the 39 traits ($p < 10^{-7}$) and resolved to 547 quantitative trait loci QTLs. We identified six multi-environment QTLs and 76 pleiotropic QTLs associated with two to five different traits. By analyzing the candidate genes for the assayed traits, we retrieved 48 potential genes containing significant functional loci. Several homologs of these candidate genes in *Arabidopsis* are described to be involved in seed or biomass formation. However, we also identified novel candidate genes, such as *SiLPT3* and *SiACS8*, which may control capsule length and capsule number traits. Altogether, we provided the highly-anticipated basis for research on genetics and functional genomics towards seed yield improvement in sesame.

Keywords: sesame; genome-wide association study; yield; QTL; candidate gene

1. Introduction

The use of high-quality oil in human daily food intake is an important part of overall well-being. Sesame (*Sesamum indicum* L.) is a source of an excellent vegetable oil rich in vital minerals, vitamins, phytosterols, polyunsaturated fatty acids, tocopherols and unique classes of lignans such as sesamin and sesamolin, which have been identified as beneficial compounds for human health [1]. Moreover, its seeds have one of the highest oil contents (55%) among major oilseed crops, as well as a high protein content [2]. The world population is growing fast and the demand for vegetable oil in quantity and high-quality is pressing. Vegetable oil consumption is expected to double by 2040 [3]. Therefore, sesame can play a significant role in satisfying this demand.

Sesame is essentially a small-scale farmer crop and its cultivation offers two main advantages: it is a very rewarding crop because of its low production cost and high sale price; and, it is also a very resilient crop, able to provide yield and generate incomes in marginal areas where many other crops cannot grow [4,5]. Over the last decade, the production of sesame seeds has doubled and the growing area has extended to more than 50 countries in the world, showing an ever-increasing interest in this crop [6]. However, sesame has a very low seed yield capacity compared to other oilseed crops [7]. According to the Food and Agriculture Organization, the average seed yield of sesame was only 578 kg/ha in 2016, ranked as the second lowest among the major oil crops [6]. Therefore, understanding the genetic basis of seed yield-related traits and applying that knowledge in sesame breeding programs might be instrumental in developing stable high-yielding sesame varieties.

The yield of any crop is a complex character, which depends upon many independent contributing components. Deep understanding of the relationship between yield and its components is crucial to the selection process and to crop improvement [8]. Sesame seed yield per plant is considered to mainly have three components, namely, the number of capsules per plant, the number of seeds per capsule and seed weight. Some other factors, including plant height, capsule dimensions, the first capsule axis height and the number of internodes, were found to be strongly associated with seed yield in sesame [9,10]. In addition, the plant growth habit, branching type, capsule shattering, management practices, and biotic and environmental factors can significantly affect sesame yield [11]. Beside the variation among cultivars for seed yield components, the within-plant variation is extremely important. For example, some sesame cultivars can have three or more capsules per leaf axil. Mosjidis and Yermanos [12] observed that seed weight from medial capsules is higher than that from lateral capsules. Moreover, Tashiro et al. [13] and later Kumazaki et al. [14] confirmed the significant differences between seed weight between capsules from nodes located at different positions along the main stem within the same plant. Accordingly, dissecting the genetic basis of the seed yield components in sesame may be challenging and will need meticulous analysis of the multiple and complex seed yield components.

Thirteen quantitative trait loci (QTL) were detected for seven seed yield-related traits using the linkage mapping approach in sesame [10]. Genome-wide association study (GWAS) has proven to be advantageous over bi-parental QTL mapping as it captures greater diversity and offers higher resolution for gene and favorable allele discovery in several plant species [15]. Recently, GWAS was also successfully applied to sesame to unravel the genetic basis of the oil production and quality traits, yield related traits, important agronomic traits, as well as salt and drought tolerance [16,17]. The objective of the hereby study was to employ the GWAS approach to comprehensively decipher the genetic basis of 39 seed yield-related traits in sesame and unlock potential alleles and genes for seed yield improvement based on a large and diverse sample phenotyped in three different environments.

2. Results

2.1. Variability and Correlation of the Seed Yield-Related Traits in the Sesame Association Panel

A total of 39 direct and indirect seed yield-related traits were studied and classified into five main trait groups: yield index, seed traits, capsule number, capsule size, and capsule pericarp (Table S1). Ten yield-related traits that were investigated in the previous research of Wei et al. [16] were also included in this study. Descriptive statistics for the traits across the 705 accessions included in this study are listed in the Table S2. Overall, the sesame diversity panel exhibited extensive trait variation across the three environments analyzed (Figure 1 and Figure S1). We selected three contrasting environments for phenotyping (Nanning (NN), Wuhan (WH) and Sanya (SY)) because they represent natural sesame growing areas in China and also cover different geographical regions of China: Central China (WH), South China (SY), Southwest China (NN). The traits appeared to be slightly higher at NN environment compared with WH and SY, but overall the yields are similar among the three locations. Some traits, especially those related to the capsule number and capsule size groups, were stable across environments; however, the traits belonging to the yield index group displayed a high variation. This observation was further confirmed with the broad-sense heritability estimates (Table S2). Generally, a large portion of the phenotypic variance in seed yield components could be attributed to the genotypic effects in sesame.

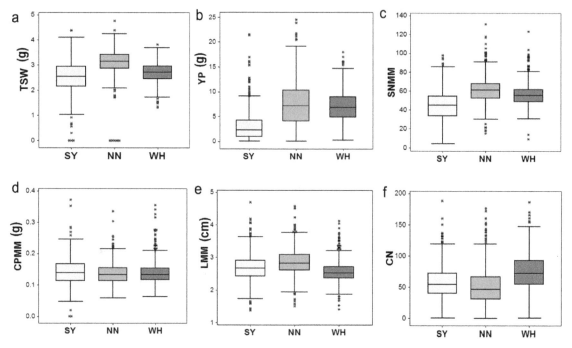

Figure 1. Boxplots displaying variation of six traits across three different environments (SY = Sanya, NN = Nanning and WH = Wuhan). Definition of the labels can be found at the end of this article.

To gain insight into the relationship between the seed yield-related traits, a clustering and correlation analysis was performed (Figure 2). It can be obviously observed that traits from the same group clustered closely, indicating strong correlations with each other. Furthermore, clustering analysis of the phenotype data highlighted three main groups (A, B and C). Group A comprised capsule number (MCNM, CN, MCNB, CNB and LCNB) and yield index (YMB and YB) related traits, which were strongly and positively correlated. This result shows that a high capsule number in a sesame plant leads to a high yield. The second group (B) was composed of mixed traits in relation to yield index, seed traits, and capsule size. From such a cluster, we inferred that accessions with high ratios of seed weight/capsule weight are likely to have a high yield. In addition, we found that high values of seed number and seed weight-traits are favorable for seed yield in sesame. Finally, Group C clustered some capsule pericarp and capsule size-related traits with moderate correlation values. Since no yield index trait was observed in this group, we concluded that it may not directly contribute to seed yield in sesame. More importantly, we found that traits from this group were negatively correlated with traits contributing to a high seed yield in sesame. For example, accessions with high capsule pericarp thickness have lower yield indexes.

2.2. Genetic Variants Associated with Seed Yield-Related Traits in Sesame

To predict significant marker-trait associations for seed yield-related traits, the mixed model was implemented in this study of the phenotype data from each environment. Genome wide association studies (GWAS) revealed 646 statistically significant loci ($p < 10^{-7}$) across the three environments associated with the 39 traits. A total of 6% of the loci were in line with the previous identified yield-related loci [16]. Significant loci were found on all of the 16 linkage groups (LG) of the genome, justifying the complex genetic architecture of the seed yield in sesame. The highest number of significant loci (86) was detected on the LG5, while the LG14 harbored only six significant loci (Table S3, Figure S2). The phenotypic variation explained by the lead loci ranged from 6.01 (SNP2372143) to 17.9% (SNP6737753 and SNP5479753), suggesting a moderate contribution to the traits (Table 1). We defined as a QTL the 88 kb region (corresponding to the linkage disequilibrium (LD) window) surrounding the peak loci and containing at least three significant loci [17]. By combining peak single nucleotide polymorphism (SNP)-trait-environment, a total of 547 QTLs were identified (Figure 3). Furthermore, by comparing peak loci through environments and traits, we uncovered six stable QTLs (detected in different environments for the same trait) and 76 pleiotropic QTLs associated with two to five various traits (Table 1). We compared the detected pleiotropic QTLs between the five groups of traits defined in this study. The results showed that most of the pleiotropic QTLs principally controlled traits from the same group (Figure 4). Few common QTLs could be observed between pairs of trait groups and there was no shared QTL for more than three traits groups. Overall, these results corroborate the phenotypic relationships observed in Figure 2. For example, there is no common QTL for the capsule pericarp and yield index groups; similarly for the capsule size and yield index groups. Conversely, the trait groups related to the yield index and capsule number exhibited the highest number of common QTLs (6), demonstrating that these groups shared similar genetic architectures. The examples presented in Figures 5 and 6, related to the trait-association for the effective capsule number in the main stem (CNM) and length of medial capsule in the main stem (LMM) of the three environments, highlight two stable QTLs detected on LG5 for CNM and LG11 for LMM. Overall, more significant loci were discovered in SY compared to the other environments.

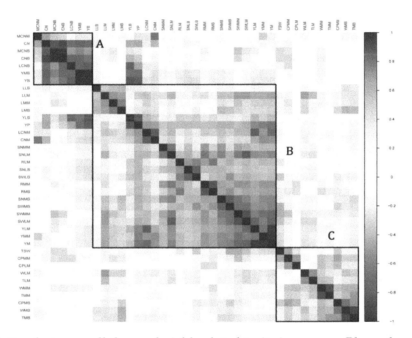

Figure 2. Correlation between all the seed yield-related traits in sesame. Blue color depicts positive correlation while red color means negative correlation. A, B and C correspond to the clusters of traits. Definition of the labels can be found at the end of this article.

Table 1. SNPs stably detected in different environments and for various traits.

LG	Position (bp)	Env.	Traits	PVE (%)	LG	Position (bp)	Env.	Traits	PVE (%)
1	1,700,170	SY	CNB	7.00	5	17,411,684	SY	SNMM	7.29
			CN	7.10				SWMM	6.79
1	1,994,183	SY	YB	8.68	6	3,404,764	SY	YB	10.65
			YMB	7.41				MCNB	7.33
1	4,450,107	SY	YB	7.55	6	3,790,583	SY	YMB	11.15
			YMB	6.99				TWB	7.13
1	6,149,415	SY	TMM	7.61				WMB	8.28
			WMM	6.58				YB	10.91
1	8,185,969	SY	YB	7.03	6	5,995,560	SY	CNB	6.99
			YMB	7.26				YMB	11.32

Table 1. *Cont.*

LG	Position (bp)	Env.	Traits	PVE (%)	LG	Position (bp)	Env.	Traits	PVE (%)
1	9,906,190	SY	YB	6.67	6	9,021,538	SY	YB	8.12
			YMB	6.88				YMB	8.34
1	11,118,941	SY	SNMB	6.32	6	9,971,772	NN	TMM	6.29
			SWMB	6.50				WMM	6.39
1	17,291,730	SY	YB	8.73	6	14,154,329	SY	YB	7.29
			YMB	9.10				YMB	6.59
2	1,201,448	SY	YB	8.76	6	14,581,641	SY	LCNM	10.29
			YMB	9.15				CNM	6.41
			YB	6.41	6	14,701,957	WH	TMM	7.09
2	5,260,400	SY	YB	7.08				WMM	6.96
			YMB	7.87	6	15,551,496	SY	RMM	7.63
2	6,057,670	NN	TMM	9.46				SNMM	6.13
			WMM	11.33	6	21,992,131	SY	YB	10.89
2	7,236,995	SY	YB	6.09				YMB	10.77
			YMB	6.11	7	6,763,527	SY	RMB	6.69
2	8,388,879	NN	TWB	6.32				SWMB	7.65
			TWB	7.24	7	1,702,826	NN	CNM	9.42
2	9,244,103	SY	TMM	6.04			WH	CNM	6.66
		WH	WMM	9.85	8	1,398,196	SY	YB	11.12
			TMM	9.25				YMB	11.99
2	11,245,765	SY	TMM	7.95	8	1,668,572	SY	YB	7.38
			WMM	7.56				SNMM	9.89
2	15,016,082	SY	YB	7.51				YM	6.66
			YMB	7.78				SWMM	8.84
2	17,451,873	SY	SNMB	6.47	8	21,325,953	SY	YB	8.97
			RMM	9.17				YMM	7.68
			SWMM	14.89	9	1,007,867	SY	RMM	8.57
			SWMB	6.83				SNMM	8.58
			SNMM	15.13	9	954,526	WH	SNMM	7.48
3	4,840,197	SY	LCNM	6.12				WMM	8.03
			CNM	6.42				TMM	7.80
3	13,198,513	SY	YB	6.68	10	1,647,805	SY	CNB	12.64
			YMB	7.24				CN	6.14
3	14,990,430	SY	YB	8.68	10	3,823,922	SY	CNB	7.43
			CNB	9.34				MCNB	6.85
			MCNB	9.40	10	7,418,158	NN	TMM	6.24
			YMB	8.81				WMM	6.31
3	16,939,689	SY	YB	6.70	10	8,305,398	SY	YB	7.76
			YMB	6.63				YMB	7.89
3	20,410,997	SY	TMM	6.47	10	10,792,029	SY	YB	7.29
			WMM	6.79				YMB	6.60
3	20,876,555	SY	YB	10.76	10	12,008,065	SY	RMM	6.83
			YMB	11.09				SNMM	6.61

Table 1. *Cont.*

LG	Position (bp)	Env.	Traits	PVE (%)	LG	Position (bp)	Env.	Traits	PVE (%)
3	20,878,243	SY	CNB	8.06	10	14,650,964	SY	YB	6.81
			MCNB	8.44				YMM	7.50
3	24,164,350	SY	TMM	6.84	10	15,097,365	SY	TMM	7.41
			WMM	6.43				WMM	6.90
4	2,505,014	SY	YB	7.98	11	6,996,833	SY	SNMM	6.53
			YMB	7.82				SWMM	7.78
4	6,419,408	SY	WMM	6.99	11	11,923,935	SY	YB	7.10
			CPMM	6.25				YMB	7.77
4	14,211,075	SY	YB	7.47	11	14,876,966	SY	YB	6.33
			YMB	6.10				YMB	6.91
5	202,984	SY	YB	9.50	11	15,137,600	SY	TMM	7.52
			YMM	6.92				WMM	6.55
5	2,854,336	NN	YM	7.85	12	328,609	SY	YM	7.53
			CNM	8.20				YMM	7.78
5	5,479,753	SY	YB	7.61	12	2,356,955	SY	YB	6.24
			CNB	8.89				YMB	6.72
			YMB	7.81	12	4,200,237	SY	YB	6.60
			CNM	13.68				YMB	7.21
		NN	CNM	17.90	12	4,895,688	SY	SNMM	8.67
5	6,737,753	SY	CNB	8.89				SWMM	6.18
			CNM	13.68	13	2,772,629	SY	YB	8.17
		NN	CNM	17.90				YMB	8.49
5	6,738,735	NN	LCNM	7.01	14	194,410	SY	YB	6.20
								YMB	6.70
			CNM	13.30	15	2,174,040	SY	YB	7.20
		SY	LCNM	13.03				YMB	7.22
		WH	CNM	14.79	15	2,372,143	WH	YB	6.29
5	6,757,688	NN	SNMB	8.02				CNM	7.03
			SWMB	10.88				YM	6.01
5	9,869,746	NN	RMB	6.54	15	3,989,016	SY	YB	6.72
			SWMB	6.08				YMB	6.69
5	11,806,702	SY	TWB	6.29	16	555,771	WH	TMM	7.55
			TWB	6.34				WMM	6.40
5	15,855,382	WH	SNMM	6.76	16	1,633,469	SY	YB	7.97
			TMM	9.11				YMB	8.17
			WMM	8.68	16	2,989,809	SY	CNB	9.26
		NN	WMM	6.23				MCNB	8.26
5	17,340,920	SY	CNB	8.64					
			MCNB	10.45					

LG = Linkage group; Env. = Environment; PVE = Phenotypic variance explained; SY = Sanya; NN = Nanning; WH = Wuhan.

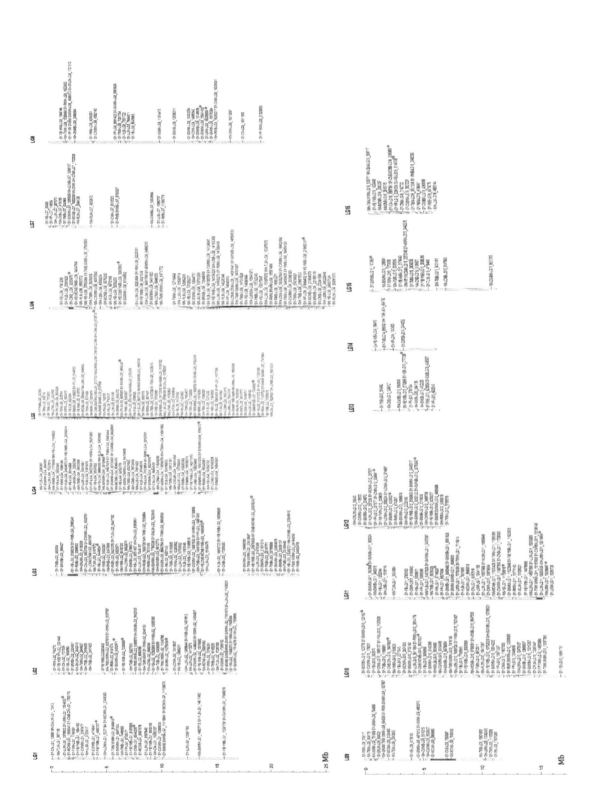

Figure 3. Genomic location of the 547 QTLs identified for seed yield-related traits in sesame. QTLs were named as follow: ENVIRONMENT-TRAIT-LINKAGEGROUP_ POSITION. Bars represent the linkage groups of sesame genome. Red portions of the bars represent the previous QTLs detected by Wu et al. [10]. Red stars represent loci previously detected by Wei et al. [16]. Definition of the labels can be found at the end of this article.

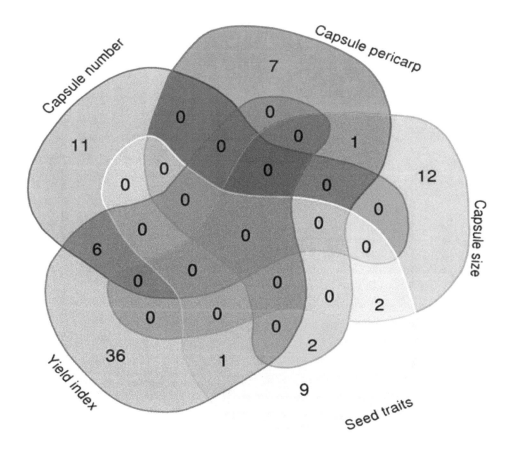

Figure 4. Venn diagram depicting the shared and common QTLs between five groups of seed yield-related traits analyzed in this study.

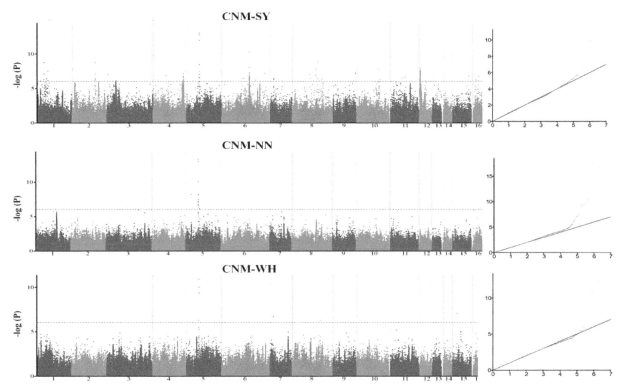

Figure 5. Genome-wide association mapping of effective capsule number in main stem (CNM) in sesame from three different environments (SY = Sanya, NN = Nanning and WH = Wuhan).

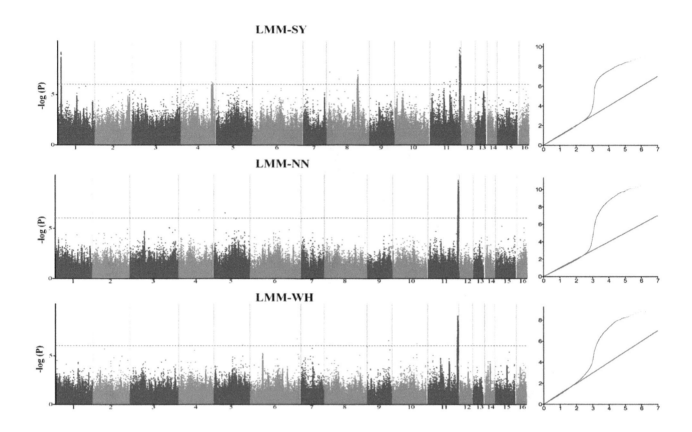

Figure 6. Genome-wide association mapping of length of medial capsule in main stem (LMM) in sesame from three different environments (SY = Sanya, NN = Nanning and WH = Wuhan).

2.3. Comparing Previous QTLs on Seed Yield-Related Traits from Bi-Parental Linkage Mapping with Our GWAS Results

In a previous study, Wu et al. [10] constructed a high-density genetic map of sesame using a population of 224 recombinant inbred lines based on the restriction-site associated DNA sequencing (RAD-seq) approach and identified several seed yield-related QTLs (plant height, first capsule height, capsule axis length, capsule number per plant, capsule length, seed number per capsule and thousand seed weight). Four similar traits, viz., capsule number per plant, capsule length, seed number per capsule and thousand seed weight, were also investigated in our study and we compared both studies to identify common genomic regions.

The physical locations of the QTLs were searched on the reference genome [18] following the descriptions of Dossa [19]. Six QTLs detected by Wu et al. [10] matched with regions around significant loci detected in this study (Table 2; Figure 3). Interestingly, we observed a good consistency between the traits related to those six QTLs and the traits associated with the corresponding significant loci. For example, the capsule length QTL (Qcl-12) from Wu et al. [10] corresponded to nine loci associated with capsule size-related traits in our study.

Also, the QTL Qcn-11 for capsule number per plant covered three significant loci identified for capsule number based on our GWAS. Another important finding is that the overlapped QTLs from Wu et al. [10] can be pleiotropic since they expanded on several significant loci which were associated with various seed yield traits in our study.

Table 2. Shared genomic regions detected for seed yield-related traits between our GWAS results and previous linkage mapping QTLs.

Traits Linkage Mapping	Code	LG	Start (bp)	End (bp)	Traits GWAS	LG	SNP Position (bp)
Grain number per capsule	Qgn-6	6	1,739,987	2,125,872	YB	6	1,741,236
					YLB	6	2,081,828
Capsule number per plant	Qcn-11	9	6,032,193	8,312,219	MCNM	9	5,988,865
					CNB	9	7,589,997
					MCNB	9	7,839,050
Capsule length	Qcl-3	3	1,566,853	2,593,783	YB	3	2,588,239
					YMB	3	2,588,241
	Qcl-4	5	9,840,981	10,961,395	YLB	5	9,857,730
					RMB	5	9,869,746
					SWMB	5	9,869,746
					TMM	5	9,895,178
					WMB	5	9,974,401
					TMB	5	10,197,769
					TMB	5	10,208,013
					YLB	5	10,705,889
					SWMB	5	10,773,145
					WMB	5	10,781,532
					LLM	5	10,786,506
					CN	5	10,786,597
					LCNM	5	10,790,853
					SWLM	5	10,958,834
	Qcl-8	4	11,220,208	11,670,895	LCNM	4	11,649,295
					WMM	4	11,658,278
					TSW	4	11,661,092
	Qcl-12	11	14,935,946	15,400,039	LLM	11	14,957,580
					LLM	11	15,003,280
					TMM	11	15,137,600
					WMM	11	15,137,600
					CPMM	11	15,138,140
					LLM	11	15,200,435
					LMM	11	15,219,964
					LMM	11	15,239,947
					LMM	11	15,289,738

2.4. Important Candidate Genes Associated with Seed Yield in Sesame

To identify the candidate genes controlling the seed yield-related traits in sesame, all the genes in 88 kb around the peak loci were retrieved [17]. In total, 7149 genes were identified and the number of genes in the LD window ranged between 8 and 42 (Table S4). Within these genes, 48 contained significant loci (Table S5). We particularly focused on these SNP-containing genes as they are more likely to modulate seed yield in sesame. Their homologs in *Arabidopsis* were identified and their functions predicted. Gene ontology analysis of these genes indicated that they are involved in developmental process, DNA and protein metabolism, response to stress, signal transduction, cell organization and biogenesis, transport and transcription (Figure 7a). Several homolog genes in *Arabidopsis* are well known to be directly or indirectly implicated in seed yield and biomass production. For example, the gene *AGL20* (AGAMOUS-like 20) plays an important role in flowering time [20], hence is directly associated with seed yield in *Arabidopsis*. In this study, we detected an intronic SNP located in the gene *SIN_1013997* (homolog of *AGL20*) strongly associated with the branch per plant seed yield and with the medial capsules in branch seed yield. Another important illustration concerns the gene *SIN_1006338* (*SiACS8*), which is located in the pleiotropic QTL associated with four various traits and was detected in all the three environments. A non-synonymous polymorphism (T/C) at the position 6,738,735 bp in this gene modulates the capsule number related traits (LCNM, CNM and CNB). An in-depth analysis suggests that the thymine allele is the favorable allele as it increases the capsule number on the stem and, therefore, leads to a higher yield (Figure 7b). Furthermore, the frequency of the T allele was rapidly increased by recent breeding, from 57% in landraces to 92% in modern cultivars. The gene *SiACS8* was previously identified as being associated with the capsule number per axil, particularly controlling the 1:3 capsules per axil in sesame [16]. These results further support our findings, indicating that *SiACS8* is indeed the causative gene controlling the capsule number trait in sesame. The homolog of *SiACS8* in *Arabidopsis* AT4G37770 (*AtACS8*) was reported to be an auxin-induced gene involved in ethylene biosynthesis, suggesting that the number of capsules on sesame stem is under the regulation of plant hormones [21].

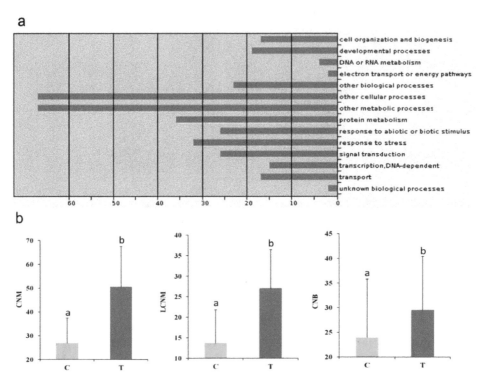

Figure 7. Functional analysis of 48 candidate gene-containing significant SNPs. (**a**) Biological function of the SNP-containing genes. (**b**) Identification of the favorable allele for the gene *SiACS8*. 262 genotypes harboring the C allele and 420 harboring the T allele were used. Different letters above bars represent significant difference ($p < 0.05$) between genotypes. The error bar indicates the standard error of the mean. Definition of the labels can be found at the end of this article.

A total of seven genes (*SIN_1017946*, *SIN_1017109*, *SIN_1021838*, *SIN_1019958*, *SIN_1011780*, *SIN_1019747* and *SIN_1014519*) involved in nutrient assimilation, carbohydrate metabolism, repression of early auxin response and kinase activity contain significant loci strongly associated with the total seed yield per plant (YP). These genes appear to be important in an effective source/sink relationship favorable for a high yield in sesame.

Some strongly associated loci were not located in the genic region; hence, gene expression analysis can give clues to pinpoint the probable candidate genes. As a proof of concept, we focused on the trait LMM and investigated the associated candidate gene. The strongest significant loci (A/G) ($-\log_{10}(p) = 9.06$) for LMM was located on the LG11 at the position 15,219,964 bp. Accessions with the guanine allele have a long capsule size as opposed to accessions with the adenosine allele. Interestingly, the frequency of the G allele in modern cultivars (20%) is comparable with that of landraces (37%), implying that this allele has not yet been intensively selected. Three genes *SIN_1011000*, *SIN_1010995* and *SIN_1010983* were found in the linkage disequilibrium window. Judging from the quantitative real time PCR (qRT-PCR) expression analysis of these genes, only *SIN_1010995* displayed a conspicuous discrepancy between the short and long capsule size accessions at different developmental stages

(Figure 8). The expression level of *SIN_1010995* (*SiLPT3*), a lipid transfer protein, was striking in the short capsule size accession but weakly expressed in the long capsule size accession. LPT3 proteins are described to be involved in cell wall edification, and more precisely in biosynthesis of cutin, which has been proposed to regulate cell adhesion during plant development [22]. The homolog gene of *SiLPT3* in *Arabidopsis AT5G59320.1* (*AtLPT3*) exhibited higher expression in the silique than other organs of *Arabidopsis*, indicating an active role in silique development [23]. Based on these observations, we speculate that *SiLPT3* regulates cell adhesion in the sesame capsule that contributes to the capsule length.

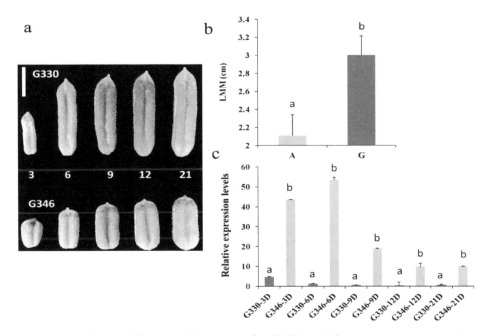

Figure 8. Expression analysis of the candidate gene for LMM trait between two contrasting accessions. (**a**) Phenotypes of G330 and G346 displaying long and short capsule length, respectively, at 3, 6, 9, 12 and 21 days after pollination. (**b**) Identification of the favorable allele at the locus 15,219,964 bp on the LG11. A total of 427 genotypes harboring the A allele and 175 harboring the G allele were used. (**c**) qRT-PCR relative expression level of the gene *SIN_1010995* between G330 and G346 at different days after pollination. Different letters above bars represent significant difference ($p < 0.05$) between genotypes. The error bar indicates the standard error of the mean. The sesame *Actin* gene (*SIN_1006268*) was used as the internal reference and 3 biological replicates and 3 technical replicates were used.

3. Discussion

The seed yield improvement of sesame is a prerequisite for the rapid expansion of the crop. Although sesame has being cultivated for a long time (~5000 years), few efforts have been made for its improvement [5]. In fact, the lack of basic information on the genetics of important agronomical traits, especially the traits complexly inherited, are causing hindrance for the breeders to achieve higher yields [24]. In this study, we observed a high variability for the assayed seed yield related traits, suggesting that our association panel harbors a large diversity necessary for genome wide association studies (GWAS). In a previous comprehensive GWAS for seed quality traits, Wei et al. [16], using the same association panel, found a low population structure, a moderate linkage disequilibrium (LD) decay (88 kb) and recommended that a high marker density, as employed in our study, could give ample power for association analyses. Several authors have studied traits that contribute to the seed yield formation in sesame. Distinctly, the capsule number per plant is a primary determinant for high seed yield in sesame [7,9,10,25,26]. In fact, sesame seeds grow in a capsule; therefore, more capsules on the plant are likely to yield more seeds [4]. Moreover, the number of seeds per capsule and the seed weight are also largely reported as important contributors to seed yield [10,27,28]. Our results match well with those of the literature, as we found that capsule size, capsule number and seed size-related traits are strongly correlated with yield indexes.

Our GWAS results revealed several clusters of significant loci, highlighting important genomic regions associated with seed yield-related traits. Interestingly, many pleiotropic QTLs were identified but an in-depth analysis indicates that very few QTLs were associated with traits from the different groups (Table S1). These results suggest that seed yield component traits from the same group have a similar genetic architecture but traits from different groups may be manipulated independently to increase the seed yield in sesame. Boyles et al. [29] also reported similar observation in sorghum with no overlapping loci for grain yield components.

The GWAS approach is recognized as a powerful tool to reconnect traits back to the underlying genetics and offers higher resolution than classical linkage mapping [30]. Previously, only one study was performed on the genetics of the sesame seed yield by employing the linkage mapping approach [10]. Comparing our results with the previous QTLs, we identified several overlapping loci associated with similar traits. Our study substantially narrows down these QTL regions which will facilitate the identification of the causal genes. In addition, several loci previously identified by Wei et al. [16] in different environments were also detected in this study, implying that these trait-associations are highly stable and could be very useful to accelerate sesame seed yield improvement efforts.

Transcriptome sequencing has been widely used to estimate gene expression changes and enables the efficiency and accuracy of candidate gene discovery in GWAS [31]. In this study, several candidate genes were retrieved from the genomic regions significantly associated with the assayed traits. To effectively pinpoint the causal genes for seed yield-related traits, additional RNA-seq data could be exploited as demonstrated in *Brassica napus*, maize, cotton, sorghum, etc. [31–33]. Nonetheless, genes containing associated SNPs which were detected in this work represent potential candidates for further functional analysis using the transgenic approach [34] and genome-editing technologies using CRISPR/Cas system. Meanwhile, the peak loci could be transformed into allele-specific markers for applications in breeding programs to design sesame varieties with improved seed yield. In fact, Asian, American and European sesame producing countries present higher yields than in Africa [6]. This can be, inter alia, related to the use of elite cultivars. For example, the modern cultivars in our panel have, on average, 70 capsules on the main stem, which is approximately double of the capsule number in landraces, and thus have a higher yield potential. Since several favorable alleles detected in this study have not yet been intensively selected, our GWAS results will undoubtedly assist in incorporating further useful alleles into the elite sesame germplasm for a seed yield increase in the future.

4. Materials and Methods

4.1. Plant Materials

In the present study, 705 cultivated sesame (*Sesamum indicum* L.) accessions were obtained from the germplasm preserved at the China National Gene Bank, Oil Crops Research Institute, Chinese Academy of Agricultural Sciences (Table S6). The panel is composed of 405 traditional landraces and 95 modern cultivars from China, as well as 205 accessions collected from 28 other countries [16]. All the accessions have been self-pollinated for four generations in Sanya, Hainan province, China (109.187° E, 18.38° N, altitude 11 m).

4.2. Field Growth Conditions

Three field trials were set in three environments in China during the years 2013 to 2014 at normal planting seasons [16]. All the accessions were grown at experiment stations in Wuhan (WH), the Hubei province (30.57° N, 114.30° E), Nanning (NN), the Guangxi province (23.17° N, 107.55° E) and Sanya (SY), the Hainan province, (109.187° E, 18.38° N). We recorded ranges of temperature (32–38/25–27 °C, day/night), relative humidity (45–72%) and rainfall (125–210 mm) during the experiment in Wuhan. In Nanning, we recorded ranges of temperature (31–34/25–26 °C, day/night), relative humidity (42–58%) and rainfall (205–235 mm) during our experiment. In Sanya, ranges of temperature (30–33/24–26 °C, day/night), relative humidity (50–75%) and rainfall (159–219 mm) were recorded during our experiment. These data show that Wuhan was the hottest location with the lowest rainfall among the 3 locations. Sanya and Nanning experimental fields have a sandy loam soil while Wuhan experimental field is characterized by a loam soil. The field trials were conducted using a randomized block design with three replications. Each plot had four rows of 2 m long spaced 0.4 m apart. At the four-leaf stage, seedlings were thinned down and eight evenly distributed plants in each row were retained for further analyses. Five uniform plants for each genotype were randomly selected to collect phenotypic data.

4.3. Trait Evaluation

Plants at the two ends of each row were not selected to avoid edge effects. Traits evaluated included (1) weight (g), length (cm), width (cm) and thickness (cm) of the dry capsule pericarp and the seed selected from different parts of the plant: medial or lateral position on the main stem or branch; (2) the seed number was counted in capsules from different parts of the plant: medial or lateral position on the main stem or branch; (3) the seed yield (g) was recorded from different parts of the plant: the capsules at medial or lateral position on the main stem or branch, total yields of the main stem, the branch and the whole plant. Based on the seed and capsule pericarp dry weights recorded from different parts of the plant, the ratio seed weight and pericarp weight were also computed. In total, 39 traits were investigated in this study and categorized into five major trait groups: yield index, seed traits, capsule number, capsule size and capsule pericarp (Table S1).

4.4. Statistical Analysis

All the statistical analyses were performed using R2.3.0 [35]. For each trait, the least square mean and descriptive statistics such as the minimum, maximum, skewness and kurtosis were estimated based on five replicates in each environment. Variation of the different traits in the different environments was represented as boxplot employing the "ggplot2" package [36]. The broad-sense heritability (H^2) was calculated as follow: $H^2 = \sigma^2{}_a/(\sigma^2{}_a + \sigma^2{}_{ae}/E + \sigma^2{}_\varepsilon/ER)$, where $\sigma^2{}_a$, $\sigma^2{}_{ay}$, and $\sigma^2{}_\varepsilon$ are estimates of the variances of accession, accession × environment interaction, and error, respectively, estimated by analysis of variance (ANOVA). E represents Environment, and R is the number of replications. Correlation among the seed yield related traits was estimated by Pearson's method at a significance level of $p < 0.05$ using the "corrplot" package [37]. For the correlation analysis, we used the best linear unbiased estimator (BLUE) values of phenotype data from the three environments.

4.5. Genome Wide Association Study Implementation

The association panel used in the present study was previously fully re-sequenced [16]. A total of 1.8 M common single nucleotide polymorphisms (SNPs) covering the whole genome with minor allele frequency >0.03 were retained for the genome wide association studies (GWAS). Phenotype-genotype association was implemented with the EMMAX model [38]. The matrix of pair-wise genetic distance derived from simple matching coefficients was used as the variance–covariance matrix of the random effect. Using the Genetic type 1 Error Calculator, version 0.2 [39], the effective number of independent SNPs were estimated to be 469,175 and the threshold to declare significant associated loci was approximately $p = 10^{-7}$ [16]. Significant associations were also selected on the threshold of $p \leq 0.01$, corrected for multiple comparisons according to the false discovery rate procedure reported by Benjamini and Hochberg [40].

4.6. Candidate Gene Mining

Based on the reference genome [18], all the genes in the 88 kb region corresponding to the average linkage disequilibrium window [16] around the peak associated loci were retrieved. Their homologs in *Arabidopsis thaliana* were predicted and their functions annotated from the database Sinbase 2.0 [18] with a cut off *E*-value of $\leq 1 \times 10^{-40}$. All the genes containing significant associated loci were prioritized. Moreover, for genomic regions where we did not find any associated SNP-containing genes, the putative candidate genes were retained if the homolog genes in *Arabidopsis thaliana* were described to be involved in seed yield or biomass formation. Gene ontology analysis of the candidate genes was performed using the Blast2GO tool v.3.1.3 [41] and plotted with the WEGO tool [42].

4.7. Gene Expression Analysis Based on Quantitative Real-Time PCR

We performed the qRT-PCR expression analysis for all the genes around the strongest associated loci with the capsule length (LMM) trait in order to pinpoint the potential candidate gene. Accession G330 with a long capsule size (~3.65 cm, at maturity stage) and accession G346 with a short capsule size (~1.90 cm, at maturity stage) were selected for this experiment. Capsules from the middle of the main stem were collected from 3 different plants (biological replicates) in Wuhan on 3, 6, 9, 12 and 21 days after pollination. RNA was extracted from fresh capsule tissues and reverse transcribed according to descriptions of Mmadi et al. [43]. In total, three genes were investigated and their gene-specific primers designed using the Primer5.0 tool [44] (Table S7). The qRT-PCR was conducted in triplicate (technical replicates) on a Roche Lightcyler® 480 instrument (Roche Molecular Systems, Inc, Basel, Switzerland) using SYBR Green Master Mix (Vazyme), according to the manufacturer's protocol. Reaction and PCR conditions are the same as the descriptions of Mmadi et al. [43]. The sesame *Actin* gene (*SIN_1006268*) was used as the internal reference and the relative gene expression values were calculated using the $2^{-\Delta Ct}$ method [45].

Supplementary Materials
Figure S1. Boxplots displaying variation of 33 traits across three different environments (SY = Sanya, NN = Nanning and WH = Wuhan). Figure S2. Manhattan plots for SNP association of all traits in the three environments (SY = Sanya, NN = Nanning and WH = Wuhan). Table S1. Full name of the 39 assayed traits. Table S2. Summary of descriptive statistics of the 39 traits in three environments. Table S3. List and position of the significant loci detected in this study. Table S4. List and functional annotation of genes around peak loci associated with the assayed traits in this study. Table S5. Candidate gene-containing significant SNPs detected in this study and their homologs in *Arabidopsis thaliana*. Table S6. Full list of the 705 accessions used in this study, their origin and their breeding status. Table S7. Primer sequences for qRT-PCR gene expression analysis.

Author Contributions: R.Z., X.W., K.D., J.Y., J.Y., D.L. participated in data collection and analysis; K.D., R.Z. wrote the manuscript; X.W. and X.Z. conceived and supervised the study. All authors have read and approved the final version of the manuscript.

Abbreviations

CN	effective capsule number in plant
CNB	effective capsule number in branch
CNM	effective capsule number in main stem
CPLM	dry capsule pericarp weight of lateral capsule in main stem
CPMB	dry capsule pericarp weight of medial capsule in branch
CPMM	dry capsule pericarp weight of medial capsule in main stem
DNA	deoxyribonucleic acid
GWAS	genome wide association study
LCNB	effective lateral capsule number in branch
LCNM	effective lateral capsule number in main stem

LD	linkage disequilibrium
LG	linkage group
LLB	length of lateral capsule in branch
LLM	length of lateral capsule in main stem
LMB	length of medial capsule in branch
LMM	length of medial capsule in main stem
MAF	minor allele frequency
MCNB	effective medial capsule number in branch
MCNM	effective medial capsule number in main stem
NN	Nanning
qRT-PCR	quantitative real-time polymerase chain reaction
QTL	quantitative trait loci
RLM	ratio of seed weight and capsule pericarp weight for lateral capsule in main stem
RMB	ratio of seed weight and capsule pericarp weight for medial capsule in branch
RMM	ratio of seed weight and capsule pericarp weight for medial capsule in main stem
RNA	ribonucleic acid
SNLB	seed number per lateral capsule in branch
SNLM	seed number per lateral capsule in main stem
SNMB	seed number per medial capsule in branch
SNMM	seed number per medial capsule in main stem
SNP	single nucleotide polymorphism
SY	Sanya
SWLB	dry seed weight of per lateral capsule in branch
SWLM	dry seed weight of per lateral capsule in main stem
SWMB	dry seed weight of per medial capsule in branch
SWMM	dry seed weight of per medial capsule in main stem
TLM	thickness of lateral capsule in main stem
TMB	thickness of medial capsule in branch
TMM	thickness of medial capsule in main stem
TSW	thousand seeds weight
WH	Wuhan
WLM	width of lateral capsule in main stem
WMB	width of medial capsule in branch
WMM	width of medial capsule in main stem
YB	yield of branch per plant
YLB	yield of lateral capsules in branch
YLM	yield of lateral capsules in main stem
YM	yield of main stem per plant
YMB	yield of medial capsules in branch
YMM	yield of medial capsules in main stem
YP	yield per plant

References

1. Anilakumar, K.R.; Pal, A.; Khanum, F.; Bawa, A.S. Nutritional, medicinal and industrial uses of sesame (*Sesamum indicum* L.) seeds: An overview. *Agric. Conspec. Sci.* **2010**, *75*, 159–168.
2. Dossa, K.; Wei, X.; Niang, M.; Liu, P.; Zhang, Y.; Wang, L.; Liao, B.; Cissé, N.; Zhang, X.; Diouf, D. Near-infrared reflectance spectroscopy reveals wide variation in major components of sesame seeds from Africa and Asia. *Crop J.* **2018**, *6*, 202–206. [CrossRef]
3. Ingersent, K.A. World agriculture: Towards 2015/2030–An FAO perspective. *J. Agric. Econ.* **2003**, *54*, 513–515.
4. Langham, D.R. Phenology of Sesame. In *Issues in New Crops and New Uses*; Janick, J., Whipley, A., Eds.; ASHS Press: Alexandria, VA, USA, 2007; p. 39.

5. Dossa, K.; Diouf, D.; Wang, L.; Wei, X.; Zhang, Y.; Niang, M.; Fonceka, D.; Yu, J.; Mmadi, M.A.; Yehouessi, L.W.; et al. The emerging oilseed crop *Sesamum indicum* enters the "Omics" era. *Front. Plant Sci.* **2017**, *8*, 1154. [CrossRef] [PubMed]

6. Food and Agriculture Organization Statistical Databases (FAOSTAT). 2017. Available online: http://faostat. fao.org/ (accessed on 19 March 2018).

7. Akhtar, K.P.; Sarwar, G.; Dickinson, M.; Ahmad, M.; Haq, M.A.; Hameed, S.; Iqbal, M.J. Sesame phyllody disease: Its symptomatology, etiology, and transmission in Pakistan. *Turk. J. Agric. For.* **2009**, *33*, 477–486.

8. Yol, E.; Uzun, B. Geographical patterns of sesame accessions grown under Mediterranean environmental conditions, and establishment of a core collection. *Crop Sci.* **2012**, *52*, 2206–2214. [CrossRef]

9. Biabani, A.R.; Pakniyat, H. Evaluation of seed yield-related characters in sesame (*Sesamum indicum* L.) using factor and path analysis. *Pak. J. Biol. Sci.* **2008**, *11*, 1157–1160. [PubMed]

10. Wu, K.; Liu, H.; Yang, M.; Tao, Y.; Ma, H.; Wu, W.; Zuo, Y.; Zhao, Y. High-density genetic map construction and QTLs analysis of grain yield-related traits in sesame (*Sesamum indicum* L.) based on RAD-Seq technology. *BMC Plant. Biol.* **2014**, *14*, 274. [CrossRef] [PubMed]

11. Diouf, M.; Boureima, S.; Diop, T.; Çagirgan, M. Gamma rays-induced mutant spectrum and frequency in sesame. *Turk. J. Field Crops* **2010**, *15*, 99–105.

12. Mosjidis, J.A.; Yermanos, D.M. Plant position effect on seed weight, oil content, and oil composition in sesame. *Euphytica* **1985**, *34*, 193–199. [CrossRef]

13. Tashiro, T.; Fukuda, Y.; Osawa, T. Oil contents of seeds and minor components in the oil of sesame, *Sesamum indicum* L.; as affected by capsule position. *Jpn. J. Crop Sci.* **1991**, *60*, 116–121. [CrossRef]

14. Kumazaki, T.; Yamada, Y.; Karaya, S.; Kawamura, M.; Hirano, T.; Yasumoto, S.; Katsuta, M.; Michiyama, H. Effects of day length and air and soil temperatures on sesamin and sesamolin contents of sesame seed. *Plant Prod. Sci.* **2009**, *12*, 481–491. [CrossRef]

15. Huang, X.; Han, B. Natural variations and genome-wide association studies in crop plants. *Annu. Rev. Plant Biol.* **2014**, *65*, 531–551. [CrossRef] [PubMed]

16. Wei, X.; Liu, K.; Zhang, Y.; Feng, Q.; Wang, L.; Zhao, Y.; Li, D.; Zhao, Q.; Zhu, X.; Zhu, X.; et al. Genetic discovery for oil production and quality in sesame. *Nat. Commun.* **2015**, *6*, 8609. [CrossRef] [PubMed]

17. Li, D.; Dossa, K.; Zhang, Y.; Wei, X.; Wang, L.; Zhang, Y.; Liu, A.; Zhou, R.; Zhang, X. GWAS uncovers differential genetic bases for drought and salt tolerances in sesame at the germination stage. *Genes* **2018**, *9*, 87. [CrossRef] [PubMed]

18. Wang, L.; Yu, S.; Tong, C.; Zhao, Y.; Liu, Y.; Song, C.; Zhang, Y.; Zhang, X.; Wang, Y.; Hua, W.; et al. Genome sequencing of the high oil crop sesame provides insight into oil biosynthesis. *Genome Biol.* **2014**, *15*, R39. [CrossRef] [PubMed]

19. Dossa, K. A physical map of important QTLs, functional markers and genes available for sesame breeding programs. *Physiol. Mol. Biol. Plants* **2016**, *22*, 613–619. [CrossRef] [PubMed]

20. Lee, H.; Suh, S.S.; Park, E.; Cho, E.; Ahn, J.H.; Kim, S.G.; Lee, J.S.; Kwon, Y.M.; Lee, I. The AGAMOUS-LIKE 20 MADS domain protein integrates floral inductive pathways in *Arabidopsis*. *Gene Dev.* **2000**, *14*, 2366–2376. [CrossRef] [PubMed]

21. Tsuchisaka, A.; Theologis, A. Heterodimeric interactions among the 1-amino-cyclopropane-1-carboxylate synthase polypeptides encoded by the *Arabidopsis* gene family. *Proc. Natl. Acad. Sci. USA* **2004**, *101*, 2275–2280. [CrossRef] [PubMed]

22. Shi, J.X.; Malitsky, S.; De Oliveira, S.; Branigan, C.; Franke, R.B.; Schreiber, L.; Aharoni, A. SHINE transcription factors act redundantly to pattern the archetypal surface of Arabidopsis flower organs. *PLoS Genet.* **2011**, *7*, e1001388. [CrossRef] [PubMed]

23. Klepikova, A.V.; Kasianov, A.S.; Gerasimov, E.S.; Logacheva, M.D.; Penin, A.A. A high resolution map of the *Arabidopsis thaliana* developmental transcriptome based on RNA-seq profiling. *Plant J.* **2016**, *88*, 1058–1070. [CrossRef] [PubMed]

24. Rao, P.V.R.; Prasuna, K.; Anuradha, G.; Srividya, A.; Vemireddy, L.R.; Shankar, V.G.; Sridhar, S.; Jayaprada, M.; Reddy, K.R.; Reddy, N.E.; et al. Molecular mapping of important agro-botanic traits in sesame. *Electron. J. Plant Breed.* **2014**, *5*, 475–488.

25. Shim, K.B.; Shin, S.H.; Shon, J.Y.; Kang, S.G.; Yang, W.H.; Heu, S.G. Classification of a collection of sesame germplasm using multivariate analysis. *J. Crop Sci. Biotechnol.* **2006**, *19*, 151–155. [CrossRef]

26. Monpara, B.A.; Khairnar, S.S. Heritability and expected genetic gain from selection in components of crop duration and seed yield in sesame (*Sesamum indicum* L.). *Plant Gene and Trait* **2016**, *7*, 1–5.

27. Emamgholizadeh, S.; Parsaeian, M.; Baradaran, M. Seed yield prediction of sesame using artificial neural network. *Eur. J. Agron.* **2015**, *68*, 89–96. [CrossRef]

28. Ramazani, S.H.R. Surveying the relations among traits affecting seed yield in sesame (*Sesamum indicum* L.). *J. Crop Sci. Biotechnol.* **2016**, *19*, 303–309. [CrossRef]

29. Boyles, R.E.; Cooper, E.A.; Myers, M.T.; Brenton, Z.; Rauh, B.L.; Morris, G.P.; Kresovich, S. Genome-wide association studies of grain yield components in diverse sorghum germplasm. *Plant Genome* **2016**, *9*, 1–17. [CrossRef] [PubMed]

30. Korte, A.; Farlow, A. The advantages and limitations of trait analysis with GWAS: A review. *Plant Meth.* **2013**, *9*, 29. [CrossRef] [PubMed]

31. Lu, K.; Peng, L.; Zhang, C.; Lu, J.; Yang, B.; Xiao, Z.; Liang, Y.; Xu, X.; Qu, C.; Zhang, K.; et al. Genome-wide association and transcriptome analyses reveal candidate genes underlying yield-determining traits in *Brassica napus*. *Front. Plant Sci.* **2017**, *8*, 206. [CrossRef] [PubMed]

32. Mao, H.; Wang, H.; Liu, S.; Li, Z.; Yang, X.; Yan, J.; Li, J.; Tran, L.S.P.; Qin, F. A transposable element in a NAC gene is associated with drought tolerance in maize seedlings. *Nat. Com.* **2015**, *6*, 8326. [CrossRef] [PubMed]

33. Sun, Z.; Wang, X.; Liu, Z.; Gu, Q.; Zhang, Y.; Li, Z.; Ke, H.; Yang, J.; Wu, J.; Wu, L.; et al. Genome-wide association study discovered genetic variation and candidate genes of fibre quality traits in *Gossypium hirsutum* L. *Plant Biotechnol. J.* **2017**, *15*, 982–996. [CrossRef] [PubMed]

34. Chowdhury, S.; Basu, A.; Kundu, S. Overexpression of a new osmotin- like protein gene (*SindOLP*) confers tolerance against biotic and abiotic stresses in sesame. *Front. Plant Sci.* **2017**, *8*, 410. [CrossRef] [PubMed]

35. R Development Core Team. *R: A Language and Environment for Statistical Computing*; R Foundation for Statistical Computing: Vienna, Austria, 2008; ISBN 3-900051-07-0.

36. Wickham, H. *Ggplot2: Elegant Graphics for Data Analysis*; Springer: New York, NY, USA, 2009.

37. Wei, T.; Simko, V. Corrplot: Visualization of a Correlation Matrix, R Package Version 0.77. 2016. Available online: http://CRAN.R-project.org/package=corrplot (accessed on 15 March 2018).

38. Kang, H.; Sul, J.; Service, S.; Zaitlen, N.; Kong, S.; Freimer, N.; Sabatti, C.; Eskin, E. Variance component model to account for sample structure in genome-wide association studies. *Nat. Genet.* **2010**, *42*, 348–354. [CrossRef] [PubMed]

39. Li, M.X.; Yeung, J.M.; Cherny, S.S.; Sham, P.C. Evaluating the effective numbers of independent tests and significant p-value thresholds in commercial genotyping arrays and public imputation reference datasets. *Hum. Genet.* **2012**, *131*, 747–756. [CrossRef] [PubMed]

40. Benjamini, Y.; Hochberg, Y. Controlling the false discovery rate: A practical and powerful approach to multiple testing. *J. R. Stat. Soc. Ser. B* **1995**, *57*, 289–300.

41. Conesa, A.; Götz, S. Blast2GO: A comprehensive suite for functional analysis in plant genomics. *Int. J. Plant Genom.* **2008**, *2008*, 619832. [CrossRef] [PubMed]

42. Ye, J.; Fang, L.; Zheng, H.; Zhang, Y.; Chen, J.; Zhang, Z.; Wang, J.; Li, S.; Li, R.; Bolund, L.; et al. WEGO: A web tool for plotting GO annotations. *Nucleic Acids Res.* **2006**, *34*, 293–297. [CrossRef] [PubMed]

43. Mmadi, M.A.; Dossa, K.; Wang, L.; Zhou, R.; Wang, Y.; Cisse, N.; Sy, M.O.; Zhang, X. Functional characterization of the versatile MYB gene family uncovered their important roles in plant development and responses to drought and waterlogging in sesame. *Genes* **2017**, *8*, 362. [CrossRef] [PubMed]

44. Lalitha, S. Primer premier 5. *Biotechnol. Softw. Int. Rep.* **2000**, *1*, 270–272. [CrossRef]

45. Livak, K.J.; Schmittgen, T.D. Analysis of relative gene expression data using real-time quantitative PCR and the $2^{-\Delta\Delta CT}$ Method. *Methods* **2001**, *25*, 402–408. [CrossRef] [PubMed]

11

Transcriptome and Hormone Comparison of Three Cytoplasmic Male Sterile Systems in *Brassica napus*

Bingli Ding, Mengyu Hao, Desheng Mei, Qamar U Zaman, Shifei Sang, Hui Wang, Wenxiang Wang, Li Fu, Hongtao Cheng * and Qiong Hu *

Key Laboratory for Biological Sciences and Genetic Improvement of Oil Crops, Ministry of Agriculture, Oil Crops Research Institute, Chinese Academy of Agricultural Sciences, Wuhan 430062, China; dingbl91@163.com (B.D.); haomengyu@caas.cn (M.H.); deshengmei@caas.cn (D.M.); qamaruzamanch@gmail.com (Q.U.Z.); 15652142445@163.com (S.S.); wanghui06@caas.cn (H.W.); wangwenxiang@caas.cn (W.W.); fuli@caas.cn (L.F.)
* Correspondence: chenghongtao@caas.cn (H.C.); huqiong01@caas.cn (Q.H.)

Abstract: The interaction between plant mitochondria and the nucleus markedly influences stress responses and morphological features, including growth and development. An important example of this interaction is cytoplasmic male sterility (CMS), which results in plants producing non-functional pollen. In current research work, we compared the phenotypic differences in floral buds of different *Brassica napus* CMS (*Polima, Ogura, Nsa*) lines with their corresponding maintainer lines. By comparing anther developmental stages between CMS and maintainer lines, we identified that in the *Nsa* CMS line abnormality occurred at the tetrad stage of pollen development. Phytohormone assays demonstrated that IAA content decreased in sterile lines as compared to maintainer lines, while the total hormone content was increased two-fold in the S_2 stage compared with the S_1 stage. ABA content was higher in the S_1 stage and exhibited a two-fold decreasing trend in S_2 stage. Sterile lines however, had increased ABA content at both stages compared with the corresponding maintainer lines. Through transcriptome sequencing, we compared differentially expressed unigenes in sterile and maintainer lines at both (S_1 and S_2) developmental stages. We also explored the co-expressed genes of the three sterile lines in the two stages and classified these genes by gene function. By analyzing transcriptome data and validating by RT-PCR, it was shown that some transcription factors (TFs) and hormone-related genes were weakly or not expressed in the sterile lines. This research work provides preliminary identification of the pollen abortion stage in *Nsa* CMS line. Our focus on genes specifically expressed in sterile lines may be useful to understand the regulation of CMS.

Keywords: cytoplasmic male sterility (CMS); phytohormones; differentially expressed genes; pollen development; *Brassica napus*

1. Introduction

Oilseed rape is one of most important oil crops worldwide, producing food, biofuel, and industrial compounds, including lubricants and surfactants. Hybrid breeding is a key technique to enhance crop production [1–3], in which cytoplasmic male sterility (CMS) plays an important role in seed production [4]. CMS is a maternally inherited trait and is beneficial for the production of F_1 hybrid seeds by generating infertile pollen without changing vegetative growth and female fertility [5]. CMS systems are not only a useful component for studying pollen development, but also an important way to utilize hybrid vigor [6]. The existence of CMS systems in plants eliminates the laborious and painstaking work of sterilization and manual emasculation in a broad range of crops. CMS can arise spontaneously in breeding lines after wide crosses, interspecific exchange of nuclear or cytoplasmic genomes, and mutagenesis [7]. Initially, it was thought that sterility was caused by mutation within the mitochondrial genome [8], however, further research has revealed that a major cause of CMS is mitochondrial DNA rearrangement, which results in plants unable to generate functional pollen [9]. Mitochondria are important cellular components for energy (ATP, NADH, $FADH_2$)-dependent metabolic pathways, including oxidative phosphorylation, respiratory electron transfer, biosynthesis of amino acids, vitamin cofactors, the Krebs cycle, and programmed cell death [10–12]. Therefore, CMS proteins were hypothesized to cause mitochondrial energy deficiency and failure to meet energy requirements during male reproductive development [13].

Currently, 10 types of CMS systems have been reported in *Brassica napus*, including the natural mutation *pol* CMS [14] and *shan2A* CMS [15], and intergeneric hybridization CMS *nap* CMS [16] and *Nsa* CMS [17]. *Nsa* CMS [17], *Ogu* CMS [18] and *tour* CMS [19] were generated by protoplast fusion of different species, resulting in a source of genetic variation within the cytoplasmic organelles [20]. Both *Pol* CMS and *Ogu* CMS are commonly used as CMS systems for *B. napus* hybrid breeding. CMS is sensitive to harsh environmental factors, including air temperature and exposure time to sunlight [21–23]. However, the *Nsa* CMS system has demonstrated stable male sterility under different environmental conditions, ensuring seed purity during hybrid seed production.

Preliminary work has demonstrated significant differences in plant endogenous hormones between CMS lines and their maintainer lines in different species [24,25]. In sugar beet, it was found the level of endogenous IAA (indole-3-acetic acid), GA3 (gibberellic acid), and ZR (zeatin-riboside), in relation to ABA (abscisic acid), differed at three developmental stages (vegetative, early flowering, and bud development) [26]. It was also demonstrated that pepper CMS line 'Bei-A' and maintainer line 'Bei-B' showed significant hormonal differences [27], with a higher IAA and ABA content and lower ZR_5 and GA_3 content observed within the CMS line [27]. The relationship between phytohormones and CMS has been widely investigated in many species, including *B. napus* [28,29], flax [30], and rice [31]. It has been shown that phytohormones ABA and IAA may be major contributors for CMS. The concentration of ABA and IAA changes at different stages of bud development between male sterile lines and their maintainer lines [29,30]. These studies collectively provide evidence for the importance of determining the endogenous level of ABA and IAA in CMS and maintainer lines when studying cytoplasmic male sterility.

Most recently, attention has focused on the provision of next-generation sequencing (NGS) technology [32–34] and the use of NGS to make studies on expressed genes and genomes in higher plants more feasible [35–37]. Currently, RNA-Seq has been used in higher plants with CMS systems in many species, including tomato [37], rice [38], and *B. napus* [36,39]. A large and growing body of literature has investigated floral buds of CMS and maintainer lines using RNA sequencing and comparative gene expression. In *Pol* CMS, unigenes related to pollen development were analyzed through transcriptome sequencing [36]. These high-throughput results will be useful for understanding the sterility mechanism of *pol* CMS in detail. Another transcriptome study of SaNa-1A CMS was also conducted in *B. napus* [40]. By comparing the sterile line and the maintainer line, many differentially expressed genes (DEGs) involved in metabolic, protein synthesis, and other pathways were identified. These results provide a basis for future research on the CMS mechanism in SaNa-1A. The existence of various CMS lines with different mitochondrial patterns offer new opportunities to explore the genetic regulation of CMS and its associated developmental effects [41].

In the current study, *Pol* CMS, *Ogu* CMS, *Nsa* CMS, and their corresponding maintainer lines (with the same nuclear genome but fertile cytoplasm) were used to carry out transcriptomic and DEG analysis. Simultaneously, we compared the morphological differences in sterile and fertile lines, and analyzed the IAA and ABA contents. We investigated the pollen abortion stage of the *Nsa* CMS line by semi-thin sectioning. This study confirms the stage of pollen abortion in *Nsa* CMS, and illustrates the mode of regulation of the different CMS systems during pollen development at the transcriptomic level.

2. Results

2.1. Phenotypic Characterization of CMS Lines and Maintainer Lines

The flower structure of rapeseed includes four sepals, four petals, six stamens (four long and two short), and one pistil from outwards to inwards. When a flower blooms, mature pollen sticks to the pistil. The pistil is almost the same height as the long stamens, allowing pollination to occur easily. In this study, we obtained three CMS systems (*Nsa* CMS, *Pol* CMS, and *Ogu* CMS) with corresponding maintainer lines. All sterile lines and their maintainer line harbor the same nuclear genome but different cytoplasm. We found that all sterile floral petals were visually wrinkled and smaller than fertile flowers in three CMS systems (Figure 1). Degeneration of stamens and shorter stamen length was observed in sterile lines as compared with the normal fertile flowers. Among the three CMS systems, the stamens of the *pol* CMS sterile line were more seriously degenerated (Figure 1F). However, the pistils of all the sterile floral buds were the same as fertile lines (Figure 1).

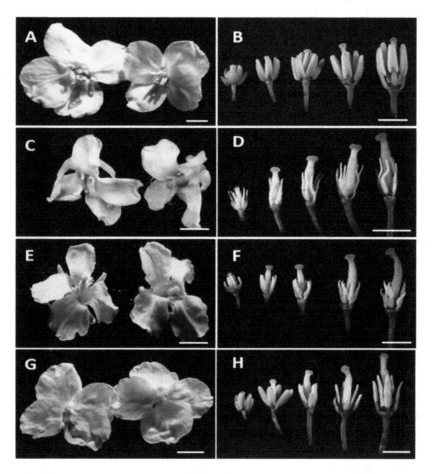

Figure 1. Flower morphology of maintainer and sterile lines of the *Pol*, *Nsa*, and *Ogu* cytoplasmic male sterility (CMS) systems. (**A–B**) Maintainer line; (**C–D**) *Nsa* CMS line; (**E–F**) *Pol* CMS line; (**G–H**) *Ogu* CMS line; Bar = 0.5 cm.

The stage at which pollen abortion occurs within the *Nsa* CMS system has not been determined clearly. For detailed characterization of the developing pollen, ultrathin specimens were observed under a microscope. By observing a semi-thin section of anthers, we conclude that the abortion period of *Nsa* CMS occurred during the tetrad period (Figure 2F). After the tetrad stage, *Nsa* CMS could not produce normal spores at the uni-nuclear stage (Figure 2G). Normal anthers form mature pollen, as shown in Figure 2D. The sterile line did not produce mature pollen but formed a large number of abnormal spores (Figure 2H).

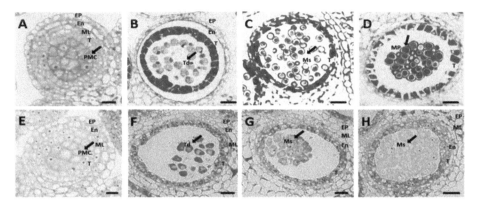

Figure 2. Comparison of maintainer line "ZS4" (**A–D**) and sterile line (**E–H**) anthers of *Nsa* CMS with toluidine blue staining. Bar = 10 μm, Ep, epidermis; En, endothecium; ML, middle layer; T, tapetum; Ms, microspore; MP, mature pollen; PMC, primary mother cells.

2.2. IAA and ABA Concentration in CMS and Maintainer Lines

Plant hormones were assessed in CMS and maintainer lines to clarify how plant hormones are altered in the three CMS systems (Figure 3). The ABA and IAA contents in flower buds were detected at S_1 (<2.5 mm size of floral buds) and S_2 stages (>2.5 mm size of floral buds) in CMS and maintainer lines, respectively. We found that ABA levels were significantly higher in all three CMS lines as compared to maintainer lines at both stages. Conversely, IAA content was significantly lower in the *Nsa* CMS line than its maintainer line at the S_1 stage, while *Ogu* and *pol* CMS lines showed no significant difference with their maintainer lines. However, IAA content was significantly lower in all CMS lines as compared to the maintainer lines at the S_2 stage. These results indicate that a significantly higher content of endogenous ABA and lower content of IAA may enhance pollen abortion in sterile lines. ABA content showed increasing and IAA decreasing trends at the S_1 stage compared to the S_2 stage. The ABA content was significantly higher at both stages in CMS lines than in their corresponding maintainer line.

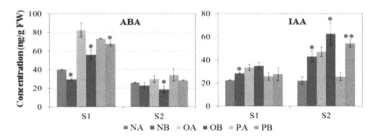

Figure 3. ABA and IAA contents of developing buds in maintainer and male sterile lines of the *Pol*, *Nsa*, and *Ogu* systems. NA, *Nsa* sterile line; NB, *Nsa* maintainer line; OA, *Ogu* sterile line; OB, *Ogu* maintainer line; PA, *Pol* sterile line; PB, *Pol* maintainer line. Asterisks indicate a significant difference was detected between CMS line and maintainer line in S_1 and S_2 stage by t-test at *p<0.05, **p<0.01.

2.3. Differentially Expressed Genes in CMS and Maintainer Lines

Using high-throughput sequencing, differentially expressed genes were detected in the sterile and corresponding maintainer lines. The flower buds used to determine phytohormone levels were also subjected to transcriptome sequence analysis. Three biological replicates were performed with the reproducibility between replicates being ≥90%. In total, 222.15 Gb of clean data were generated (with all samples Q30 ≥ 90%). Differentially expressed genes (DEGs) were identified in Biocloud (Biomarker Technologies). For each CMS system, DEGs were found between the male sterile line and the corresponding maintainer line. DEGs exhibiting a two-fold change or greater were selected according to the *q*-values [39]. At the S_1 stage, we identified 1306, 1262, and 4127 DEGs in the *Nsa*, *Pol*, and *Ogu* systems, respectively. More DEGs (2369, 1690, and 3035) were discovered at the S_2 stage in the three CMS systems. Among the three CMS systems, the largest number of DEGs were observed in the *Ogu* CMS system at the S_1 stage. Among the total 4127 DEGs, 2158 genes were upregulated and 1969 genes were downregulated. The smallest number of DEGs was observed in the *Pol* CMS system, in which 806 genes were upregulated and 456 genes were downregulated at the S_1 stage (Figure 4). Many more upregulated DEGs with high-fold change (>5-fold) were found at the S_2 stage compared to the S_1 stage in all three systems (Figure 4). Furthermore, only the *Ogu* CMS system exhibited more DEGs, including upregulated and downregulated genes, in the S_1 stage than the S_2 stage. More DEGs were observed in the *Pol* CMS, and especially in the *Nsa* CMS system at the S_2 stage than the S_1 stage.

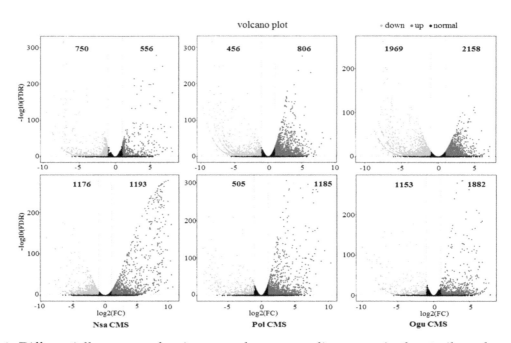

Figure 4. Differentially expressed unigenes and corresponding genes in the sterile and maintainer lines. The genes were selected with "$p \leq 0.01$" and "fold change ≥ 2". The X-axis is the log of 2-fold change in expression between the sterile and maintainer lines at two stages. Y-axis shows the statistical significance of the differences with the value of log10 (FDR). The spots in different colors are representing expression of different genes. Black spots represent genes without significant expression. Red spots mean 2-fold upregulated genes from maintainer lines to sterile lines. Green spots represent significantly 2-fold down-expressed genes from maintainer lines to sterile lines.

2.4. Gene Ontology and Classification of Three CMS Lines

At the S_1 stage, we observed that only 156 unigenes were co-differentially expressed in the three CMS lines, compared to 581 unigenes at the S_2 stage (Figure 5A,B). KEGG classification and functional enrichment was performed for DEGs at both stages (Figure 5C,D). At the S_1 stage, five categories were identified, including environmental information processing, genetic information processing, organismal systems, cellular processes, and metabolism (Figure 5C). At the S_2 stage, genes were divided into four categories, including metabolism, genetic information processing, cellular processes, and environmental information processing (Figure 5D). At the S_1 stage, in the environmental information processing category, 3% of DEGs were associated with plant hormone signal transduction. Only 1% of the DEGs were relative to plant–pathogen interaction. Within the cellular processes category, the highest number of DEGs was related to the peroxisome. Significantly enriched DEGs were identified as being involved in pentose–glucuronate interconversions and starch–sucrose metabolism among the metabolic components category. At the S_2 stage, metabolic components were significantly enriched, including starch–sucrose, arginine–proline, glycerophospholipid, alanine–aspartate–glutamate, and amino–nucleotide sugar metabolism. From these analyses, we can determine starch and sucrose play an important role in metabolism at this stage. Transcriptomic data also revealed that many DEGs were enriched in plant hormonal signal transduction pathways.

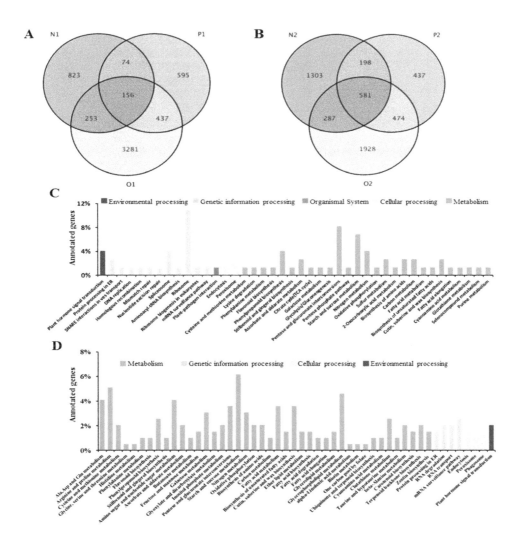

Figure 5. Co-differentially expressed unigenes (**A**, **B**) and GO annotations (**C**, **D**) of differentially expressed genes (DEGs). Venn diagrams of differentially expressed genes at (**A**) the S1 stage and (**B**) the S2 stage in *Nsa* (N), *Ogu* (O), and *Pol* (P) male sterile lines. N1, S1 stage of *Nsa* CMS system; O1, S1 stage of *Ogu* CMS system; P1, S1 stage of *Pol* CMS system; N2, S2 stage of *Nsa* CMS system; O2, S2 stage of *Ogu* CMS system; P2, S2 stage of *Pol* CMS system; (**C**) and (**D**), the X-axis indicates the percentage of genes in each categories, and y-axis showed classification of unigenes.

2.5. Verification of DEGs by RT-PCR

We conducted RT-PCR to validate the results generated by RNA-Seq. To determine whether transcription factors and hormone-related genes were differentially expressed, we quantified expression of these genes in the three CMS lines by semi-quantitative polymerase chain reaction (RT-PCR). Expression of genes encoding transcription factors or involved in ABA or IAA signaling were enriched in maintainer lines compared to male sterile lines (Figure 6). From the RT-PCR results, we found that almost all selected genes were highly expressed in maintainer lines compared to sterile lines. Most of the genes showed higher expression levels at the S_2 stage compared to the S_1 stage in all CMS systems (*Pol*, *Ogu*, and *Nsa*). This result was consistent with the RNA-Seq results.

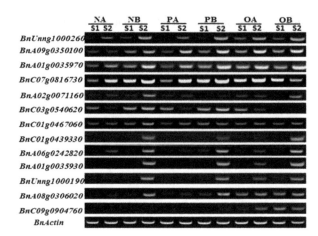

Figure 6. Gene expression difference in three cytoplasmic male sterile materials of S1 and S2 stages. NA, *Nsa* sterile line; NB, *Nsa* maintainer line; OA, *Ogu* sterile line; OB, *Ogu* maintainer line; PA, *Pol* sterile line; PB, *Pol* maintainer line. The *BnActin* gene was used as the control.

3. Discussion

The widespread existence of CMS in plant species may be related to the potential to promote outcrossing and prevent inbreeding depression. Independent CMS lines differ not only in their sequences and origins [42], but also in their phenotype, including changes in microspore development [43] and breakdown pattern of tapetum structure [44]. Pollen development comprises a series of defined physiological events. A large volume of published studies describe the role of energy (ATP, NADH, $FADH_2$) in the pollen abortion process [45]. Our results also show that many energy production or conversion related genes are differentially expressed between male sterile and maintainer lines. In the semi-thin sections of the *Nsa* CMS and maintainer lines (Figure 2), we identified differences in the epidermis, endothecium, tapetum, microspore, and mature pollen. Our results revealed that pollen abortion occurred at the tetrad stage in the *Nsa* CMS line. Following the tetrad stage, *Nsa* CMS plants could not produce normal spores at the uni-nuclear stage (Figure 2H). Pollen abortion occurred due to the breakdown of tapetum and premature or delayed degeneracy [46,47]. In *pol* CMS, abortion was started at stage 4 (pollen development period). Anthers of the sterile line could not differentiate sporogenous cells, with the middle layer, endothecium, and tapetum being indistinguishable. The results in sterile anthers filled with numerous, highly vacuolated cells [36]. Due to abnormal development in the early stage, *pol* CMS lines cannot produce normal tetrads. In *Ogu* CMS, abortion was also identified to start at the tetrad stage by comparing the cell morphology of three central stages (the tetrad, mid-microspore, vacuolated microspore) of pollen development [44]. It was found that the tapetal cells developed a large vacuole at tetrad stage in the *Ogu* CMS line. The anther development of sterile line SaNa-1A, a line with CMS derived from somatic hybrids between *B. napus* and *Sinapis alba*, is also abnormal from the tetrad stage [40]. The abortion phenotype of *Nsa* CMS is the same as the SaNa-1 sterile line in all four stages of pollen development. Similar to SaNa-1 CMS, *Nsa* CMS was derived from somatic hybrids between *B. napus* and *Sinapis arvensis*. Together with *Ogu* CMS, which is derived from intergeneric hybridization between *B. napus* and *Raphanus sativa*, all the alloplasmic CMS systems have the same pollen development abortion stage.

Phytohormones (IAA and ABA) are generally known for their specific role in the induction and promotion of DNA synthesis, and play a role in metabolic pathways [48,49]. It was observed that high and exogenous application of ABA induced pollen abortion by specifically suppressing apoplastic sugar transport in pollen [22,50,51]. The ABA content in younger floral buds was higher than that of elder ones in all three CMS systems, whereas the IAA content showed a converse trend. However, similar to previous studies in other species, sterile lines of the three CMS systems showed some differences in IAA and ABA content in both male sterile and their corresponding maintainer lines [52,53].

Transcriptomic analysis detected a total of 5619 DEGs at the S_1 stage, with 156 co-differentially expressed in all three CMS lines, and 5208 DEGs at the S_2 stage, with 581 co-differentially expressed in all three CMS lines. KEGG analysis divided these co-differentially expressed genes into 42 and 50 categories at the S_1 and S_2 stages, respectively. At both the S_1 and S_2 stages, half of the genes were involved in metabolism. Many genes for mitochondrial energy metabolism and pollen development were also differentially expressed in multiple CMS systems [36,54,55]. It has been shown that the presence of infertility genes affects the transcription of genes involved in the energy metabolism of the mitochondria, resulting in impairment of the normal physiological functions of the mitochondria, which leads to infertility [13].

Previous studies identified a link between plant hormones and cytoplasmic male infertility [22,24]. In addition, transcription factors have also been implicated in pollen infertility [56,57]. Therefore, the expression of some transcription factors and hormone-related genes were selected to verify the results generated by RNA-Seq. RT-PCR results indicated that the expression level of selected genes in maintainer lines was significantly higher than that of the male sterile lines. This provides further support that levels of phytohormone precursor genes and some transcription factors may be correlated with cytoplasmic male sterility. Increased IAA content was detected in maintainer lines compared to male sterile lines in all CMS systems. Coincident with this result, we observed IAA signaling-related genes, including two *IAA19* genes, were significantly enriched in maintainer lines (Figure 6).

4. Materials and Methods

4.1. Plant Materials

Pol CMS, *Ogu* CMS, *Nsa* CMS and their corresponding maintainer lines were used in this study. Materials were cultivated in the field of Oil Crops Research Institute, Chinese Academy of Agricultural Sciences (OCRI-CAAS), Wuhan, China. Anthers at different stages were collected for morphological study, and the abortion stage was studied by semi-thin sectioning of floral buds. Samples of floral buds (<2.5 mm and >2.5 mm) were collected and stored at −70 °C for further RNA-sequencing and hormonal quantification.

4.2. Morphology and Semi-Thin Sections

The floral buds of sterile and fertile lines were examined under the microscope (Olympus: CX31RTSF). At different stages, samples collected from the sterile and fertile line were fixed in FAA solution [38% formaldehyde, 70% ethanol, and 100% acetic acid (1:1:18)]. A vacuum chamber was used to evacuate the air and volatiles from the sample bottles. Fixed floral buds were dehydrated by a graded series of ethanol (70, 85, 95, and 100%) for one hour. Pre-infiltration and penetration by Technovit 7100 resin steps were undertaken to produce semi-thin sections ~3 μm thick. Samples were stained by 1% toluidine blue (Sigma Aldrich, St. Louis, MO, USA) for 3 min, and 5 specimens of each stage were observed to take images under an optical microscope (Olympus: CX31RTSF, Tokyo, Japen).

4.3. Phytohormone (ABA and IAA) Quantification

About 60 floral buds (smaller than 2.5 mm and larger than 2.5 mm) of *Pol* CMS, *Ogu* CMS, *Nsa* CMS, and their corresponding maintainer lines were quantified for ABA and IAA phytohormones. Samples were collected and extracted using methanol compounds [58]. The extraction was carried out by adding 1 mL of MeOH (methyl alcohol) with water (8:2) into each tube containing fresh plant material. Samples were shaken for 30 min before centrifugation at 12000 rpm at 4 °C for 10 min. The supernatant was transferred to a new microcentrifuge tube and dried in a speed vacuum. After drying, 100 μL of MeOH was added to each sample. Each sample was homogenized using a vortex mixer and centrifuged at 12000 RPM at 4 °C for 10 min. Phytohormones within the supernatant were separated by HPLC (Agilent 1200) and analyzed by a hybrid triple quadrupole/linear ion trap mass spectrometry (ABI 4000 Q-Trap, Applied Biosystems, Foster City, CA, USA).

4.4. Illumina Sequencing and Analysis of DEGs

About 60 floral buds were harvested from the plants of each line (*Ogu* CMS, *Nsa* CMS, and their corresponding maintainer lines) at the same time. Samples collected from each line were pooled, frozen in liquid nitrogen, and stored at −70 °C for RNA preparation. Total RNA from two stages of floral buds (<2.5 mm and >2.5 mm) of *pol* CMS, *Ogu* CMS, *Nsa* CMS, and their corresponding maintainer lines were extracted by using RNA kits (Tiangen, Beijing, China) in accordance with the manufacturer's protocol. The integrity of the total RNA was checked by 1% agarose gel electrophoresis. The concentration was detected by Nano-Drop (Thermo Scientific, Madison, WI, USA) and purity of RNA was determined by Agilent 2100 Bio-analyzer (Agilent, Waldbronn, Germany). RNA (10 μL) was sequenced using the Illumina HiSeq 2000 (Illumina, San Diego, CA, USA) and 150 bp of data collected per run. After removing adapters and low-quality data, the resulting clean data was aligned to the *B. napus* reference genome [59]. Potential duplicate molecules were removed from the aligned BAM/SAM format records. FPKM (fragments per kilobase of exon per million fragments mapped) values were used to analyze gene expression by the software Cufflinks [60]. Three biological replicates were performed for each sample.

4.5. Semi-Quantitative (RT-PCR) Analysis of DEGs

The DEG results were confirmed by RT-PCR using the same RNA samples which were used for RNA library construction. Complementary DNA was generated from the RNA template by using the reverse transcription kit (Vazyme, Nanjing, China). Specific primers for differentially expressed genes were designed to amplify 600–750 bp sequences (Table S1). RT-PCR was carried out by using a program of 95 °C for 5 min (initial hot start), 30 cycles of 95 °C for 30 s, 56 °C for 35 s, and 72 °C for 5 min. Three biological replicates were analyzed for each sample.

5. Conclusions

Considerable effort has been taken to identify the pollen abortion stage in *Pol* and *Ogu* CMS lines. Conversely, the pollen abortion stage in the *Nsa* CMS line had not been determined clearly. From this study, we identified the tetrad stage of *Nsa* CMS for pollen abortion by using the semi-thin sectioning of floral buds. This information will help support the application of *Nsa* CMS in plant breeding. Higher content of ABA and lower content of IAA was observed in sterile lines when compared to maintainer lines in all male sterile systems. This result may reveal that ABA and IAA play different roles in fertile pollen development. During the two stages, genes involved in energy production were enriched in maintainer lines in comparison to sterile lines for all CMS systems investigated.

Author Contributions: Data curation, M.H.; Formal analysis, B.D.; Funding acquisition, H.C. and Q.H.; Investigation, S.S.; Project administration, D.M. and Q.H.; Resources, H.W. and L.F.; Software, W.W.; Writing—original draft, B.D., Q.U.Z. and H.C.; Writing—review & editing, Q.H. All authors read and approved the final manuscript.

Acknowledgments: We thank Rachel Wells in John Innes Center for proof-reading and English correction.

References

1.   Wang, H.Z. Review and future development of rapeseed industry in China. *Chin. J. Oil Crop Sci.* **2010**, 2, 300–302.

2.   Fu, T.D. Breeding and utilization of rapeseed hybrid. *Hubei Sci. Technol.* **2000**, 167–169.

3.   Allender, C.J.; King, G.J. Origins of the amphiploid species *Brassica napus L*. Investigated by chloroplast and nuclear molecular markers. *BMC Plant Biol.* **2010**, *10*, 54. [CrossRef] [PubMed]

4.   Yamagishi, H.; Bhat, S.R. Cytoplasmic male sterility in brassicaceae crops. *Breed Sci.* **2014**, *64*, 38–47. [CrossRef] [PubMed]

5.   Shinada, T.; Kikuchi, Y.; Fujimoto, R.; Kishitani, S. An alloplasmic male-sterile line of *Brassica oleracea* harboring the mitochondria from Diplotaxis muralis expresses a novel chimeric open reading frame, *orf72*. *Plant Cell Physiol.* **2006**, *47*, 549–553. [CrossRef] [PubMed]

6.   Shiga, T.; Baba, S. Cytoplasmic male sterility in oil seed rape. *Brassica napus* L., and its utilization to breeding. *Japan J. Breed.* **1973**, *23*, 187–197.

7.   Hanson, M.R.; Bentolila, S. Interactions of mitochondrial and nuclear genes that affect male gametophyte development. *Plant Cell* **2004**, *16*, 154–169. [CrossRef]

8.   Schnable, P.S.; Wise, R.P. The molecular basis of cytoplasmic male sterility and fertility restoration. *Trends Plant Sci.* **1998**, *3*, 175–180. [CrossRef]

9.   Horn, R.; Guptac, K.J.; Colombo, N. Mitochondrion role in molecular basis of cytoplasmic male sterility. *Mitochondrion* **2014**, *19*, 198–205. [CrossRef]

10.   Siedow, J.N.; Umbach, A.L. Plant mitochondrial electron transfer and molecular biology. *Plant Cell* **1995**, *7*, 821–831. [CrossRef]

11.   Logan, D.C. The mitochondrial compartment. *J. Exp. Bot.* **2006**, *57*, 1225–1243. [CrossRef] [PubMed]

12.   Youle, R.J.; Karbowski, M. Mitochondrial fission in apoptosis. *Nat. Rev. Mol. Cell Biol.* **2005**, *6*, 657. [CrossRef] [PubMed]

13.   Chen, L.; Liu, Y.G. Male sterility and fertility restoration in crops. *Annu. Rev. Plant Biol.* **2014**, *65*, 579–606. [CrossRef] [PubMed]

14.   Fu, T.D. Production and research of rapeseed in the People's Republic of China. *Eucarpia. Crucif. News* **1981**, *6*, 6–7.

15.   Shen, J.X.; Wang, H.Z.; Fu, T.D.; Tian, B.M. Cytoplasmic male sterility with self-incompatibility, a novel approach to utilizing heterosis in rapeseed (*Brassica napus* L.). *Euphytica* **2008**, *162*, 109–115. [CrossRef]

16.   Thompson, K.F. Cytoplasmic male-sterility in oil-seed rape. *Heredity* **1972**, *29*, 253–257. [CrossRef]

17.   Hu, Q.; Andersen, S.; Dixelius, C.; Hansen, L. Production of fertile intergeneric somatic hybrids between brassica napus and sinapis arvensis for the enrichment of the rapeseed gene pool. *Plant Cell Rep.* **2002**, *21*, 147–152.

18.   Ogura, H. Studies on the new male-sterility in japanese radish, with special reference to the utilization of this sterility towerds the practical raising of hybrid seeds. *Mém. Fac. Agric. Kagoshima Univ.* **1967**, *6*, 1446–1459.

19.   Rawat, D.S.; Anand, I.J. Male sterility in indian mustard. *Indian J. Genet. Plant Breed.* **1979**, *39*, 412–414.

20.   Kemble, R.J.; Barsby, T.L. Use of protoplast fusion systems to study organelle genetics in a commercially important crop. *Biochem. Cell Biol.* **1988**, *66*, 665–676. [CrossRef]

21.   Burns, D.R.; Scarth, R.; McVetty, P.B.E. Temperature and genotypic effects on the expression of *pol* cytoplasmic male sterility in summer rape. *Can. J. Plant Sci.* **1991**, *71*, 655–661. [CrossRef]

22.   Zhang, J.K.; Zong, X.F.; Yu, G.D.; Li, J.N.; Zhang, W. Relationship between phytohormones and male sterility in thermo-photo-sensitive genic male sterile (TGMS) wheat. *Euphytica* **2006**, *150*, 241–248. [CrossRef]

23.   Zhou, Y.M.; Fu, T.D. Genetic improvement of rapeseed in China. In Proceedings of the 12th International Rapeseed Congress, Wuhan, China, 26–30 March 2007.

24.   Sawhney, V.K.; Shukla, A. Male sterility in flowering plants: Are plant growth substances involved? *Am. J. Bot.* **1994**, *81*, 1640–1647. [CrossRef]

25.   Tian, C.G.; Zhang, M.Y.; Duan, J. Preliminary study on the changes of phytohormones at different development stage in cytoplasmic male sterility line and its maintainer of rape. *Sci. Agric. Sin.* **1998**, *31*, 20–25.

26.   Wang, H.Z.; Wu, Z.D.; Han, Y. Relationships between endogenous hormone contents and cytoplasmic male sterility in sugarbeet. *Sci. Agric. Sin.* **2008**, *41*, 1134–1141.

27.   Wu, Z.M.; Hu, K.L.; Fu, J.Q.; Qiao, A.M. Relationships between cytoplasmic male sterility and endogenous hormone content of pepper bud. *J. South China Agric. Univ.* **2010**, *31*, 1–4.

28.   Dubas, E.; Janowiak, F.; Krzewska, M.; Hura, T.; Żur, I. Endogenous aba concentration and cytoplasmic membrane fluidity in microspores of oilseed rape (*Brassica napus* L.) genotypes differing in responsiveness to androgenesis induction. *Plant Cell Rep.* **2013**, *32*, 1465–1475. [CrossRef] [PubMed]

29. Shukla, A.; Sawhney, V.K. Abscisic acid: One of the factors affecting male sterility in *Brassica napus*. *Physiol. Plant.* **2010**, *91*, 522–528. [CrossRef]

30. Guan, T.; Dang, Z.; Zhang, J. Studies on the changes of phytohormones during bud development stage in thermo-sensitivity genic male-sterile flax. *Chin. J. Oil Crop Sci.* **2007**, *3*, 248–253.

31. Huang, S.; Zhou, X. Relationship between rice cytoplasmic male sterility and contents of GA (1+4) and IAA. *Acta Agric. Bor-Sin.* **1994**, *9*, 16–20.

32. Wang, Z.; Gerstein, M.; Snyder, M. RNA-seq: A revolutionary tool for transcriptomics. *Nat. Rev. Genet.* **2009**, *10*, 57–63. [CrossRef] [PubMed]

33. Grabherr, M.G.; Haas, B.J.; Yassour, M.; Levin, J.Z.; Thompson, D.A.; Amit, I.; Adiconis, X.; Fan, L.; Raychowdhury, R.; Zeng, Q.; et al. Full-length transcriptome assembly from RNA-Seq data without a reference genome. *Nat. Biotechnol.* **2011**, *29*, 644–652. [CrossRef] [PubMed]

34. Qi, Y.X.; Liu, Y.B.; Rong, W.H. RNA-Seq and its applications: A new technology for transcriptomics. *Hereditas* **2011**, *33*, 1191. [CrossRef] [PubMed]

35. Torti, S.; Fornara, F.; Vincent, C.; Andrés, F.; Nordström, K.; Göbel, U.; Knoll, D.; Schoof, H.; Coupland, G. Analysis of the *Arabidopsis* shoot meristem transcriptome during floral transition identifies distinct regulatory patterns and a leucine-rich repeat protein that promotes flowering. *Plant Cell.* **2012**, *24*, 444–462. [CrossRef] [PubMed]

36. An, H.; Yang, Z.; Yi, B.; Wen, J.; Shen, J.; Tu, J.; Ma, C.; Fu, T. Comparative transcript profiling of the fertile and sterile flower buds of *pol CMS* in *B. napus*. *BMC Genom.* **2014**, *15*, 258. [CrossRef] [PubMed]

37. Jeong, H.J.; Kang, J.H.; Zhao, M.; Kwon, J.K.; Choi, H.S.; Bae, J.H.; Lee, H.A.; Joung, Y.H.; Choi, D.; Kang, B.C. Tomato male sterile 1035 is essential for pollen development and meiosis in anthers. *J. Exp. Bot.* **2014**, *65*, 6693–6709. [CrossRef] [PubMed]

38. Yan, J.; Zhang, H.; Zheng, Y.; Ding, Y. Comparative expression profiling of mirnas between the cytoplasmic male sterile line meixianga and its maintainer line meixiangb during rice anther development. *Planta* **2015**, *241*, 109–123. [CrossRef] [PubMed]

39. Qu, C.; Fu, F.; Liu, M.; Zhao, H.; Liu, C.; Li, J.; Tang, Z.; Xu, X.; Qiu, X.; Wang, R.; et al. Comparative transcriptome analysis of recessive male sterility (RGMS) in sterile and fertile *Brassica napus* lines. *PLoS ONE* **2015**, *10*, e0144118. [CrossRef]

40. Du, K.; Liu, Q.; Wu, X.Y.; Jiang, J.J.; Wu, J.; Fang, Y.J.; Li, A.M.; Wang, Y. Morphological structure and transcriptome comparison of the cytoplasmic male sterility line in *Brassica napus* (SaNa-1A) derived from somatic hybridization and its maintainer line SaNa-1B. *Front. Plant Sci.* **2016**, *7*, 1313. [CrossRef]

41. Leino, M.; Teixeira, R.; Landgren, M.; Glimelius, K. Brassica napus lines with rearranged arabidopsis mitochondria display *CMS* and a range of developmental aberrations. *Theor. Appl. Genet.* **2003**, *106*, 1156–1163. [CrossRef]

42. Mackenzie, S. Male sterility and hybrid seed production. *Plant Biotechnol. Agric.* **2012**, 185–194.

43. Datta, R.; Chamusco, K.C.; Chourey, P.S. Starch biosynthesis during pollen maturation is associated with altered patterns of gene expression in maize. *Plant Physiol.* **2002**, *130*, 1645–1656. [CrossRef] [PubMed]

44. González-Melendi, P.; Uyttewaal, M.; Morcillo, C.N.; Hernández Mora, J.R.; Fajardo, S.; Budar, F.; Lucas, M.M. A light and electron microscopy analysis of the events leading to male sterility in Ogu-INRA *CMS* of rapeseed (*Brassica napus*). *J. Exp. Bot.* **2008**, *59*, 827–838. [CrossRef] [PubMed]

45. Yang, J.H.; Huai, Y.; Zhang, M.F. Mitochondrial atpa gene is altered in a new orf220-type cytoplasmic male-sterile line of stem mustard (*Brassica juncea*). *Mol. Biol. Rep.* **2009**, *36*, 273–280. [CrossRef] [PubMed]

46. Luo, X.D.; Dai, L.F.; Wang, S.B.; Wolukau, J.; Jahn, M.; Chen, J.F. Male gamete development and early tapetal degeneration in cytoplasmic male sterile pepper investigated by meiotic, anatomical and ultrastructural analyses. *Plant Breed.* **2006**, *125*, 395–399. [CrossRef]

47. Li, N.; Zhang, D.S.; Liu, H.S.; Yin, C.S.; Li, X.X.; Liang, W.Q.; Yuan, Z.; Xu, B.; Chu, H.W.; Wang, J. The rice tapetum degeneration retardation gene is required for tapetum degradation and anther development. *Plant Cell* **2006**, *18*, 2999–3014. [CrossRef]

48. Minocha, S.C. The role of auxin and abscisic acid in the induction of cell division in jerusalem artichoke tuber tissue cultured in vitro. *Z. Pflanzenphysiol.* **1979**, *92*, 431–441. [CrossRef]

49. Rook, F.; Hadingham, S.A.; Li, Y.; Bevan, M.W. Sugar and aba response pathways and the control of gene expression. *Plant Cell Environ.* **2006**, *29*, 426–434. [CrossRef]

50. Ji, X.; Dong, B.; Shiran, B.; Talbot, M.J.; Edlington, J.E.; Hughes, T.; White, R.G.; Gubler, F.; Dolferus, R. Control of abscisic acid catabolism and abscisic acid homeostasis is important for reproductive stage stress tolerance in cereals. *Plant Physiol.* **2011**, *156*, 647–662. [CrossRef]

51. De Storme, N.; Geelen, D. The impact of environmental stress on male reproductive development in plants: Biological processes and molecular mechanisms. *Plant Cell Environ.* **2014**, *37*, 1–18.

52. Duca, M. Genetic-phytohormonal interactions in male fertility and male sterility phenotype expression in sunflower (*Helianthus annuus* L.)/interacciones genético-fitohormonales en la expresión fenotípica de la androfertilidad y androesterilidad en girasol (helianthus annuus l.)/interactions génétiques phytohormonales dans l'expression phénotypique d'un male fertile et male sterile du tournesol (helianthus annuus L.). *Helia* **2008**, *31*, 27–38.

53. Singh, S.; Sawhney, V. Abscisic acid in a male sterile tomato mutant and its regulation by low temperature. *J. Exp. Bot.* **1998**, *49*, 199–203. [CrossRef]

54. Yang, J.H.; Zhang, M.F.; Yu, J.Q. Mitochondrial *nad2* gene is co-transcripted with CMS-associated *orfB* gene in cytoplasmic male-sterile stem mustard (*Brassica juncea*). *Mol. Biol. Rep.* **2009**, *36*, 345–351. [CrossRef] [PubMed]

55. Heng, S.; Liu, S.; Xia, C.; Tang, H.; Xie, F.; Fu, T.; Wan, Z. Morphological and genetic characterization of a new cytoplasmic male sterility system (oxa CMS) in stem mustard (*Brassica juncea*). *Theor. Appl. Genet.* **2018**, *131*, 59–66. [CrossRef] [PubMed]

56. Xing, M.; Sun, C.; Li, H.; Hu, S.; Lei, L.; Kang, J. Integrated analysis of transcriptome and proteome changes related to the Ogura cytoplasmic male sterility in cabbage. *PLoS ONE* **2018**, *13*, e0193462. [CrossRef] [PubMed]

57. Li, Y.; Ding, X.; Wang, X.; He, T.; Zhang, H.; Yang, L.; Wang, T.; Chen, L.; Gai, J.; Yang, S. Genome-wide comparative analysis of DNA methylation between soybean cytoplasmic male-sterile line NJCMS5A and its maintainer NJCMS5B. *BMC Genom.* **2017**, *18*, 596. [CrossRef] [PubMed]

58. Wang, H.; Cheng, H.; Wang, W.; Liu, J.; Hao, M.; Mei, D.; Zhou, R.; Fu, L.; Hu, Q. Identification of *BnaYUCCA6* as a candidate gene for branch angle in *Brassica napus* by QTL-seq. *Sci. Rep.* **2016**, *6*, 38493. [CrossRef] [PubMed]

59. Chalhoub, B.; Denoeud, F.; Liu, S.; Parkin, I.A.; Tang, H.; Wang, X.; Chiquet, J.; Belcram, H.; Tong, C.; Samans, B.; et al. Early allopolyploid evolution in the post-Neolithic Brassica napus oilseed genome. *Science* **2014**, *345*, 950–953. [CrossRef] [PubMed]

60. Cheng, H.; Hao, M.; Wang, W.; Mei, D.; Wells, R.; Liu, J.; Wang, H.; Sang, S.; Tang, M.; Zhou, R.; et al. Integrative RNA- and miRNA-Profile Analysis Reveals a Likely Role of BR and Auxin Signaling in Branch Angle Regulation of *B. napus*. *Int. J. Mol. Sci.* **2017**, *18*, 887. [CrossRef] [PubMed]

12

Constitutive Expression of *Aechmea fasciata SPL14* (*AfSPL14*) Accelerates Flowering and Changes the Plant Architecture in *Arabidopsis*

Ming Lei [1,2,3,4], **Zhi-ying Li** [1,2,3,4], **Jia-bin Wang** [1,2,3,4], **Yun-liu Fu** [1,2,3,4], **Meng-fei Ao** [1,2,3,4] **and Li Xu** [1,2,3,4,*]

1   Institute of Tropical Crops Genetic Resources, Chinese Academy of Tropical Agricultural Sciences, Danzhou 571737, China; leiming_catas@126.com (M.L.); xllizhiying@vip.163.com (Z.-y.L.); jiabinwangfuhu@sina.com (J.-b.W.); fyljj_2007@126.com (Y.-l.F.); 17889981612@163.com (M.-f.A.)
2   Key Laboratory of Crop Gene Resources and Germplasm Enhancement in Southern China, Danzhou 571737, China
3   Key Laboratory of Tropical Crops Germplasm Resources Genetic Improvement and Innovation, Danzhou 571737, China
4   Mid Tropical Crop Gene Bank of National Crop Resources, Danzhou 571737, China
*   Correspondence: xllzy@263.net

Abstract: Variations in flowering time and plant architecture have a crucial impact on crop biomass and yield, as well as the aesthetic value of ornamental plants. *Aechmea fasciata*, a member of the Bromeliaceae family, is a bromeliad variety that is commonly cultivated worldwide. Here, we report the characterization of *AfSPL14*, a squamosa promoter binding protein-like gene in *A. fasciata*. *AfSPL14* was predominantly expressed in the young vegetative organs of adult plants. The expression of *AfSPL14* could be upregulated within 1 h by exogenous ethephon treatment. The constitutive expression of *AfSPL14* in *Arabidopsis thaliana* caused early flowering and variations in plant architecture, including smaller rosette leaves and thicker and increased numbers of main inflorescences. Our findings suggest that AfSPL14 may help facilitate the molecular breeding of *A. fasciata*, other ornamental and edible bromeliads (e.g., pineapple), and even cereal crops.

Keywords: *Aechmea fasciata*; squamosa promoter binding protein-like; flowering time; plant architecture; bromeliad

1. Introduction

The squamosa promoter binding protein (SBP)-like (SPL) proteins are plant-specific transcription factors (TFs) that play essential roles in the regulation networks of plant growth and development [1]. The genes encoding SPL proteins were first identified in snapdragon (*Antirrhinum majus*), and were then found in almost all other green plants [2–5]. All SPL proteins contain a highly conserved DNA-binding domain termed the SBP domain, which consists of approximately 76 amino acid residues and features two zinc-binding sites and a bipartite nuclear localization signal (NLS) [6]. Many studies of various species have revealed the diverse functions of SPLs, which are involved in a broad range of important biological processes including the leaf development [7–10], embryonic development [11], fertility controlling [12,13], copper homeostasis [14,15], as well as the biosynthesis of phenylpropanoids and sesquiterpene. In addition to affecting these developmental aspects, several SPL factors which can be regulated by miR156, an evolutionary highly conserved microRNA (miRNA), also play crucial roles in the control of flowering time. The overexpression of *AtSPL3, AtSPL4, AtSPL5, AtSPL9, AtSPL15,* and *OsSPL16* can significantly promote flowering [3,9,16–19]. AtSPL9, together with AtSPL3 and the AtSPL2/10/11 group promote the floral meristem identity by directly regulating the same or different target genes [9,19]. Interestingly, compared with the positive regulation of accelerated flowering by the SPLs described above, AtSPL14 appears to be a negative regulator of vegetative-phase changes and floral transitions [20]. Another function of miR156-regulated SPL factors is in plant architecture formation and yield. *Teosinte Glume Architecture* (*TGA1*), an *SPL* gene, is responsible for the liberation of the kernel during domestication and evolution in *Z. mays* [21]. In *Triticum aestivum*, TaSPL3/17 play important roles in reducing the number of tillers and the outgrowth rate of axillary buds [22]. Two SPL homologs, TaSPL20 and TaSPL21, together reduce plant height and increase the thousand-grain weight [23]. In switchgrass (*Panicum virgatum*), miR156-regulated SPL4 suppresses the formation of both aerial and basal buds and controls the shoot architecture [24]. In *O. sativa*, higher expression of *OsSPL14* can reduce the tiller number, increase the lodging resistance, promote panicle branching and enhance grain yields [25,26]. The multifaceted functions of SPLs demonstrate complex and interesting regulation networks underlying plant lifestyles.

Bromeliaceae is one of the most morphologically diverse families and is widely distributed in tropical and subtropical areas [27]. Although certain cultivated species of bromeliads are appreciated for their edible fruits (e.g., pineapple: *Ananas comosus*) or medicinal properties (e.g., *Bromelia antiacantha*), the vast majority are appreciated for their ornamental value [28]. However, the unsynchronized natural flowering time of cultivated bromeliads always results in increased cultivation and harvesting costs and decreased economic value of fruits and ornamental flowers [29]. To date, several efforts were made to uncover the mechanism of flowering of bromeliads induced by age, photoperiod, autonomous and exogenous ethylene, or ethephon [29–33], but the precise molecular mechanism remained unknown.

Here, we characterized the *SPL* gene *AfSPL14*, from *Aechmea fasciata*, a popular ornamental flowering bromeliad. Phylogenetic analyses showed that AfSPL14 is closely related to OsSPL17, OsSPL14, AtSPL9, and AtSPL15. Furthermore, the expression of *AfSPL14* transcripts responded to plant age and exogenous ethephon treatment. The constitutive expression of *AfSPL14* in *Arabidopsis* promotes branching and accelerates flowering under long-day (LD) conditions. These results suggested that AfSPL14, a TF of the SPL family, might be involved in the process of flowering and in plant architecture variations of *A. fasciata*.

2. Results

2.1. Isolation and Sequence Analysis of AfSPL14 in A. fasciata

The *SPL* cDNA was isolated using the rapid amplification of cDNA ends (RACE) technique, and then named *AfSPL14*. The cDNA of *AfSPL14* was 1504-bp long and presented a 123-bp 5′ untranslated region (UTR), a 331-bp 3′ UTR, and a 1050-bp open reading frame (ORF), which was predicted to encode a 349-amino acid protein with a molecular weight (MW) and an isoelectric point (pI) of 37.53 kDa and 9.08, respectively.

To investigate the evolutionary relationships between the AfSPL14 and SPL proteins of other species, a phylogenetic tree was constructed using the neighbor-joining method with 1000 bootstrap replicates with 13 SBPs of *Physcomitrella patens*, 16 SPLs of *Arabidopsis*, and 19 SPLs of *O. sativa* (Table S1). Because the alignment of the full-length protein sequences showed no consensus sequences except for SBP domains (data not shown), only the highly-conserved SBP domains were used for the phylogenetic analysis. The unrooted phylogenetic tree classified all SBP domains into seven groups (I-VII), and AfSPL14 was clustered into group III, with AtSPL9, AtSPL15, OsSPL7, OsSPL14, and OsSPL17; however, this group did not contain SBP domains of *P. patens* (Figure 1), which was similar to the results obtained by others [34]. The fact that AfSPL14 has been classified with three SPL genes in *O. sativa* and two in *Arabidopsis* suggests that the SBP domains of these six SPLs might have undergone species-specific evolutionary processes after speciation.

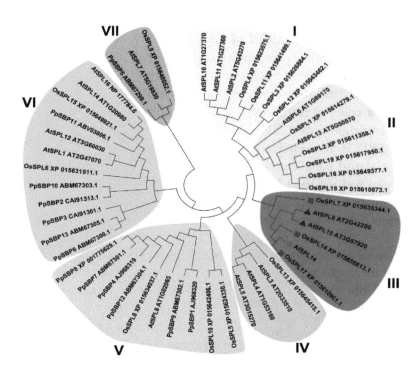

Figure 1. Phylogenetic analysis of AfSPL14 and SPLs of *Arabidopsis*, *O. sativa* and *P. patens* based on the conserved SBP domains. The unrooted tree was created using the neighbor-joining method with 1000 bootstrap replicates with 13 SBPs of *P. patens*, 16 SPLs of *Arabidopsis*, and 19 SPLs of *O. sativa*. The sequences of all these SBP domains are listed in Table S1.

The multiple sequence alignment of the AfSPL14 protein with SPL homologs of other species indicated that the SBP domain was highly conserved among the species (Figure 2a). All SBP domains could be divided into four motifs, which were zinc finger-like structure 1 (Zn1), Zn2, joint peptide (Jp) of Zn1 and Zn2, and NLS. The first zinc finger was C4H, and the second zinc finger-like structure was C2HC. The Jp plays a crucial role in modifying the protein-DNA interaction process [6], and was also highly conserved in all aligned sequences (Figure 2a). In addition, the bipartite NLS motif, which partially overlapped Zn2, was highly conserved (Figure 2a,b).

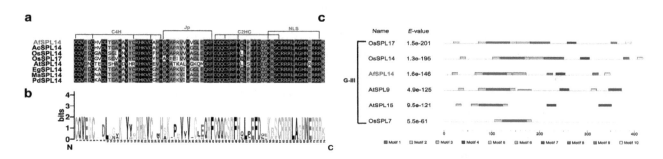

Figure 2. Sequence alignment and logo view of variable SBP domains, and putative motifs of variable SPLs in group III. (**a**) Multiple alignment of the SBP domains using DNAMAN software. The four conserved motifs, which include two zinc finger-like structures (C4H, C2HC), Jp and NLS, are indicated; (**b**) Sequence logo view of the consensus SBP domains. The overall height of the stack and the height of each letter represent the sequence conservation at that position and the relative frequency of the corresponding amino acid at that position, respectively; (**c**) putative motifs of variable SPLs in group III identified by MEME software online (http://meme-suite.org/tools/meme). The color boxes represent different putative motifs for which the sequences are listed in Table S2 online. G-III indicates group III from Figure 1.

To further examine conserved sequences other than the SBP domain, the online Multiple EM for Motif Elicitation (MEME) tool was used to identify putative motifs in the SPL proteins in group III [35]. As shown in Figure 2c, all members of group III contained motifs 1, 2, and 5, and these motifs had similar distributions. Actually, motif 2 belongs to C4H, and motif 5 belongs to NLS; motif 1 contains Jp, C2HC, and partials of C4H and NLS (Figure 2c, Table S2). Compared with OsSPL14 and OsSPL17, which contained all motifs except motif 7, AfSPL14 had a similar motif distribution but lacked motifs 7 and 8 (Figure 2c). These results suggested that AfSPL14 might have conserved functions with OsSPL14 and OsSPL17.

Exon-intron organization of all members of group III genes were generated based on genome sequences and the corresponding CDSs (Figure 3). As shown in Figure 3, each member of these genes had two introns and three exons; thus, they shared a similar exon-intron composition. All members of rice and *Arabidopsis* in group III were targets of miR156 [9,36], and a putative miR156 target site was also observed in *AfSPL14* (Figure 3). The consistency of the motif investigation, exon-intron organization, and phylogenetic analysis indicates putative similarities in functional regions and sites among the genes in group III.

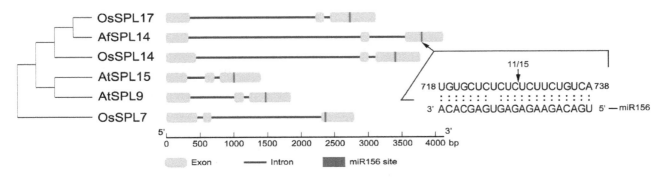

Figure 3. Exon-intron structures of *SPL* genes in group III from Figure 2 and AfmiR156 cleavage site in *AfSPL14* determined by 5′ RLM-RACE. For the determination of the AfmiR156 cleavage site in *AfSPL14*, 15 clones were selected randomly for sequencing, and 11 of them were cleaved in the position indicated by the arrow towards the base interval of *AfSPL14* RNA sequence.

2.2. AfSPL14 Was a Target of miR156 of A. fasciata (AfmiR156)

As all members of rice and *Arabidopsis* in group III were targets of miR156, there was also a putative miR156 target site in *AfSPL14* (Figure 3). To test whether the mRNA of *AfSPL14* was indeed targeted for degradation and was cleaved at the predicted position by AfmiR156, 5′ RNA ligase mediated rapid amplification of cDNA ends (RLM-RACE) was carried out to map the 5′ terminus of the cleavage fragment. DNA sequencing results of the amplified product demonstrated that *AfSPL14* could be indeed cleaved by AfmiR156 (Figure 3).

2.3. Transcript Profiling of AfSPL14 in A. fasciata

To gain insights into the role of AfSPL14 in *A. fasciata*, we determined the gene's expression profiles in various organs at different developmental stages via reverse transcription followed by quantitative real-time PCR (RT-qPCR). The transcripts of *AfSPL14* could be detected in almost all tested tissues except the roots of the adult plant prior to flower bud differentiation (Figure 4a). *AfSPL14* mRNA was more abundant in the central leaves and stems regardless of the developmental stage (Figure 4a,b). The accumulation of *AfSPL14* transcripts in the central leaves and stems showed significant changes during development, with the highest level observed in adult plants prior to flower bud differentiation, a relatively lower level observed in juvenile plants, and the lowest level observed in 39-day-after-flowering (DAF) adult plants (Figure 4a,b), suggesting that AfSPL14 might be involved in phase transitions.

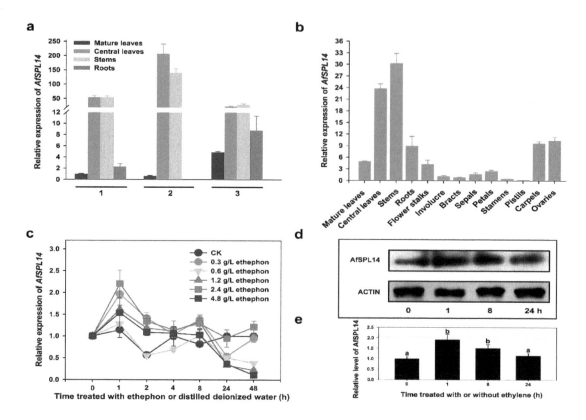

Figure 4. Expression of *AfSPL14* transcripts in various tissues of *A. fasciata* and immunoblot analysis of AfSPL14 in central leaves treated with or without ethephon. (**a**) Expression level of *AfSPL14* transcripts in various tissues of juvenile and adult plants. (1) juvenile plants; (2) adult plants prior to flower bud differentiation; (3) 39-DAF flowering adult plants. Samples were collected at 10:00 am. (**b**) Expression level of *AfSPL14* transcripts in the vegetative and reproductive organs of 39-DAF flowering adult plants. Samples were collected at 10:00 am. (**c**) Expression level of *AfSPL14* transcripts in the central leaves of *A. fasciata* in response to exogenous ethephon treatment at different concentrations for different time. In the panels, 0, 1, 2, 4, and 8 h represents the samples collected at 10:00, 11:00, 12:00, 14:00, and 18:00, respectively; 24 h and 48 h represent the treated samples collected at 10:00 am at the next day and the next two days, respectively. For CK, 10 mL of distilled deionized H_2O was poured into the cylinder shapes of *A. fasciata*. 0 h represents the samples treated without ethephon or distilled deionized H_2O. (**d**) Immunoblot analysis of the AfSPL14 protein level in the central leaves of *A. fasciata* treated with 10 mL of 0.6 g·L^{-1} exogenous ethephon for 1, 8, and 24 h, or without ethephon (0 h). The total proteins were separated using SDS-PAGE, and the transferred proteins were then probed with a rabbit polyclonal AfSPL14 antibody or a rabbit polyclonal Actin antibody, respectively. (**e**) Relative level of AfSPL14 protein in the central leaves of *A. fasciata* treated with 10 mL of 0.6 g·L^{-1} exogenous ethephon for 1, 8 and 24 h, or without ethephon (0 h). Three independent experiments were performed, the values are shown as the means and error bars indicate the standard deviation ($n = 3$). ANOVA was conducted, and means were separated by DNMRT.

2.4. Response to Exogenous Ethephon Treatment

To induce bromeliad flowering, ethylene or ethephon is widely used [29]. In fact, flowering induction by ethylene or ethephon is age-dependent. Plants of *A. comosus* which were somewhat less than about 1.0 kg fresh weight in subtropical regions respond only minimally to ethylene or ethephon [29]. Similar to 'Smooth Cayenne' and other variations of *A. comosus*, adult plants (but not juveniles) of *A. fasciata* could be induced by ethephon. Our previous investigation also showed that above 96% of 12-month-old adult plants could be induced to flower by 10 mL of exogenous ethephon treatment at 0.6 g·L^{-1} within two weeks, but that none of 6-month-old juvenile plants flowered under the same condition [37]. Here, we investigated the possible response of *AfSPL14* in adult plants of *A. fasciata* to exogenous ethephon treatment at different concentrations. As shown in Figure 4c, the expression of *AfSPL14* transcripts in the central leaves of adult plants prior to flower bud differentiation increased transiently after treatment for 1 h. Interestingly, a rapid decrease of the expression level of *AfSPL14* transcripts was observed after treatment for 2 h, almost reaching lower levels than in control plants after 24 h (Figure 4c), suggesting the remarkable effect of ethylene on the expression of *AfSPL14*.

To investigate the effect of ethylene on the level of AfSPL14 protein, we extracted the total proteins from the central leaves of *A. fasciata* treated with or without 10 mL of 0.6 g·L^{-1} ethephon. An immunoblot analysis was performed using a specific antibody against the AfSPL14 protein (Figure 4d). The level of AfSPL14 after treatment for 1 h was ~200% higher than the level in the untreated central leaves (Figure 4d,e). Consistent with the changes in the relative expression of *AfSPL14* mRNA, the level of AfSPL14 also gradually decreased after continuous treatment for 8 and 24 h (Figure 4d,e).

2.5. AfSPL14 Does Not Exhibit Transactivation Activity in Yeast

To test whether AfSPL14 is a transcriptional activator, the ORF (1-349 amino acids), the N terminus containing the SBP domain (1–141 amino acids) (AfSPL14N), and the C terminus (142–349 amino acids) (AfSPL14C) of AfSPL14 were fused with the GAL4 binding domain carried by the pGBKT7 (pBD) vector, respectively. The expression vectors pBD-AfSPL14, pBD-AfSPL14N, and pBD-AfSPL14C were then transformed into the yeast strain Y2HGold carrying the dual reporter genes AUR1-C and MEL1, respectively. As shown in Figure 5, similar to the negative control pBD, but not the positive control pGAL4, all the yeast cells carrying the three tested vectors could not grow on a medium containing SD/−Trp/+AbA/+X-α-Gal, indicating that AfSPL14 could not activate the transcription of the dual reporter genes in yeast.

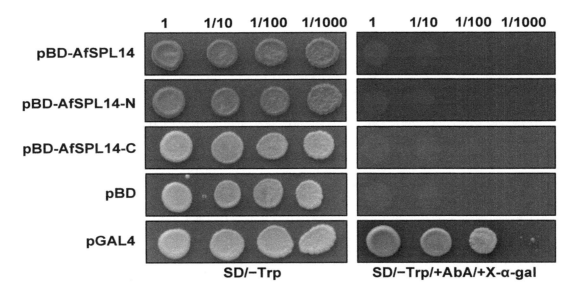

Figure 5. Transactivation activity assay of AfSPL14 in yeast cells. The pGBKT7 (pBD) vectors were fused with the full-length of AfSPL14 (pBD-AfSPL14), the N terminus of AfSPL14 (pBD-AfSPL14-N) and the C terminus of AfSPL14 (pBD-AfSPL14-C), respectively. Each kind of these constructs was then transformed into Y2HGold cells which contained the reporter genes *AUR1-C* and *MEL1*. pBD and pGAL4 plasmids were transformed into Y2HGold cells and used as negative and positive controls, respectively. Yeast clones containing the right constructs grew on SD/−Trp medium at dilutions of 1, 1/10, 1, 100, and 1/1000 for three to five days, and were then transferred onto SD/−Trp/+AbA/+X-α-Gal medium for continuous growth for three further days to test their transactivation activities. SD: synthetic dropout; AbA: Aureobasidin A; SD/−Trp: SD medium without Trp; SD/−Trp/+AbA/+X-α-gal: SD medium without Trp, but with 40 mg/L X-α-gal and 200 µg/L AbA.

2.6. Constitutive Expression of AfSPL14 in Arabidopsis

To assess the function of AfSPL14 in flowering, we induced the ectopic expression of *AfSPL14* with the 35S CaMV promoter (*Pro35S::AfSPL14*) in *Arabidopsis* ecotype Columbia (Col-0) (WT) (Figure S1). Under LD conditions, the flowering time of *Pro35S::AfSPL14* transgenic plants was significantly earlier ($p = 6.26 \times 10^{-8}$) than that of the WT and the WT transformed with the empty vector (Vector) (Figure 6a–c). Although the difference of the number of rosette leaves was minor between the *Pro35S::AfSPL14* transgenic plants and WT, the statistical analysis indicated that the number was significantly lower ($p = 4.69 \times 10^{-5}$) in the *Pro35S::AfSPL14* transgenic plants (Figure 6c). The *Pro35S::AfSPL14* transgenic lines were also smaller than the WT (Figure 6a–c). In addition, the *Pro35S::AfSPL14* transgenic plants showed morphological changes in the reproductive phase. A comparison between the WT and Vector, which only has one main inflorescence per plant, showed that a majority of the transformants of *Pro35S::AfSPL14* developed two main inflorescences (Figure 6b,d). Interestingly, a second inflorescence could be developed from the base of the main inflorescence, and it could also develop from the node of the main inflorescence or even be divided randomly from the non-node position of the main inflorescence (Figure 6b,d). Another change in the reproductive phase was the thickening of the main inflorescence in the *Pro35S::AfSPL14* transgenic plants compared with that of the WT (Figure 6e,f).

To further confirm whether the expression of *AfSPL14* in the *Pro35S::AfSPL14* transgenic plants altered the expression of downstream flowering genes, RT-qPCR analysis was performed with the *Arabidopsis* shoot apices grown under LD conditions as materials. The *Arabidopsis* shoot apices were harvested from the central parts of *Arabidopsis* seedlings, and contain the youngest rosette leaves. As expected, compared to WT, the expression level of the genes *SUPPRESSOR OF OVEREXPRESSION OF CONSTANS 1* (*SOC1*), *FRUITFULL* (*FUL*) and *APETALA1* (*AP1*), which encode floral inductive factors, was substantially upregulated at the shoot apex of *Pro35S::AfSPL14* transgenic plants (Figure 7a,b,e). However, the expression level of another gene encoding plant-specific transcription factor LEAFY (LFY), which is also a positive regulator inducing flowering at the shoot apex, showed no clear difference between WT and *Pro35S::AfSPL14* transgenic plants (Figure 7c). The expression level of *Flowering Locus T* (*FT*), an integrator of flowering pathways and defined as a florigen, was also considerably upregulated at the shoot apex of *Pro35S::AfSPL14* transgenic plants (Figure 7d). In addition, the expression of floral organ identity genes, such as *AtAP2* and *AtAP3*, was also upregulated (Figure 7f,g).

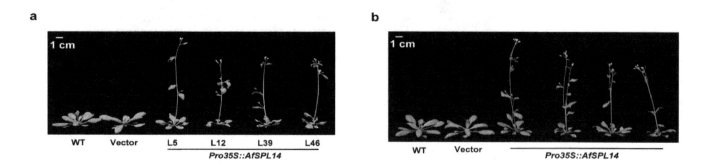

Figure 6. *Cont.*

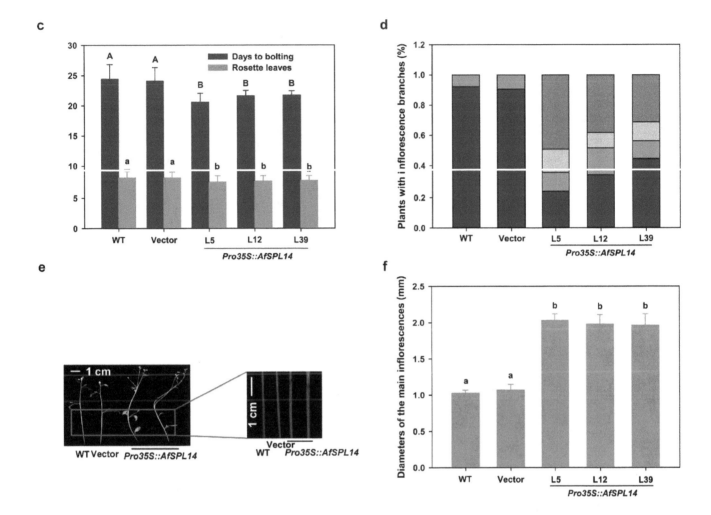

Figure 6. Phenotype analysis of *Pro35S::AfSPL14* transgenic plants. (**a**) Flowering *Pro35S::AfSPL14* transgenic plants shown next to WT and WT transformed with the empty vector (Vector) under LD conditions. L5, L12, L39, and L46 indicate the different lines. (**b**) Flowering *Pro35S::AfSPL14* transgenic plants that had two main inflorescences under LD conditions. (**c**) Days and number of rosette leaves to bolting of the WT, Vector and *Pro35S::AfSPL14* transgenic plants grown under LD conditions. Values are the means ± standard deviation. Seventy-nine plants were scored for each line. Difference letters indicate statistical differences. (**d**) Percentages of plants which have two main inflorescences grown under LD conditions. The number of plants with the second main stem developed from the base (orange), the node (ginger), and the non-node (dark green) position of the main inflorescences and plants with only one inflorescence (dark red) was calculated. One hundred and twenty-eight 38-day-old, long-day-grown plants and ninety-six 55-day-old, short-day-grown plants were counted for each line. (**e**) Bending and thicker main inflorescences of *Pro35S::AfSPL14* transgenic plants under LD conditions. (**f**) The diameter of main inflorescences of WT, Vector and transgenic plants. Forty-eight 38-day-old, long-day-grown plants were counted for each line. The diameter of the positions which were 5 cm distance from the basal of the main inflorescences was measured. ANOVA was conducted, and means were separated by DNMRT.

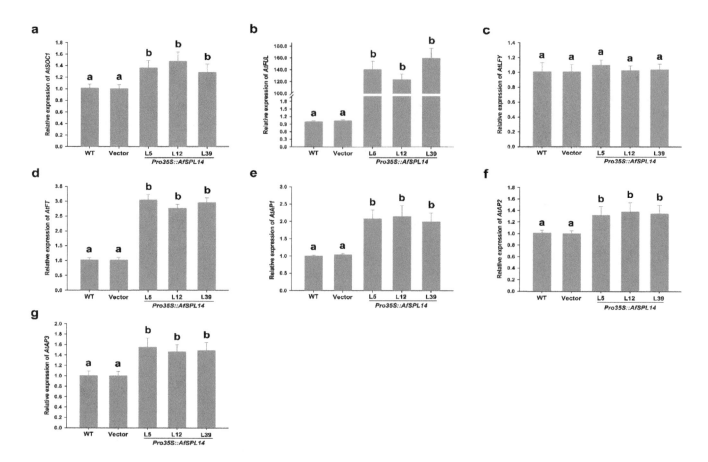

Figure 7. RT-qPCR analysis of flowering related genes at the shoot apex of WT, Vector and *Pro35S::AfSPL14* transgenic plants. Relative expression of three flowering promoting genes, *SUPPRESSOR OF OVEREXPRESSION OF CONSTANS 1 (SOC1)* (**a**), *FRUITFULL (FUL)* (**b**), *LEAFY* (**c**), and one florigen *Flowering Locus T (FT)* (**d**), and three flowering organ identify genes, *APETALA1 (AP1)* (**e**), *AP2* (**f**) and *AP3* (**g**) was performed. Fourteen-day-old long-day-grown seedlings were used. Three biological replicates and three technical replicates were performed. Transcript levels were normalized using *AtACTB* gene as a reference. All primers used here are listed in Table S3 online. ANOVA was conducted, and means were separated by DNMRT.

3. Discussion

SPL proteins are plant-specific TFs, and have been reported in many plants, including *Antirrhinum majus* [2], *Arabidopsis thaliana* [3], *Chlamydomonas* [14], *O. sativa* [36], *P. patens* [5], tomato [38], *Triticum aestivum* [39], *Castor Bean* [40], *Prunus mume* [41], *Citrus* [42], pepper [43], Petunia [44], *Brassica napus* [45] and *Chrysanthemum* [46]. In the present study, we identified the *SPL* gene *AfSPL14* in *A. fasciata*, an economically valuable, short-day ornamental plant that exhibits crassulacean acid metabolism (CAM). We discussed the correlation between this gene and the plant hormone ethylene, which has been widely used to induce flowering of members in numbers of the Bromeliaceae family. We also suggested that AfSPL14 was a putative flowering inducer and ideal plant architecture generator, based on its heterologous constitutive expression in *Arabidopsis*.

Compared with many other plant species in which the role of ethylene in the regulation of flowering appears complicated, in a majority of bromeliads including pineapple and *A. fasciata*, flowering can be triggered by a small burst of ethylene production in the meristem in response to exogenous ethylene or ethephon treatment [30,33]. In previous studies, several ethylene biosynthesis, signaling and responsive genes were identified and characterized [30–33,47]. Here, we found that exogenous ethephon induced the expression of *AfSPL14* transcripts rapidly and dramatically within

1 h (Figure 4c). Interestingly, the expression level of *AfSPL14* transcripts gradually declined after continuous treatment for 8 h (Figure 4c). In fact, several *SPLs* in some other species also could be transiently upregulated and then downregulated by ethylene, for example, *MdSBP20* and *MdSBP27* in the leaves of apple (*Malus* × *domestica* Borkh.) cv. 'Fuji'34, and *SPL7* and *SPL9* in the fruit of *Cavendish banana* [48]. A more precise identification of the changes in the translational level of AfSPL14 in response to the exogenous ethephon treatment showed a consistence with the changes at the transcriptional level. After treatment for 1 h, the expression of the AfSPL14 protein was also dramatically induced to a higher level compared with that in the untreated central leaves (Figure 4d,e). However, after treatment for 8 h and 24 h, the amount of AfSPL14 also decreased gradually (Figure 4d,e). Furthermore, three 5′-ATGTA-3′ core sequences were enclosed in the nearly 3000-bp-length promoter sequence of *AfSPL14* promoter (Figure S2). The 5′-ATGTA-3′ core sequence might interact with ethylene insensitive 3 (EIN3), a crucial factor in the ethylene signaling pathway that could activate or inhibit the expression of downstream genes at the transcriptional level. Further investigation should be performed regarding the regulation of *AfSPL14* by exogenous ethephon at the transcriptional and post-transcriptional levels.

Previous studies of the molecular regulation of the model species *Arabidopsis* have identified at least five genetic pathways relevant to flowering, namely: the photoperiod, vernalization, gibberellic acid (GA), and the autonomous and aging pathways [49]. During this process, at least 180 genes were involved [50]. *SPLs* are indispensable among these genes, and are involved in several signaling pathways. For example, AtSPL3 and AtASPL9 act independently of *FT*, and directly activate flower-promoting *MADS box* genes, thus defining a separate endogenous flowering pathway [18]. AtSPL9 could also acts upstream of *FT* and promotes *FT* expression [51,52]. AtSPL15 integrates the GA pathway and the aging pathway to promote flowering [53]. In addition, AtSPL3/4/5 link developmental aging and photoperiodic flowering [54]. Moreover, the enhancement of the miR156 site-mutated *OsSPL14* gene could also accelerate flowering [55]. Phylogenetic and motif analyses of AfSPL14 and the SPLs of *Arabidopsis* and OsSPL14 showed that the former was similar to AtASPL9, AtSPL15, and OsSPL14 (Figures 1–3), implying a putative conserved function, such as, flowering promotion. Recently, an age-dependent flowering pathway was identified by the regulation of CmNF-YB8, a nuclear factor, through directly triggering miR156-SPL-regulated processes in the short day plant chrysanthemum (*Chrysanthemum morifolium*) [56]. The flowering of pineapple and *A. fasciata* is also age dependent, and the juvenile plants cannot flower naturally, even when treated with exogenous ethylene [30,33]. The expression of *AfSPL14* transcripts was higher in the central leaves and stems of adult plants prior to flower bud differentiation compared with that of the juvenile plants (Figure 4a). This fact is similar to the increasing pattern of accumulation of *AtSPL9* and *AtSPL15* in the meristem with age [9,17,53], and inconsistent with the expression profile of *AfAP2-1*, a putative flowering TF encoding gene identified in *A. fasciata* [33]. These results suggest that AfSPL14 might act positively in the juvenile-to-vegetative phase transition and flowering pathway regulated by developmental age.

The constitutive expression of *AfSPL14* in *Arabidopsis* significantly promoted flowering under LD conditions (Figure 6a–c), which was inconsistent with the flowering-delayed phenotype caused by the constitutive expression of *AfAP2-1* in *Arabidopsis* [33], thus suggesting that AfSPL14 is an activator of flowering integrator and floral inductive genes such as *AtFT*, *AtAP1*, *AtSOC1*, and *AtFUL* (Figure 7a,b,d,e). A previous study demonstrated that the overexpression of AtSPL3 could strongly induce *AtFUL*, but has a weaker effect, or no effect at all, on *AtSOC1* in the shoot apex in *Arabidopsis* [18]. Interestingly, compared with that of WT, the expression level of *AtFUL* at the shoot apex of *Pro35S::AfSPL14* transgenic plants was upregulated dramatically, while *AtSOC1* was slightly induced (Figure 7a,b), suggesting that similar to AtSPL3, AfSPL14 might also induce flowering via an endogenous pathway.

Phylogenetic and motif analyses of AfSPL14 with variable SPLs suggested that it was closer and more similar to OsSPL14 than to AtSPL9 and AtSPL15 (Figures 2 and 3); this is consistent with the evolutionary distances among *A. fasciata*, rice and *Arabidopsis*. In addition, the genes have similar exon-intron structures (Figure 3). Higher expression of *OsSPL14* could reduce the tiller number, increase the lodging resistance, promote panicle branching, and enhance the grain yield [25,26]. Importantly, in addition to the acceleration of flowering, the constitutive expression of *AfSPL14* in *Arabidopsis* also promotes the number of main inflorescences and produces thicker and sturdier culms (Figure 6a–f). However, we did not find the transactivator activity of AfSPL14 in yeast cells (Figure 5), in opposition to the results reported on OsSPL14 [26]. These results suggested functional conservation and diversification in AfSPL14 and OsSPL14. Interestingly, the repression of *AtSPL10* caused reduced apical dominance, and increased the number of main inflorescences [10]. A loss-of-function mutation of *AtSPL9* and *AtSPL15* resulted in altered main stem architecture and enhanced branching [17]. The main inflorescence-changed phenotypes of constitutive expressed *AfSPL14* in *Arabidopsis* and loss-of-function *AtSPL9*, *AtSPL10,* and *AtSPL15* mutants appeared to be similar.

Certain SPLs can be regulated by miR156, two miRNAs that can regulate the expression of SPL proteins at the post-transcriptional level [57]. Similar to *OsSPL14*, *AfSPL14* also had a miR156 cleavage site in its CDS sequence (Figure 3). Because of a point mutation in the OsmiR156-directed site of *OsSPL14*, grain yield was enhanced [25,26]. Many SPLs positively regulate grain yield [23,58–60]. The thicker main inflorescence phenotype in *AfSPL14*-constitutive expressed *Arabidopsis* implied that this gene might act positively in the regulation of flower stalk diameter in *A. fasciata*. Further investigation should focus on the morphological changes of flowers in *AfSPL14*-overexpressed and/or *AfSPL14*-silenced *A. fasciata*, the morphological changes of flowers in *AfSPL14*-overexpressed and/or *AfSPL14*-silenced pineapple, and the cloning and functional characterization of possible homologs of *AfSPL14* in pineapple.

4. Materials and Methods

4.1. Plant Materials and Sample Preparation

The *A. fasciata* specimens used in this study were planted in a greenhouse (ambient temperature of 30–32 °C) located in the experimental area of the Institute of Tropical Crop Genetic Resources, Chinese Academy of Tropical Agricultural Sciences (CATAS). For the tissue-specific expression and western blot analyses, different tissue samples, including mature leaves, central leaves, stems, roots, and various flower organs, were collected.

The wild-type (WT) and transgenic plants of *Arabidopsis* used in this study were of the Columbia ecotype (Col-0). Seeds were surface sterilized in 0.1% $HgCl_2$ for 10 min and then washed with sterilized distilled water five times. The washed seeds were then plated on MS medium containing sugar (2%) and agar (0.8%) and incubated in the dark at 4 °C for 2 days. The plates were then moved to a chamber at 23 °C under LD (16 h light) conditions, with a photon flux density (120 $\mu mol\ m^{-2}\ s^{-1}$) for continuous growth.

4.2. Isolation and Sequencing of the AfSPL14 Gene

Total RNA was extracted from the central leaves of *A. fasciata* using the hexadecyl trimethyl ammonium bromide (CTAB) method [33], and then used for the RACE at the 5′ and 3′ ends according to the manufacturer's instructions for the SMARTer™ RACE cDNA Amplification Kit (Clontech, Tokyo, Japan). The specific 5′ and 3′ fragments were cloned into pEASY-blunt vectors (Transgen, Beijing, China), and then sequenced by Thermo Fisher Scientific (Guangzhou, China). The gene accession number of *AfSPL14* is MF114304. The primers used here are listed in Table S3 online.

4.3. Bioinformatic Analysis

The ORF of *AfSPL14* was predicted using the ORF Finder (https://www.ncbi.nlm.nih.gov/orffinder/). The sequence logo was generated by the online WebLogo 3 platform (http://weblogo.threeplusone.com/). A phylogenetic tree was constructed with MEGA version 6.0 using the neighbor-joining method with 1000 bootstrap replications [61]. The scheme of exon-intron structures was generated by Gene Structure Display Server 2.0 (http://gsds.cbi.pku.edu.cn/index.php). Putative motifs of variable SPLs were identified by MEME software online with default settings (http://meme-suite.org/tools/meme).

4.4. 5′ RLM-RACE

5′ RLM-RACE was performed according to the manufacturer's instructions of FirstChoice® RLM-RACE Kit (Thermo Fisher Scientific, New York, NY, USA). For the next amplification, 10 μg of total RNA, which was isolated from central leaves of 12-month-old *A. fasciata* plants using the CTAB method [33], was used. The gene specific primers of *AfSPL14* for the first and second PCR products are *AfSPL14*-5outer and *AfSPL14*-5inner, respectively. The second PCR products were gel purified and subcloned into pEASY-T3 Vector (Transgen, Beijing, China) for sequencing. Primers used for 5′ modified RACE are listed in Table S3 online.

4.5. RT-qPCR

First-strand cDNA was synthesized using the TransScript One-Step gDNA Removal and cDNA Synthesis SuperMix (Transgen, Beijing, China) according to the manufacturer's instructions. Quantitative real-time PCR (qPCR) was conducted on a Therma PikoReal 96™ Real-Time PCR System (Thermo Fisher Scientific, Waltham, MA, USA) using the TransStart Tip Green qPCR SuperMix Kit (Transgen, Beijing, China). The total reactions (20 μL) described in this protocol converted total RNA (500 ng ~ 5 μg) into the first-strand cDNA. The first-strand reaction products (20 μL) were diluted with sterilized distilled H_2O 5 times, and diluted products (1 μL) were used for total qPCR reactions (10 μL). Three biological replicates and three technical replicates were performed. The relative expression levels of specific genes were calculated using the $2^{-\Delta\Delta Ct}$ method with the *β-actin* gene (*ACTB*) of *A. fasciata* or *Arabidopsis* as the internal control [62]. All primers used for qPCR are listed in Table S3 online.

4.6. Transgenic Plants

For the transgenic constructs, the coding sequence (CDS) of *AfSPL14* was cloned into the KpnI-SalI sites of the binary vector Cam35S-gfp under the control of the cauliflower mosaic virus (CaMV) 35S promoter. The constructs were then delivered into *Agrobacterium tumefaciens* strain EHA105 by the freeze-thaw method [63]. Col-0 background *Arabidopsis* was transformed using the floral dipping method [64]. For the selection of transgenic plants, the seeds were planted on MS agar medium supplemented with hygromycin (25 mg/L). Seedlings conferring resistance to hygromycin were then transplanted in a chamber at 23 °C under LD conditions. Transgenic plants were verified by genomic PCR and RT-PCR using primers *AfSPL14-OX* F and *AfSPL14-OX* R, which were listed in Table S3 online. T3 transgenic plants were used for next experiments.

4.7. Transactivation Analysis of AfSPL14 in Yeast Cells

The yeast strain Y2HGold was transformed with plasmids containing the pGBKT7 (pBD) vector with the ORF or fragments of AfSPL14 fused in frame with GAL4 DNA binding domain. The primers used are listed in Table S3 online. pBD and pGAL4 were used as negative and positive controls, respectively. Transformants were selected on synthetic dropout (SD) medium lacking tryptophan (SD/−Trp) (Clontech, Tokyo, Japan) and then dripped onto SD/−Trp/+AbA/+X-α-gal to determine the transactivation activity.

4.8. Exogenous Ethephon Treatment of A. fasciata

To test the response to ethylene, adult (12-month-old) *A. fasciata* plants which were grown in pots in our greenhouse were treated with ethephon (10 mL) at 0.3 g·L^{-1}, 0.6 g·L^{-1}, 1.2 g·L^{-1}, 2.4 g·L^{-1}, 4.8 g·L^{-1} for 1, 2, 4, 8, 24, or 48 h. All treatments were applied by pouring the specific concentration of ethephon solution into the leaf whorl of each plant, with the same quantity of water as control. The central leaves were then physically isolated and immediately frozen in liquid nitrogen for further research.

4.9. SDS-PAGE and Immunoblot Analysis

Total proteins were extracted from the central leaves of adult *A. fasciata* plants. The physically isolated and immediately frozen central leaves (0.5 g) were homogenized with extraction buffer (1 mL) (Tris (1 mol/L, pH 6.8); DL-dithiothreitol (0.2 mol/L); sodium dodecyl sulfate (4% (g/mL)); glycerol (20%)) by using pestle and mortar. After centrifugation at 12,000 rotation per minute (rpm) for 15 min at 4 °C, the supernatants were transferred into new tubes and 4 times volume of acetones were added. After being vortexed for 2 min and then placed on ice for 1 h, the mixture was centrifuged again, as above. The supernatants were discarded and 4 times volume of acetone was added. The mixture was then vortexed and centrifuged again; the extracted proteins were diluted with 0.5× extraction buffer, boiled at 100 °C for 10 min, and then centrifuged at 13,000 rpm for 5 min. The supernatants (total central leaf proteins) were separated using 15% sodium dodecyl sulfate polyacrylamide gel electrophoresis (SDS-PAGE) containing urea (6 mol/L). After electrophoresis, the proteins were transferred onto nitrocellulose membranes (Amersham Biosciences, Pittsburgh, PA, USA) and probed with a rabbit polyclonal AfSPL14 antibody (Jiaxuan Biotech, Beijing, China) or a rabbit polyclonal Actin antibody (Agrisera, Vännäs, Sweden). After incubation with horseradish peroxidase conjugated goat anti-rabbit IgG (Jiaxuan Biotech, Beijing, China), the signals were detected by enhanced chemiluminescence (Jiaxuan Biotech, Beijing, China). X-ray films were scanned and analyzed using ImageMaster™ 2D Platinum software (GE Healthcare, Pittsburgh, PA, USA). Protein concentration of each extract was determined by using a protein assay kit (Bio-Rad, Hercules, CA, USA) with BSA as the standard.

4.10. Data Analysis

Values represent means ± standard deviation of two or three biological replicates. ANOVA was conducted, and the means were separated by Duncan's New Multiple Range Test (DNMRT).

Author Contributions: Conceptualization, L.X. and M.L.; Methodology, L.X.; Software, J.-b.W.; Validation, M.L., and L.X.; Formal Analysis, J.-b.W.; Investigation, M.L. and M.-f.A.; Resources, Z.-y.L., Y.-l.F. and L.X.; Data Curation, M.L., J.-b.W. and M.-f.A.; Writing-Original Draft Preparation, M.L.; Writing-Review & Editing, L.X.

Acknowledgments: This work was supported by grants from the National Natural Science Foundation of China (31372106, 31601793), the National Science Foundation of Hainan Province (20163127) and the Fundamental Scientific Research Funds for CATAS-TCGRI (1630032016006).

Abbreviations

SBP	SQUAMOSA PROMOTER BINDING PROTEIN
SPL	SBP-LIKE
miRNA	microRNA
TGA1	*TEOSINTE GLUME ARCHITECTURE*
LD	long day
RACE	Rapid amplification of cDNA ends
UTR	Untranslated region
ORF	Open reading frame
MW	Molecular weight
pI	Isoelectric point
Jp	Joint peptide
NLS	Nuclear localization signal

RT-qPCR	Reverse transcription followed by quantitative real-time PCR
RLM-RACE	RNA ligase mediated rapid amplification of cDNA ends
DAF	Day after flowering
WT	Wild Type
SOC1	*SUPPRESSOR OF OVEREXPRESSION OF CONSTANS 1*
FUL	*FRUITFULL*
AP1	*APETALA1*
LFY	*LEAFY*
FT	*Flowering Locus T*
AP2	*APETALA2*
AP3	*APETALA3*
CAM	Crassulacean acid metabolism
EIN3	ETHYLENE INSENSITIVE 3
GA	Gibberellic acid
CTAB	Hexadecyl trimethyl ammonium bromide
CDS	the coding sequence
SDS-PAGE	Sodium dodecyl sulfate polyacrylamide gel electrophoresis

References

1. Preston, J.C.; Hileman, L.C. Functional evolution in the plant *SQUAMOSA-PROMOTER BINDING PROTEIN-LIKE (SPL)* gene family. *Front. Plant Sci.* **2013**, *4*, 80. [CrossRef] [PubMed]

2. Klein, J.; Saedler, H.; Huijser, P. A new family of DNA binding proteins includes putative transcriptional regulators of the *Antirrhinum majus* floral meristem identity gene *SQUAMOSA. Mol. Gen. Genet.* **1996**, *250*, 7–16. [PubMed]

3. Cardon, G.H.; Hohmann, S.; Nettesheim, K.; Saedler, H.; Huijser, P. Functional analysis of the *Arabidopsis thaliana SBP-box* gene *SPL3*: A novel gene involved in the floral transition. *Plant J.* **1997**, *12*, 367–377. [CrossRef] [PubMed]

4. Arazi, T.; Talmor-Neiman, M.; Stav, R.; Riese, M.; Huijser, P.; Baulcombe, D.C. Cloning and characterization of micro-RNAs from moss. *Plant J.* **2005**, *43*, 837–848. [CrossRef] [PubMed]

5. Riese, M.; Höhmann, S.; Saedler, H.; Münster, T.; Huijser, P. Comparative analysis of the *SBP-box* gene families in *P. patens* and seed plants. *Gene* **2007**, *401*, 28–37. [CrossRef] [PubMed]

6. Yamasaki, K.; Kigawa, T.; Inoue, M.; Tateno, M.; Yamasaki, T.; Yabuki, T.; Aoki, M.; Seki, E.; Matsuda, T.; Nunokawa, E.; Ishizuka, Y.; et al. A novel zinc-binding motif revealed by solution structures of DNA-binding domains of *Arabidopsis* SBP-family transcription factors. *J. Mol. Biol.* **2004**, *337*, 49–63. [CrossRef] [PubMed]

7. Moreno, M.A.; Harper, L.C.; Krueger, R.W.; Dellaporta, S.L.; Freeling, M. *Ligulelessl* encodes a nuclear-localized protein required for induction of ligules and auricles during maize leaf organogenesis. *Genes Dev.* **1997**, *11*, 616–628. [CrossRef] [PubMed]

8. Lee, J.; Park, J.J.; Kim, S.L.; Yim, J.; An, G. Mutations in the rice *liguleless* gene result in a complete loss of the auricle, ligule, and laminar joint. *Plant Mol. Biol.* **2007**, *65*, 487–499. [CrossRef] [PubMed]

9. Wang, J.W.; Schwab, R.; Czech, B.; Mica, E.; Weigel, D. Dual effects of miR156-targeted *SPL* genes and *CYP78A5/KLUH* on plastochron length and organ size in *Arabidopsis thaliana. Plant Cell* **2008**, *20*, 1231–1243. [CrossRef] [PubMed]

10. Shikata, M.; Koyama, T.; Mitsuda, N.; Ohme-Takagi, M. *Arabidopsis SBP-box* genes *SPL10, SPL11* and *SPL2* control morphological change in association with shoot maturation in the reproductive phase. *Plant Cell Physiol.* **2009**, *50*, 2133–2145. [CrossRef] [PubMed]

11. Nodine, M.D.; Bartel, D.P. MicroRNAs prevent precocious gene expression and enable pattern formation during plant embryogenesis. *Genes Dev.* **2010**, *24*, 2678–2692. [CrossRef] [PubMed]

12. Xing, S.; Salinas, M.; Hohmann, S.; Berndtgen, R.; Huijser, P. miR156-targeted and nontargeted SBP-box transcription factors act in concert to secure male fertility in *Arabidopsis. Plant Cell* **2010**, *22*, 3935–3950. [CrossRef] [PubMed]

13. Xing, S.; Salinas, M.; Garcia-Molina, A.; Hohmann, S.; Berndtgen, R.; Huijser, P. *SPL8* and miR156-targeted *SPL* genes redundantly regulate *Arabidopsis* gynoecium differential patterning. *Plant J.* **2013**, *75*, 566–577. [CrossRef] [PubMed]

14. Kropat, J.; Tottey, S.; Birkenbihl, R.P.; Depege, N.; Huijser, P.; Merchant, S. A regulator of nutritional copper signaling in *Chlamydomonas* is an SBP domain protein that recognizes the GTAC core of copper response element. *Proc. Natl. Acad. Sci. USA* **2005**, *102*, 18730–18735. [CrossRef] [PubMed]

15. Yamasaki, H.; Hayashi, M.; Fukazawa, M.; Kobayashi, Y.; Shikanai, T. SQUAMOSA promoter binding protein-like7 is a central regulator for copper homeostasis in *Arabidopsis*. *Plant Cell* **2009**, *21*, 347–361. [CrossRef] [PubMed]

16. Wu, G.; Poethig, R.S. Temporal regulation of shoot development in *Arabidopsis thaliana* by miR156 and its target *SPL3*. *Development* **2006**, *133*, 3539–3547. [CrossRef] [PubMed]

17. Schwarz, S.; Grande, A.V.; Bujdoso, N.; Saedler, H.; Huijser, P. The microRNA regulated *SBP-box* genes *SPL9* and *SPL15* control shoot maturation in *Arabidopsis*. *Plant Mol. Biol.* **2008**, *67*, 183–195. [CrossRef] [PubMed]

18. Wang, J.W.; Czech, B.; Weigel, D. MiR156-regulated SPL transcription factors define an endogenous flowering pathway in *Arabidopsis thaliana*. *Cell* **2009**, *138*, 738–749. [CrossRef] [PubMed]

19. Yamaguchi, A.; Wu, M.F.; Yang, L.; Wu, G.; Poethig, R.S.; Wagner, D. The microRNA-regulated SBP-box transcription factor SPL3 is a direct upstream activator of *LEAFY*, *FRUITFULL*, and *APETALA1*. *Dev. Cell* **2009**, *17*, 268–278. [CrossRef] [PubMed]

20. Stone, J.M.; Liang, X.; Nekl, E.R.; Stiers, J.J. *Arabidopsis* AtSPL14, a plant-specific SBP-domain transcription factor, participates in plant development and sensitivity to fumonisin B1. *Plant J.* **2005**, *41*, 744–754. [CrossRef] [PubMed]

21. Wang, H.; Nussbaum-Wagler, T.; Li, B.; Zhao, Q.; Vigouroux, Y.; Faller, M.; Bomblies, K.; Lukens, L.; Doebley, J.F. The origin of the naked grains of maize. *Nature* **2005**, *436*, 714–719. [CrossRef] [PubMed]

22. Liu, J.; Cheng, X.; Liu, P.; Sun, J. miR156-regulated TaSPLs interact with TaD53 to regulate *TaTB1* and *TaBA1* expression in bread wheat. *Plant Physiol.* **2017**, *174*, 1931–1948. [CrossRef] [PubMed]

23. Zhang, B.; Xu, W.; Liu, X.; Mao, X.; Li, A.; Wang, J.; Chang, X.; Zhang, X.; Jing, R. Functional conservation and divergence among homoeologs of *TaSPL20* and *TaSPL21*, two *SBP-box* genes governing yield-related traits in hexaploid wheat. *Plant Physiol.* **2017**, *174*, 1177–1191. [CrossRef] [PubMed]

24. Gou, J.; Fu, C.; Liu, S.; Tang, C.; Debnath, S.; Flanagan, A.; Ge, Y.; Tang, Y.; Jiang, Q.; Larson, P.R.; Wen, J.; Wang, Z.Y. The miR156-SPL4 module predominantly regulates aerial axillary bud formation and controls shoot architecture. *New Phytol.* **2017**, *216*, 829–840. [CrossRef] [PubMed]

25. Jiao, Y.; Wang, Y.; Xue, D.; Wang, J.; Yan, M.; Liu, G.; Dong, G.; Zeng, D.; Lu, Z.; Zhu, X.; et al. Regulation of *OsSPL14* by OsmiR156 defines ideal plant architecture in rice. *Nat. Genet.* **2010**, *42*, 541–544. [CrossRef] [PubMed]

26. Miura, K.; Ikeda, M.; Matsubara, A.; Song, X.J.; Ito, M.; Asano, K.; Matsuoka, M.; Kitano, H.; Ashikari, M. OsSPL14 promotes panicle branching and higher grain productivity in rice. *Nat. Genet.* **2010**, *42*, 545–549. [CrossRef] [PubMed]

27. Givnish, T.J.; Barfuss, M.H.; Van Ee, B.; Riina, R.; Schulte, K.; Horres, R.; Gonsiska, P.A.; Jabaily, R.S.; Crayn, D.M.; Smith, J.A.; et al. Phylogeny, adaptive radiation, and historical biogeography in Bromeliaceae: Insights from an eight-locus plastid phylogeny. *Am. J. Bot.* **2011**, *98*, 872–895. [CrossRef] [PubMed]

28. Zanella, C.M.; Janke, A.; Palma-Silva, C.; Kaltchuk-Santos, E.; Pinheiro, F.G.; Paggi, G.M.; Soares, L.E.; Goetze, M.; Buttow, M.V.; Bered, F. Genetics, evolution and conservation of Bromeliaceae. *Genet. Mol. Biol.* **2012**, *35*, 1020–1026. [CrossRef] [PubMed]

29. Bartholomew, D.P.; Paull, R.E.; Rohrbach, K.G. *The Pineapple-Botany, Production and Uses*; CABI Publishing: Wallingford, UK, 2003; p. 176.

30. Trusov, Y.; Botella, J.R. Silencing of the *ACC synthase* gene *ACACS2* causes delayed flowering in pineapple [*Ananas comosus* (L.) Merr.]. *J. Exp. Bot.* **2006**, *57*, 3953–3960. [CrossRef] [PubMed]

31. Lv, L.; Duan, J.; Xie, J.; Wei, C.; Liu, Y.; Liu, S.; Sun, G. Isolation and characterization of a *FLOWERING LOCUS T* homolog from pineapple (*Ananas comosus* (L.) Merr). *Gene* **2012**, *505*, 368–373. [CrossRef] [PubMed]

32. Lv, L.L.; Duan, J.; Xie, J.H.; Liu, Y.G.; Wei, C.B.; Liu, S.H.; Zhang, J.X.; Sun, G.M. Cloning and expression analysis of a *PISTILLATA* homologous gene from pineapple (*Ananas comosus* L. Merr). *Int. J. Mol. Sci.* **2012**, *13*, 1039–1053. [CrossRef] [PubMed]

33. Lci, M.; Li, Z.Y.; Wang, J.B.; Fu, Y.L.; Ao, M.F.; Xu, L. AfAP2-1, An age-dependent gene of *Aechmea fasciata*, responds to exogenous ethylene treatment. *Int. J. Mol. Sci.* **2016**, *17*, 303. [CrossRef] [PubMed]

34. Li, J.; Hou, H.; Li, X.; Xiang, J.; Yin, X.; Gao, H.; Zheng, Y.; Bassett, C.L.; Wang, X. Genome-wide identification and analysis of the *SBP-box* family genes in apple (*Malus x domestica* Borkh.). *Plant Physiol. Biochem.* **2013**, *70*, 100–114. [CrossRef] [PubMed]

35. Bailey, T.L.; Elkan, C. Fitting a mixture model by expectation maximization to discover motifs in biopolymers. *Int. Conf. Intell. Syst. Mol. Biol.* **1994**, *2*, 28–36.

36. Xie, K.; Wu, C.; Xiong, L. Genomic organization, differential expression, and interaction of SQUAMOSA promoter-binding-like transcription factors and microRNA156 in rice. *Plant Physiol.* **2006**, *142*, 280–293. [CrossRef] [PubMed]

37. Li, Z.; Wang, J.; Zhang, X.; Lei, M.; Fu, Y.; Zhang, J.; Wang, Z.; Xu, L. Transcriptome sequencing determined flowering pathway genes in *Aechmea fasciata* treated with ethylene. *J. Plant Growth Regul.* **2016**, *35*, 316–329. [CrossRef]

38. Salinas, M.; Xing, S.; Hohmann, S.; Berndtgen, R.; Huijser, P. Genomic organization, phylogenetic comparison and differential expression of the SBP-box family of transcription factors in tomato. *Planta* **2012**, *235*, 1171–1184. [CrossRef] [PubMed]

39. Zhang, B.; Liu, X.; Zhao, G.; Mao, X.; Li, A.; Jing, R. Molecular characterization and expression analysis of *Triticum aestivum squamosa-promoter binding protein-box* genes involved in ear development. *J. Integr. Plant Biol.* **2013**, *56*, 571–581. [CrossRef] [PubMed]

40. Zhang, S.; Ling, L. Genome-wide identification and evolutionary analysis of the *SBP-box* gene family in castor bean. *PLoS ONE* **2014**, *9*, e86688. [CrossRef] [PubMed]

41. Xu, Z.; Sun, L.; Zhou, Y.; Yang, W.; Cheng, T.; Wang, J.; Zhang, Q. Identifiation and expression analysis of the *SQUAMOSA promoter-binding protein (SBP)-box* gene family in *Prunus mume*. *Mol. Genet. Genom.* **2015**, *290*, 1701–1715. [CrossRef] [PubMed]

42. Shalom, L.; Shlizerman, L.; Zur, N.; Doron-Faigenboim, A.; Blumwald, E.; Sadka, A. Molecular characterization of *SQUAMOSA PROMOTER BINDING PROTEIN-LIKE (SPL)* gene family from *Citrus* and the effect of fruit load on their expression. *Front. Plant Sci.* **2015**, *6*, 389. [CrossRef] [PubMed]

43. Zhang, H.X.; Jin, J.H.; He, Y.M.; Lu, B.Y.; Li, D.W.; Chai, W.G.; Khan, A.; Gong, Z.H. Genome-wide identification and analysis of the SBP-box family genes under *Phytophthora capsici* stress in pepper (*Capsicum annuum* L.). *Front. Plant Sci.* **2016**, *7*, 504. [CrossRef] [PubMed]

44. Preston, J.C.; Jorgensen, S.A.; Orozco, R.; Hileman, L.C. Paralogous *SQUAMOSA PROMOTER BINDING PROTEIN-LIKE (SPL)* genes differentially regulate leaf initiation and reproductive phase change in petunia. *Planta* **2016**, *243*, 429–440. [CrossRef] [PubMed]

45. Cheng, H.; Hao, M.; Wang, W.; Mei, D.; Tong, C.; Wang, H.; Liu, J.; Fu, L.; Hu, Q. Genomic identification, characterization and differential expression analysis of *SBP-box* gene family in *Brassica napus*. *BMC Plant Biol.* **2016**, *16*, 196. [CrossRef] [PubMed]

46. Song, A.; Gao, T.; Wu, D.; Xin, J.; Chen, S.; Guan, Z.; Wang, H.; Jin, L.; Chen, F. Transcriptome-wide identification and expression analysis of *chrysanthemum* SBP-like transcription factors. *Plant Physiol. Biochem.* **2016**, *102*, 10–16. [CrossRef] [PubMed]

47. Li, Y.H.; Wu, Q.S.; Huang, X.; Liu, S.H.; Zhang, H.N.; Zhang, Z.; Sun, G.M. Molecular cloning and characterization of four genes encoding ethylene receptors associated with pineapple (*Ananas comosus* L.) flowering. *Front. Plant Sci.* **2016**, *7*, 710. [CrossRef] [PubMed]

48. Bi, F.; Meng, X.; Ma, C.; Yi, G. Identification of miRNAs involved in fruit ripening in Cavendish bananas by deep sequencing. *BMC Genom.* **2015**, *16*, 776. [CrossRef] [PubMed]

49. Srikanth, A.; Schmid, M. Regulation of flowering time: All roads lead to Rome. *Cell. Mol. Life Sci.* **2011**, *68*, 2013–2037. [CrossRef] [PubMed]

50. Fornara, F.; de Montaigu, A.; Coupland, G. SnapShot: Control of flowering in *Arabidopsis*. *Cell* **2010**, *141*, 550. [CrossRef] [PubMed]

51. Jung, J.H.; Seo, Y.H.; Seo, P.J.; Reyes, J.L.; Yun, J.; Chua, N.H.; Park, C.M. The GIGANTEA-regulated microRNA172 mediates photoperiodic flowering independent of CONSTANS in *Arabidopsis*. *Plant Cell* **2007**, *19*, 2736–2748. [CrossRef] [PubMed]

52. Wu, G.; Park, M.Y.; Conway, S.R.; Wang, J.W.; Weigel, D.; Poethig, R.S. The sequential action of miR156 and miR172 regulates developmental timing in *Arabidopsis*. *Cell* **2009**, *138*, 750–759. [CrossRef] [PubMed]

53. Hyun, Y.; Richter, R.; Vincent, C.; Martinez-Gallegos, R.; Porri, A.; Coupland, G. Multi-layered regulation of *SPL15* and cooperation with *SOC1* integrate endogenous flowering pathways at the *Arabidopsis* shoot meristem. *Dev. Cell* **2016**, *37*, 254–266. [CrossRef] [PubMed]

54. Jung, J.H.; Lee, H.J.; Ryu, J.Y.; Park, C.M. SPL3/4/5 integrate developmental aging and photoperiodic signals into the FT-FD module in *Arabidopsis* flowering. *Mol. Plant* **2016**, *9*, 1647–1659. [CrossRef] [PubMed]

55. Luo, L.; Li, W.; Miura, K.; Ashikari, M.; Kyozuka, J. Control of tiller growth of rice by *OsSPL14* and Strigolactones, which work in two independent pathways. *Plant Cell Physiol.* **2012**, *53*, 1793–1801. [CrossRef] [PubMed]

56. Wei, Q.; Ma, C.; Xu, Y.; Wang, T.; Chen, Y.; Lu, J.; Zhang, L.; Jiang, C.Z.; Hong, B.; Gao, J. Control of *chrysanthemum* flowering through integration with an aging pathway. *Nat. Commun.* **2017**, *8*, 829. [CrossRef] [PubMed]

57. Ling, L.Z.; Zhang, S.D. Exploring the evolutionary differences of *SBP-box* genes targeted by miR156 and miR529 in plants. *Genetica* **2012**, *140*, 317–324. [CrossRef] [PubMed]

58. Wang, S.; Wu, K.; Yuan, Q.; Liu, X.; Liu, Z.; Lin, X.; Zeng, R.; Zhu, H.; Dong, G.; Qian, Q.; et al. Control of grain size, shape and quality by OsSPL16 in rice. *Nat. Genet.* **2012**, *44*, 950–954. [CrossRef] [PubMed]

59. Chuck, G.S.; Brown, P.J.; Meeley, R.; Hake, S. Maize SBP-box transcription factors unbranched2 and unbranched3 affect yield traits by regulating the rate of lateral primordia initiation. *Proc. Natl. Acad. Sci. USA* **2014**, *111*, 18775–18780. [CrossRef] [PubMed]

60. Si, L.; Chen, J.; Huang, X.; Gong, H.; Luo, J.; Hou, Q.; Zhou, T.; Lu, T.; Zhu, J.; Shangguan, Y.; et al. OsSPL13 controls grain size in cultivated rice. *Nat. Genet.* **2016**, *48*, 447–456. [CrossRef] [PubMed]

61. Tamura, K.; Stecher, G.; Peterson, D.; Filipski, A.; Kumar, S. MEGA6: Molecular Evolutionary Genetics Analysis version 6.0. *Mol. Biol. Evol.* **2013**, *30*, 2725–2729. [CrossRef] [PubMed]

62. Livak, K.J.; Schmittgen, T.D. Analysis of relative gene expression data using real-time quantitative PCR and the $2^{-\Delta\Delta CT}$ method. *Methods* **2001**, *25*, 402–408. [CrossRef] [PubMed]

63. Holsters, M.; de Waele, D.; Depicker, A.; Messens, E.; van Montagu, M.; Schell, J. Transfection and transformation of *Agrobacterium tumefaciens*. *Mol. Genet. Genom.* **1978**, *163*, 181–187. [CrossRef]

64. Clough, S.J.; Bent, A.F. Floral dip: A simplified method for *Agrobacterium*-mediated transformation of *Arabidopsis thaliana*. *Plant J.* **1998**, *16*, 735–743. [CrossRef] [PubMed]

13

Expression of Maize MADS Transcription Factor *ZmES22* Negatively Modulates Starch Accumulation in Rice Endosperm

Kangyong Zha [†], Haoxun Xie [†], Min Ge, Zimeng Wang, Yu Wang, Weina Si and Longjiang Gu *

National Engineering Laboratory of Crop Stress Resistance breeding, Anhui Agricultural University, Hefei 230036, China; kangyongzha929@163.com (K.Z.); xhx521xz@163.com (H.X.); 13215616815@163.com (M.G.); wzmhanchang@163.com (Z.W.); wangyu20180712@163.com (Y.W.); weinasi@ahau.edu.cn (W.S.)
* Correspondence: longjianggu@163.com
† These authors contributed equally to this work.

Abstract: As major component in cereals grains, starch has been one of the most important carbohydrate consumed by a majority of world's population. However, the molecular mechanism for regulation of biosynthesis of starch remains elusive. In the present study, *ZmES22*, encoding a MADS-type transcription factor, was modestly characterized from maize inbred line B73. *ZmES22* exhibited high expression level in endosperm at 10 days after pollination (DAP) and peaked in endosperm at 20 DAP, indicating that *ZmES22* was preferentially expressed in maize endosperm during active starch synthesis. Transient expression of *ZmES22* in tobacco leaf revealed that ZmES22 protein located in nucleus. No transactivation activity could be detected for ZmES22 protein via yeast one-hybrid assay. Transformation of overexpressing plasmid 35S::*ZmES22* into rice remarkedly reduced 1000-grain weight as well as the total starch content, while the soluble sugar was significantly higher in transgenic rice lines. Moreover, overexpressing *ZmES22* reduced fractions of long branched starch. Scanning electron microscopy images of transverse sections of rice grains revealed that altered expression of *ZmES22* also changed the morphology of starch granule from densely packed, polyhedral starch granules into loosely packed, spherical granules with larger spaces. Furthermore, RNA-seq results indicated that overexpressing *ZmES22* could significantly influence mRNA expression levels of numerous key regulatory genes in starch synthesis pathway. Y1H assay illustrated that ZmES22 protein could bind to the promoter region of *OsGIF1* and downregulate its mRNA expression during rice grain filling stages. These findings suggest that *ZmES22* was a novel regulator during starch synthesis process in rice endosperm.

Keywords: *Zea mays* L.; MADS transcription factor; *ZmES22*; starch

1. Introduction

Maize (*Zea mays* L.) is one of the most widely grown crop world-wide, as well as a critical model for various biological researches, especially for endosperm development [1]. Starch is the major component of maize grains, which accounted up to 71% on a dry weight basis. Therefore, comprehensive understanding of the molecular mechanism for regulation of starch synthesis will facilitate increase in yield to feed growing population.

Starch is composed of two major components, known as amylose and amylopectin. The process of starch biosynthesis has been reported to be under finely regulated by numerous genes, which mainly encoded multiple subunits or isoforms of four enzymes: ADP-glucose pyrophosphorylase (AGPase), starch synthase (SS), starch branching enzyme (SBE), and starch debranching enzyme (DBE) [2,3]. At the initial stage of starch synthesis, glucose-1-phosphate, together with ATP, are converted to

ADP-glucose (ADPG) via AGPase. In the developing endosperm, ADPG is mainly produced in the cytosol and transferred into amyloplast through an adenylate translocator, BT1 [4]. Afterwards, the synthesis of starch is furthered by chain elongation by transferring ADPG to the nonreducing end of a glucan primer. The amylose chain elongation is completed by granule-bound starch synthase I (GBSSI), whereas, amylopectin chains are elongated by a soluble form of starch synthase (SSI, SSII, SSIII, and SSIV). α-1,6-Glucosidic linkages is then introduced by starch branching enzyme (BEI and BEII) and finally, fine structure of amylopectin is achieved through removal of unnecessary branches by starch debranching enzymes (ISA and Pullulanase). Mutants defective in any key genes exhibited apparent abnormal characters of starch in reserve organs. Mutations in *OsAGPL2*, one of the large subunits of AGPase, caused severe defects in grain filling and starch synthesis [5]. Loss-of-function mutations occurred in *OsBT1* gene, which encoded an ADPG translocator, resulted in a remarkable reduction in grain weight than wild type [4]. Deficiency of *OsSSIIa* lead to a chalky interior appearance and the endosperm of the mutant lines are mainly consisted of loosely packed, spherical starch granules with larger air spaces [5]. *Grain Incomplete Filling 1 (OsGIF1)*, encoding a cell-wall invertase, was of great importance in regulation of sucrose unloading from phloem into cells of reserve organs. Mutant lines of *OsGIF1* showed severe defects in grain filling and in turn reduced the grain weight to 70% of wild type rice at 30 days after pollination (DAP) [6].

Since starch biosynthesis and accumulation are critical determinants for both grain quality and production, key transcriptional regulators, including several transcription factors (TFs), have also been demonstrated to play an important regulatory role in starch synthesis. Null mutants of *OsBZIP58* seeds exhibited altered starch composition as well as morphological defects with apparent white belly region [7]. SUSIBA2, a WRKY family transcription factor, could directly bind to the promoter of *pISA1* gene to regulate its expression, thus affecting the synthesis of starch in barley [8]. Additionally, one of AP2 family of transcription factors, SERF1, negatively regulates rice grain filling, and genetic mutations could enhance the starch synthesis process of rice [9]. *ZmbZIP91* was proved to be a key regulator of the starch synthesis by directly binding to ACTCAT elements in the promoters of starch synthesis genes [10]. The inhibition of *ZmDof3* led to defects of the kernel phenotype with decreased starch content and a partially patchy aleurone layer [1]. Altered expression of transcription factors, causing abnormal features in reserve organs, could provide profound implications in understanding the molecular mechanisms that control starch biosynthesis. Despite these research highlights, a comprehensive understanding of factors that regulate the expression of genes in network of starch synthesis remains largely unknown, especially in maize. Hence, screening and identification of key transcription factors involved in starch synthesis will be of great importance in breeding of high-yielding crops.

In previous studies, a total of 2298 transcription factors were identified and further examined using RNA-seq dataset from 18 representative tissues from maize [11], which provided profound clues regarding to the relationship between development and dynamic expression profiles of key transcription factors. With an emphasis on endosperm-specificity, we identified 36 transcription factors that were preferentially highly expressed in maize endosperm [12]. The mRNA expression profiles of one gene, encoding a typical MADS transcription factor (GRMZM2G159397, designated as *ZmES22*), were further confirmed via qRT-PCR assays. To test if this gene was related to starch synthesis, *ZmES22* was cloned from maize inbred line B73. Afterwards, molecular properties and biological functions were modestly comprehensively characterized in transgenic rice lines. Overexpressing *ZmES22* in rice significantly reduced 1000-grain weight as well as hindered starch accumulation. Besides, altered expression of *ZmES22* in transgenic rice also changed the starch structure and morphology of starch granules. Furthermore, RNA-seq analysis demonstrated that numerous key regulatory genes in starch synthesis were differentially expressed compared to that in WT plants. Yeast one hybrid assay revealed that ZmES22 could bind to the promoter of *OsGIF1* and downregulated its expression during grain filling process. This study illustrated that *ZmES22* could be a newfound transcription factor, which negatively regulated starch synthesis in rice endosperm.

2. Results

2.1. Sequence Analysis and Construction of Phylogenetic Tree for ZmES22 Homologues

As one of the largest transcription factor family in eukaryote, MADS-box proteins has been characterized by its important roles in a variety of aspects during plant growth and development [13,14]. To test if *ZmES22* were related to starch synthesis, this gene was firstly cloned from maize inbred line B73. *ZmES22* contained an open reading frame (ORF) of 723 bp and encoded a protein of 240 amino acids with a predicted molecular weight (Mw) of 27,903 Da and an isoelectric point (pI) of 8.92. Pfam analysis of ZmES22 revealed that the deduced protein sequence consisted of four conserved domains, namely the MADS-box domain (MADS-box), intervening (I), K-box domain, and the C terminal domain (Figure S1). In order to find homologs of *ZmES22*, blastp program was explored for protein sequence of ZmES22 to search against protein database for *Zea mays*, *Oryza sativa*, and *Arabidopsis thaliana*, respectively. Afterwards, pairwise amino acid distances were calculated using MEGA7 with Jones–Taylor–Thornton (JTT) model, and genes with diversity less than 0.8 were retained according to empirical experience. A total of 16 genes, including 6 from *Zea mays*, 5 from *Oryza sativa*, and 5 from *Arabidopsis thaliana* were identified, respectively (Tables S1 and S2). Phylogenetic tree was constructed using the conserved MADS domain, and clear orthologous relationship could be observed between *ZmES22* and *OsMADS7* (Figure 1 and Figure S2). Furthermore, the Multiple EM for Motif Elicitation (MEME) motif website search program was explored to identify the conserved motifs for all 16 homologues. Great majority homologues contained Motif 1, Motif 2, Motif 3, and Motif 4, indicating these motifs were probably evolutionary conserved (Figure 1). While, presence or absence for remained motifs was more variable.

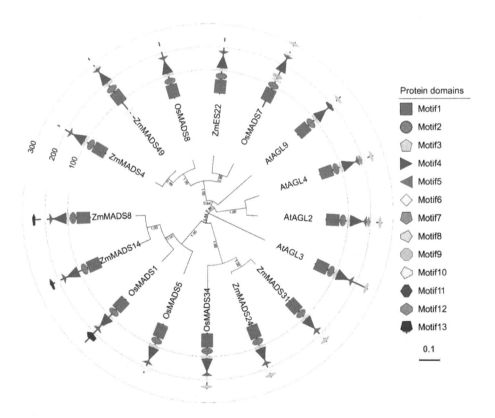

Figure 1. Phylogenetic tree of *ZmES22* homologous genes from maize, rice and Arabidopsis. Phylogenic tree of homologous genes of *ZmES22* from maize, rice and Arabidopsis MADS proteins, which was constructed using conserved MADS domain with MEGA7 software via Neighbor-joining method. Bootstrap value was indicated at each branch point. Gene IDs and predicted functions are listed in Supplementary Table S1.

2.2. Expression Profiles and Subcellular Localization of ZmES22

qRT-PCR assays were performed to investigate expression profiles of *ZmES22*. In line with previous transcriptome analysis, compared with nutritive organs, such as root, stem and leaf, *ZmES22* exhibited higher relative expression levels in reproductive organs (Figure 2). Intriguingly, significantly higher expression level of *ZmES22* was observed in endosperm than embryo at 10 DAP, and mRNA expression of *ZmES22* peaked in endosperm at 20 DAP. To ascertain the location of *ZmES22* protein, coding sequence of *ZmES22* was inserted into empty vector 35S::GFP. Afterwards, 35S::*ZmES22*-GFP construct and 35S::GFP were efficiently transfected tobacco leaf cells separately via *Agrobacterium* infiltration (Figure 3). Green fluorescence of 35S::GFP could be observed throughout the cell, whereas, green fluorescence of ZmES22-GFP fusion protein appeared only in nucleus (Figure 3), illustrating that ZmES22 protein functioned in nucleus. However, yeast one-hybrid assay demonstrated that ZmES22 protein did not have transcriptional activity in yeast cells (Figure S3). These results indicated that *ZmES22* may be involved in the regulation of development of endosperm with the help of other proteins.

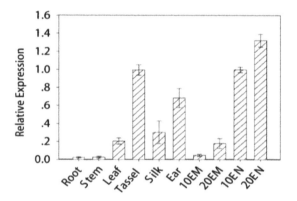

Figure 2. Expression pattern of ZmES22 across diverse tissues. Expression patterns of *ZmES22* in root, stem, leave, tassel, silk, ear, embryo and endosperm was quantified via qRT-PCR. The developmental stage of the embryo and endosperm is indicated by 10 and 20 DAP. Maize *Actin1* was used as the internal control. Error bars are standard deviations of three technical repeats and two biological repeats.

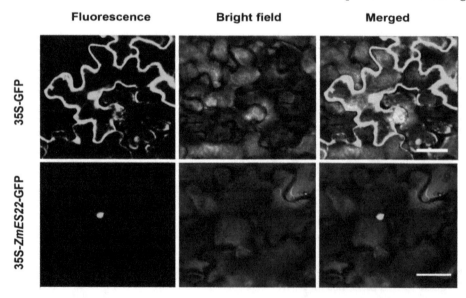

Figure 3. Subcellular localization of ZmES22 in tobacco. The 35S::*ZmES22*-GFP fusion construct and 35S::GFP vector were transiently expressed in tobacco epidermal cells and examined by a confocal laser scanning biological microscope, respectively. Bars = 50 μm.

2.3. Analysis of Agronomic Characters of ZmES22 Overexpression Transgenic Rice

To illustrate the function of *ZmES22*, twelve independent rice lines, which overexpressed *ZmES22* under the drive of CaMV 35S promoter, were obtained via *Agrobacterium* mediated transformation. qRT-PCR assays revealed that *ZmES22* expressed at distinct levels in transgenic rice lines, among which L8, L9, and L10 exhibited significantly higher expression (Figure S4). Therefore, these three transgenic rice lines was selected for further research. Compared to wild type (WT) plants, overexpression rice lines exhibited no visible difference during both the vegetative and reproductive stages, with similar plant height as well as panicle architecture (Figure S5). After maturation, agronomic traits, including grain length, grain width, grain thickness, and 1000-grain weight were minutely characterized for both transgenic rice lines and WT plants. There was no significant change in either grain length or grain width between transgenic plants and WT plants (Figure 4A,B). Nevertheless, grain thickness was dramatically decreased in overexpressed rice lines (Figure 4C, Student's *t*-test, *p*-value = 4.8×10^{-5}). Accordingly, 1000-grain weight of transgenic plants were significantly depleted by 3.88 g than that of WT plants (Figure 4D, Student's *t*-test, *p*-value = 1.8×10^{-8}). Additionally, total starch content, apparent amylose content (AAC) and soluble sugar content of both transgenic rice lines and WT plants were measured according to previously reported methods. Surprisingly, compared to WT plants, both total starch content and AAC were significantly reduced (Figure 5A,B, Student's *t*-test, *p*-value = 0.02), whereas, the content of soluble sugar in transgenic rice lines were significantly increased by 38% than that of WT plants (Figure 5C, Student's *t*-test, *p*-value = 8.4×10^{-4}). In particular, content of soluble sugar was two times larger in transgenic line L9 than that in WT plants. These results revealed that overexpression of *ZmES22* gene could significantly block starch biosynthesis process in endosperm of rice.

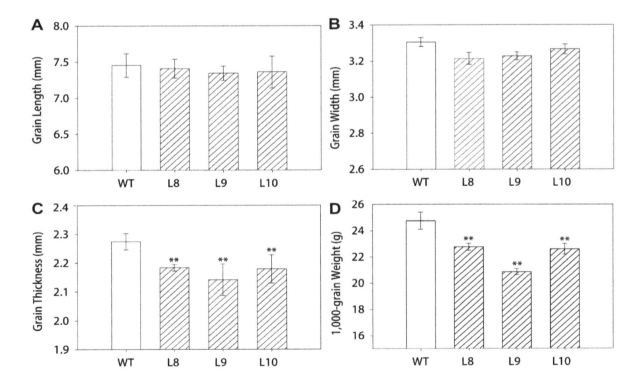

Figure 4. Agronomic characters of seeds from transgenic rice lines that overexpressed *ZmES22*. Grain agronomic characters including grain length (**A**), grain width (**B**), grain thickness (**C**), and 1000-grain weight (**D**) were minutely measured. Data are presented as mean ± SD of three replicates. L: transgenic lines of *ZmES22* seeds; WT: wild-type plants (Zhonghua 11), Student's *t*-test, ** *p*-value < 0.01.

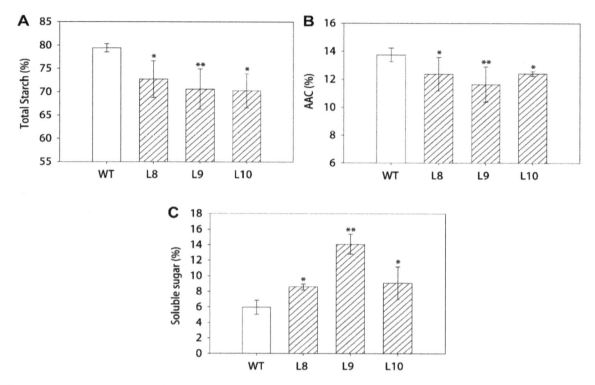

Figure 5. Overexpression of *ZmES22* in rice altered the starch composition. (**A**) Total starch content in rice endosperm. (**B**) Apparent amylose content (AAC) in rice endosperm. (**C**) Soluble sugar content in rice endosperm. Data are presented as mean ± SD of three replicates. L: transgenic lines of *ZmES22* seeds; WT: wild-type plants (Zhonghua 11), Student's *t*-test, * *p*-value < 0.05, ** *p*-value < 0.01.

2.4. Overexpression of ZmES22 Influences Starch Structure in Transgenic Rice

Both amylopectin blue value and the maximum absorption wavelength reflect the ability of amylopectin binding to iodine. Therefore, different BV and kmax can provide indicators for the basic distinction of starch structure [15]. To detect whether the relative content of amylose and amylopectin were altered by overexpression of *ZmES22*, BV and kmax of starch from both *ZmES22* overexpression rice seeds and WT plants were determined accordingly. As shown in Figure 6A,B, both BV and kmax of amylose and amylopectin in three transgenic lines were significantly smaller than that of WT plants (Student's *t*-test, *p*-value = 2.2×10^{-16}). Furthermore, morphology of starch granules was examined via scanning electron microscopy (SEM) [16]. SEM images of transverse sections of rice grains revealed that both central and dorsal endosperms were filled with densely packed, polyhedral starch granules in both transgenic rice and WT seeds, while ventral endosperm of transgenic rice seeds exhibited an apparent abnormity with a visible chalky region (Figure 6C), which was mainly consisted of loosely packed, spherical starch granules with larger air spaces. These results indicated that the overexpressing *ZmES22* could change starch structure as well as influence morphology of starch granules in transgenic rice lines.

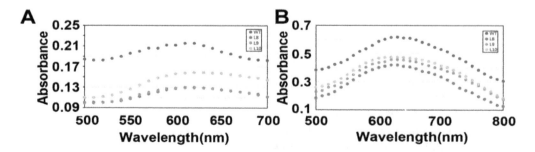

Figure 6. *Cont.*

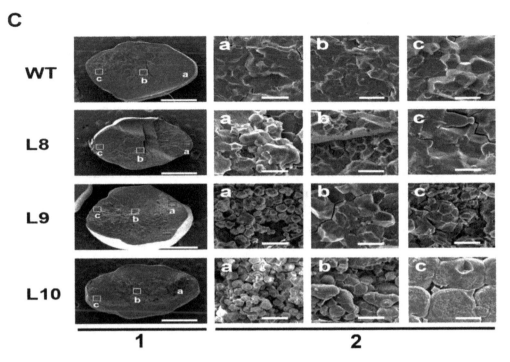

Figure 6. Blue value (BV), maximum absorbance (kmax) and scanning electron microscopy (SEM) images of the transverse sections of transgenic rice seeds. (**A**) BV at 600 nm and kmax represent the ability to combine with iodine. (**B**) BV at 680 nm and kmax represent the ability to combine with iodine. (**C**) Cross-sections of mature seeds are shown in (1). SEM of the ventral area of mature endosperm is shown in a of (2) and indicated by a red square in (1). Bars: 1 mm in (**1**); 10 μm in (2) a: dorsal; b: center; c: belly.

2.5. Overexpression of ZmES22 Influence Expression Profiles of Numerous Starch Synthesis Related Genes at 20 DAP Endosperm

To further explore the molecular basis of *ZmES22* in regulation of starch synthesis, expression profiles of 17 genes, which were preferentially expressed in developing endosperms and were demonstrated to be involved in starch synthesis, were compared between transgenic rice lines with WT plants at different developmental stages (3, 6, 10, and 20 DAP, Figure 7). The results illustrated that, except for *OsISA2*, *OsSSI*, and *OsSSIIa*, great majority of characterized starch synthesis related genes were downregulated depending on the individual genes when compared to WT plants (Figure 7). Interestingly, expression levels of *OsBEI* and *OsPUL* exhibited similar tendency that they were remarkably upregulated as grains got maturity (Figure 7). As is described previously, *ZmES22* was highly expressed in 20 DAP endosperm, therefore, we proposed that genes differentially expressed in transgenic rice plants at 20 DAP endosperm might be potentially key regulators. Nevertheless, no significant changes could be observed among all of tested genes at 20 DAP endosperms (Figure 7). In order to further investigate possible regulation by *ZmES22*, 20-DAP seeds for both overexpression rice lines and WT plants were collected for RNA-seq analysis, each was repeated with two biological replicates (Table 1, Figure 8A,B). Collectively, 1902 differentially expressed genes (DEGs), consisting of 986 upregulated and 916 downregulated genes in overexpression rice lines (Figure 8C), were determined with the following criteria:(1) the minimum fold-change of gene expression was 2.0; (2) the maximum adjusted p value was 0.05. In order to validate the RNA-seq data, 10 DEGs, including 5 upregulated and 5 downregulated genes, were randomly selected for quantitative real-time PCR analysis, and the results illustrated that RNA-seq data are of satisfactory quality (Figure 8D). To analyze the functional enrichments of the DEGs, both Gene Ontology (GO) and Kyoto Encyclopedia of Genes and Genomes (KEGG) analysis were performed using R package ClusterProfiler. DNA metabolic process, response to stress and carbohydrate metabolic process are the three mostly enriched GO items (Figure S6). Moreover, six pathways were significantly enriched in KEGG analysis (Figure 9), including starch and sucrose metabolism pathway (Figure S7), galactose metabolism, phenylalanine metabolism, plant hormone signal transduction pathway (Figure S8), etc. Interestingly, one gene, named *GIF1* (*Os04g0413500*), which was reported to be a key regulator to rice grain-filling and yield, was significantly enriched in carbohydrate metabolic process as well as starch and sucrose metabolism pathway in KEGG. The *gif1* mutant exhibited slower grain-filling rate and showed markedly more grain chalkiness than wild-type plants [6]. In the present study, *GIF1* gene was downregulated as much as 8-fold in overexpression rice lines compared with WT plants (Fisher's exact test, $p = 5.0 \times 10^{-5}$). The relative mRNA expression level of *OsGIF1* in four different developmental endosperms (3, 6, 10, and 20 DAP) were further confirmed by real-time quantitative PCR (Figure S9A). Because overexpression *ZmES22* lead to similar phenotype as *gif1* mutant, we therefore wonder if ZmES22 could bind to the promoter of *GIF1* and negatively regulate its expression?

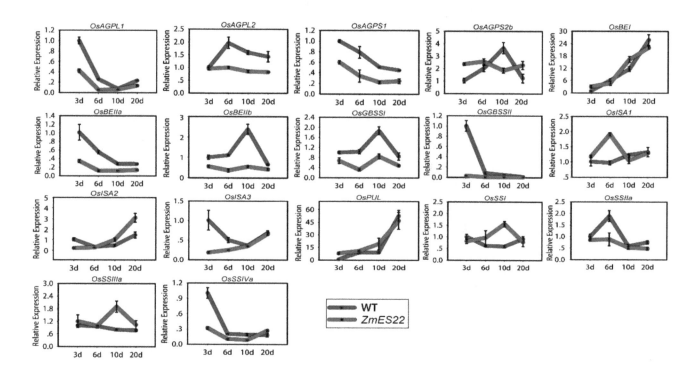

Figure 7. Expression profiles of 17 starch synthesis related genes across diverse developmental stages. Blue line represents wild type (Zhonghua 11), and red line denotes transgenic lines. d is short for days after pollination (DAP). The mRNA expression level of each gene in the three DAP seeds of wild type was used as a control. All data are shown as means ±SD from three biological replicates and two technical replicates. Primers are listed in Supplemental Table S3.

Table 1. Statistics of sequencing data

Sample	Clean Reads	Mapped Reads	Clean Base (Gb)	Mapped Base (Gb)	Mapping Rate (%)	Concordant Pair Rate (%)	Q30 (%)	GC Content (%)
WT-1	66,417,130	61,285,695	6.64	6.13	92.27	85.4	94.03	56.95
WT-2	65,482,702	60,254,372	6.55	6.03	92.02	84.7	94.36	57.31
ZmES22-1	65,615,264	60,672,105	6.56	6.07	92.47	85.7	94.15	57.22
ZmES22-2	65,985,744	60,618,187	6.6	6.06	91.9	84.4	92.45	57.01

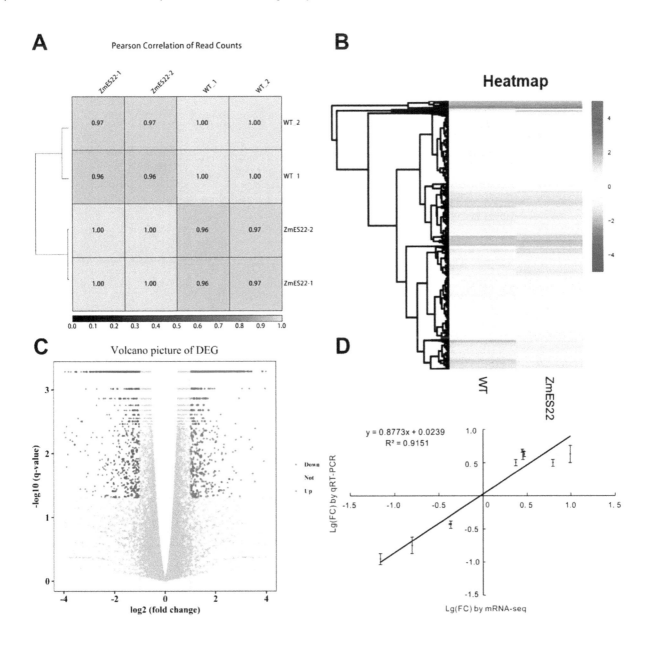

Figure 8. RNA-seq analysis of endosperm at 20 DAP for transgenic rice lines and wild-type plants. (**A**) Pearson correlation of read counts. (**B**) Heat map comparison between Zhonghua 11 and L9. (**C**) A volcano plot of differentially expressed genes (DEGs) about Zhonghua 11 and L9. (**D**) Validation of transcription group data. Expression level changes (log2 (fold change)) of 10 randomly selected DEGs analyzed by RNA-Seq (x-axis) were compared with expression data obtained by qRT-PCR (y-axis).

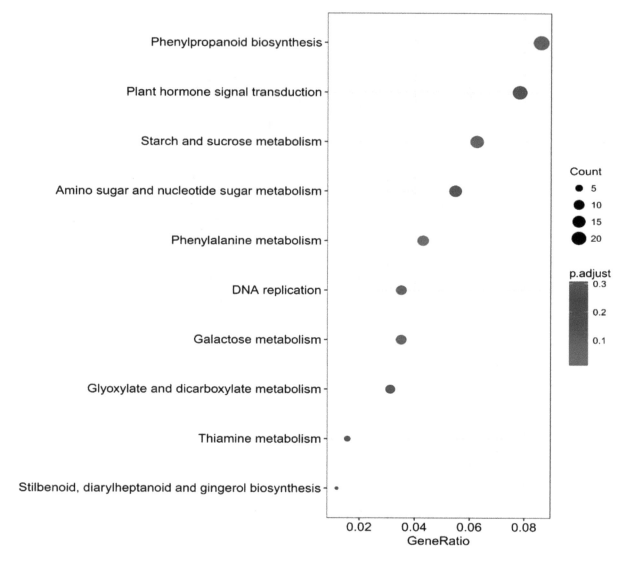

Figure 9. KEGG enrichment analysis for DEGs between transgenic rice lines and wild type plants. KEGG pathways that were enriched for DEGs between transgenic rice lines in comparison to wild type Zhonghua 11. The black circle denotes DEGs that were annotated to one KEGG pathway, and the color panel denoted the p-value for each KEGG pathway.

2.6. ZmES22 Could Bind to Promoter GIF1 Gene

To determine if ZmES22 could bind to the promoter of *GIF1*, we firstly extracted 2000 base-pair sequences from upstream of *GIF1* and subjected it web site (http://bioinformatics.psb.ugent.be/webtools/plantcare/html/) to predict if promoter of *GIF1* contained conserved element that MADS type transcription factors could bind to [17]. Surprisingly, a conserved element (CATGT) was located at minus 365 base-pair in upstream of *OsGIF1* gene [18] (Figure S9B). Afterwards, yeast one hybrid assay was explored to determine whether ZmES22 protein could bind to this element. Complete coding sequence of *ZmES22* was inserted into vector containing both activation domain (AD) and was drove by P_{T7}. Two repeated copies of CATGT was synthesized as bait (designated as pGIF1). Simultaneously, a mutant (CAGGT, designated as pmGIF1) was also used as a negative control. As was illustrated in Figure 10, both the growth of the yeast in null and negative control were obviously inhibited in SD/-Ura medium with 900 ng/mL AbA in the yeast one-hybrid assay, however, the yeast co-transformed with P_{T7}-*ZmES22* and pGIF1-AbAi grows well, indicating that ZmES22 binds to the core element of the promoter of *OsGIF1* (Figure 10).

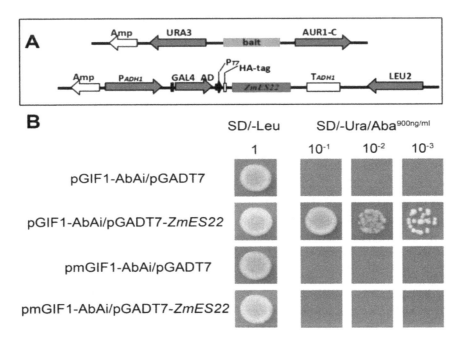

Figure 10. ZmES22 could bind to the core motif of *OsGIF1* via yeast one hybrid assay. (**A**) Schematic structure of yeast expression construct pGAD-*ZmES22* and reporter construct. (**B**) Yeast Y1HGlod was transformed with the vector pGADT7-*ZmES22* and CATGT (*pGIF1*) or mutant tandem repeats (*pmGIF1*) plasmids. The transformants were screened by plating on SD/-Leu/AbA plates to verified the interaction between ZmES22 and the core motif of *GIF1* promoter.

3. Discussion

The endosperm is the tissue that most flowering plants produce in the seeds after fertilization. Endosperm development involves the process of starch synthesis and storage protein accumulation. Recent studies revealed that process of starch synthesis was remarkably conserved ranging from green algae to extant higher plants, suggesting that genes encoding starch biosynthesis related enzymes were functionally conserved across diverse lineages [19,20]. To date, enzymes involved in starch synthesis has been soundly documented in rice. Therefore, rice endosperm is a particularly ideal model to screen and identify key transcription factors that could finely tune the process of starch synthesis in maize [21].

In the present study, transgenic rice that overexpressed one MADS type transcription factor *ZmES22* from maize, exhibited obvious defects with respect to grain characteristics, represented by loosely packed starch granules, reduced 1000-grain weight, and altered apparent amylose and total starch content, suggesting that *ZmES22* might play a key role in regulation of starch synthesis pathway in the transgenic lines. It has been reported that the process of starch biosynthesis was regulated by 17 genes, which mainly encoded multiple subunits or isoforms of four enzymes: ADP-glucose pyrophosphorylase (AGPase), starch synthase (SS), starch branching enzyme (SBE), and starch debranching enzyme (DBE) [2,3]. Interestingly, no significant expression change could be detected for all of these 17 genes in endosperm at 20 DAP between wide type and transgenic rice based on clues from both qRT-PCR and RNA-seq results. However, the mRNA expression levels of majority of starch synthesis related genes, except for *OsISA1, OsISA2, OsSSI,* and *OsSSIIa*, decreased during the early stages of endosperm development stages compared to that in wild type rice. These results indicated that overexpression of *ZmES22* could negatively affect mRNA expression of the majority of starch synthesis related genes in distinct degree during the early endosperm development stages.

KEGG analysis revealed that DEGs was significantly enriched in starch and sucrose metabolism pathway, in which the mRNA expression of one gene, named *OsGIF1*, decreased as much as 8-fold in transgenic rice. Previous result demonstrated that grain filling rate of *gif1* mutants was slower and accompanied with distinct chalkiness and loosely packed starch granules, while overexpression of *GIF1* driven by its native promoter produces larger grains [6]. The phenotype of *gif1* mutant was consistent to transgenic rice lines that overexpressed *ZmES22* gene from maize. Evidence from qRT-PCR also validated that mRNA expression level of *OsGIF1* continually decreased in overexpression rice lines in comparison to wild type plants. These results indicated that overexpression of *ZmES22* in rice might inhibit the mRNA expression of *OsGIF1*. As is reported that typic MADS type transcription factor are plant specific and often contained four functional domains, the MADS-box conserved domain (MADS-box), intervening (I), K-box domain, which is homologous to keratin (K), and the C terminal domain. MADS-box domain could bind to the promoter and regulate the expression of downstream genes. For example, *ZmMADS47* directly binds the core motif CATGT of promoter of zein genes and activated its expression [18]. Additionally, CATGT element also resided in the upstream of transcription starting sites (-365 base pair), yeast one hybrid assay demonstrated that ZmES22 could bind to the core motif of *OsGIF1* and repress its expression.

The primary results provide evidence that *ZmES22* affect starch synthesis and endosperm development through binding to and downregulating the expression of *OsGIF1*, and in turn influencing carbon distribution and transportation of sucrose on grain filling in rice plant.

In conclusion, a MADS type transcription factor from maize was modestly characterized via overexpression in rice and the molecular mechanism for its anticipant role in regulation of starch synthesis were also explored by RNA-seq and yeast one hybrid assay in the present study. Starch synthesis is a complicated and sophisticated process, which is regulated by numerous transcription factors via protein–protein and protein–DNA interactions. In order to shed light on how *ZmES22* influence the starch synthesis in rice, *ZmES22* mutant are being created via CRISPR/Cas9 system.

4. Materials and Methods

4.1. Plant Materials and Growth Conditions

Ten representative tissues, including root, stem, leaf, tassel, filament, ear, 10 DAP (days after pollination) embryo, 20 DAP embryo, 10 DAP endosperm, and 20 DAP endosperm, were collected from maize inbred line B73 plants, which were grown in a greenhouse with paddy soil at 28 °C under a 14 h light/10 h dark photoperiod. Each tissue was repeatedly sampled from three individual plants as three biological replicates. Both wild type cultivars (*Oryza sativa* L. *japonica* cv. Zhonghua 11) and transgenic rice lines were grown under natural conditions in experimental field plots for Anhui Agricultural University in Anhui Province, China. Rice endosperms at 3, 6, 10, and 20 DAP were harvested for qRT-PCR assay.

4.2. RNA Extraction and Real-Time RT-PCR Analysis

Total RNA of collected samples was extracted using RNAiso Plus Kit (Takara, Kusatsu, Japan) according to manufacturer's instructions. Afterwards, first-strand cDNA was generated using reverse transcription system (Promega, Madison, WI, USA). The qRT-PCR (quantitative real-time PCR) was performed using SYBR Green Master (Roche, Basel, Switzerland) on an ABI 7300 Real Time PCR System (Applied Biosystems, Foster City, CA, USA), and the reactions were performed according to previous report [22]. In detail, the reaction conditions were set as following: 50 °C for 2 min, then 95 °C for 10 min, followed by 40 cycles of 95 °C for 15 s and 60 °C for 1 min. Each cDNA sample was quantified in three replicates. The achieved data were calculated by 2_DDCt method as described previously [23]. Maize *Actin1* gene was used as internal control to normalize the detection threshold for each of three replicates.

4.3. Subcellular Localization

Full-length open reading frame (ORF) of *ZmES22* without the termination codon was amplified and cloned into pCAMBIA1305 vector under the drive of cauliflower mosaic virus (CaMV) 35S promoter. The fusion construct 35S:: *ZmES22*-GFP and empty vector 35S:: GFP were transformed into leaves for 35-day-old tobacco (*Nicotiana bethamiana*) via a syringe without needle, respectively. Infiltrated tobacco plants were transferred into dark condition for 12 h, followed by normal illumination for 48 h. Finally, green fluorescent signals were examined using confocal microscope (Olympus FV1000, Tokyo, Japan).

4.4. Transcriptional Activation Assay

Full-length ORF of *ZmES22* was amplified and inserted into pGBKT7 vector (Clontech, San Deigo, CA, USA), which were fused with the GAL4 DNA-binding domain beforehand. Subsequently, both negative control (null pGBKT7 vector) and positive control (co-transformation of pGBKT7-53 with pGADT7-T vectors), together with fusion construct pGBKT7-*ZmES22* was transformed into yeast strain AH109, respectively. AH109 strain carries HIS3, ADE2 and MEL1 reporter genes. Transformed yeast cells were then cultured on SD/-Trp medium for 3 days at 30 °C and then transferred into SD/-Trp/-His/-Ade/X-α-GAL medium for 3 days at 30 °C.

4.5. Generation of Transgenic Rice Lines

Full-length ORF of *ZmES22* was amplified and inserted into an overexpression vector pCAMBIA1301a under the drive of CaMV 35S promoter and a NOS terminator. Recombination construct pCAMBIA1301a-*ZmES22* also harbored a GUS reporter gene and was transformed into *japonica* rice cultivar Zhonghua 11 via *Agrobacterium* mediated transformation [24]. Both histochemical staining of GUS activity and PCR experiments followed by sanger sequencing were utilized to validate if transgenic rice lines were positive ones.

4.6. Determination of Agronomic Characters and Measurement of Grain Quality

Vernier caliper was adopted to measure the length, width and thickness for 100 uniformly mature seeds at the longest, widest, and thickest point, respectively. 1000-grain weight was determined by counting ten independent repeats of 100-grain samples on an electronic balance. Each measurement was repeated for three times. Embryos and follicles were separated from the embryo and ground into powder. The starch content was measured with total starch determination kit (K-TSTA; Megazyme, Bray, County Wicklow, Ireland) according to manufacturer's protocol. Apparent amylose content (AAC) of the samples was measured using iodine colorimetry (K-AMYL; Megazyme) [15]. Anthrone method was applied to determine soluble sugar content [7].

4.7. Measurement of Starch Blue Value (BV) and Maximum Absorption Wavelength (kmax)

The separation of amylose and amylopectin, and measurement of the blue value (BV) and maximum absorption wavelength (kmax) of starch were referred to the modified alkali impregnation method [25]. Detailedly, 5 mg isolated amylose powder was dissolved in 8 mL 90% dimethyl sulfoxide and then diluted to 100 mL double distilled water. The absorption spectra of the starch-iodine complex were examined ranging from 500 to 800 nm. The BV was A_{600}. While, with respect to amylopectin, 15 mg isolated amylopectin powder was dissolved in 100 mL double distilled water. The absorption spectra were examined ranging from 500 to 700 nm, and the BV was set to A_{680}.

4.8. Observation of Starch Granules by Scanning Electron Microscopy (SEM)

According to the methods in previous report [25], Hitachi S-3000N scanning electron microscope (SEM) (Hitachi, Tokyo, Japan) were used to observe the morphology of starch granules. SEM images were distinguished through cross-sections of mature rice seeds including ventral, central, and dorsal area of mature endosperm.

4.9. RNA-Seq and Data Analysis

Seeds for both overexpression transgenic rice lines and Zhonghua 11 at 20 DAP were collected for RNA-seq, each group were repeated twice as two biological replicates. Subsequently, RNA was isolated and then high throughput sequencing was performed on BGISEQ-500 platform in Beijing Genomics Institute (BGI; Shenzhen, China). After trimming of low-quality and adaptor sequences from raw sequencing reads, clean data were aligned to *Oryza sativa* ssp. *japonica* cv [26]. Nipponbare genome (IRGSP-1.0, http://rapdb.dna.affrc.go.jp/) [27] using TopHat2 software [28]. The resulted BAM alignment files were subject to Cufflinks to calculate gene expression levels [29]. Differentially expressed genes (DEGs) were determined by Cuffdiff with default parameters, based on the following criteria: (1) the minimum fold-change of gene expression was 2.0; (2) the maximum adjusted p value was 0.05 [30]. The RNA-seq data were validated using quantitative real-time PCR analysis for ten randomly selected DEGs. The R package ClusterProfiler were explored to conduct both Gene Ontology (GO) and Kyoto Encyclopedia of Genes and Genomes (KEGG) analysis [31]. The raw sequencing dataset has been submitted to NCBI's Gene Expression Omnibus (GEO; http://www.ncbi.nlm.nih.gov/geo/) under accession number SRP063765.

4.10. Yeast One-Hybrid Assay

Yeast one-hybrid assays were implemented originally according to the Matchmaker® Gold Yeast One-Hybrid Library Screening System User Manual (Clontech). To test the ability of ZmES22 to bind to the core motif CATGT of *GIF1* promoter, CATGT and mutant tandem repeats were cloned and inserted into the *Bam*HI and *Hind*III site of the p53/AbAi vector. Yeast Y1HGlod was transformed with the vector pGADT7-ZmES22 and CATGT or mutant tandem repeats plasmids. To evaluate interaction between ZmES22 and the core motif CATGT of *GIF1* promoter, the transformants were screened by plating on SD /-Leu/AbA plates.

5. Conclusions

Starch is one of the major components of cereal grains, providing sufficient calories for both human diet and animal feed. Therefore, comprehensive understanding molecular basis of starch synthesis process and its regulatory network is of vital importance. In the present study, we identified a gene *ZmES22*, encoding a typical MADS type transcription factor, which were exclusively highly expressed in maize endosperm, indicating its crucial role in endosperm development of maize. When *ZmES22* was overexpressed in rice, the 1000-grain weight, together with total starch content were remarkably reduced, whereas, the soluble sugar content was significantly higher when compared to wild type. Moreover, overexpression *ZmES22* altered the relative fraction of long branched starch and changed the morphology of starch granule from densely packed, polyhedral starch granules into loosely spherical granules with larger spaces. These results demonstrate that *ZmES22* is a negative regulator that could affect the starch biosynthesis process. Moreover, RNA-seq and qRT-PCR results further illustrated that overexpression of *ZmES22* could downregulate mRNA expression level of numerous key genes in starch synthesis pathway, particularly in early developmental stages in transgenic rice lines. Furthermore, ZmES22 could bind to the promoter region of the *OsGIF1* and downregulate its mRNA expression throughout the endosperm developmental stages. Therefore, we proposed that *ZmES22* might affect starch biosynthesis as well as reducing the rate of grain filling by downregulation of *OsGIF1* in rice. Whether knock-down or knockout of the *ZmES22* gene could contribute to increase of yield in maize remains to be demonstrated.

Author Contributions: L.G. designed the research; K.Z. conducted the molecular experiments and analyzed the data; H.X. performed the rice transformation; M.G., Z.M., and Y.W. analyzed agronomic characters, detected starch content; L.G. and W.S. drafted the manuscript.

References

1. Qi, X.; Li, S.; Zhu, Y.; Zhao, Q.; Zhu, D.; Yu, J. ZmDof3, a maize endosperm-specific Dof protein gene, regulates starch accumulation and aleurone development in maize endosperm. *Plant Mol. Biol.* **2017**, *93*, 7–20. [CrossRef] [PubMed]
2. Nakamura, Y. Towards a Better Understanding of the Metabolic System for Amylopectin Biosynthesis in Plants: Rice Endosperm as a Model Tissue. *Plant Cell Physiol.* **2002**, *43*, 718–725. [CrossRef]
3. James, M.G.; Denyer, K.; Myers, A.M. Starch synthesis in the cereal endosperm. *Curr. Opin. Plant Biol.* **2003**, *6*, 215–222. [CrossRef]
4. Li, S.; Wei, X.; Ren, Y.; Qiu, J.; Jiao, G.; Guo, X.; Tang, S.; Wan, J.; Hu, P. OsBT1 encodes an ADP-glucose transporter involved in starch synthesis and compound granule formation in rice endosperm. *Sci. Rep.* **2017**, *7*, 40124. [CrossRef]
5. Lee, S.-K.; Hwang, S.-K.; Han, M.; Eom, J.-S.; Kang, H.-G.; Han, Y.; Choi, S.-B.; Cho, M.-H.; Bhoo, S.H.; An, G.; et al. Identification of the ADP-glucose pyrophosphorylase isoforms essential for starch synthesis in the leaf and seed endosperm of rice (*Oryza sativa* L.). *Plant Mol. Biol.* **2007**, *65*, 531–546. [CrossRef] [PubMed]
6. Wang, E.; Wang, J.; Zhu, X.; Hao, W.; Wang, L.; Li, Q.; Zhang, L.; He, W.; Lu, B.; Lin, H.; et al. Control of rice grain-filling and yield by a gene with a potential signature of domestication. *Nat. Genet.* **2008**, *40*, 1370–1374. [CrossRef]
7. Wang, J.-C.; Xu, H.; Zhu, Y.; Liu, Q.-Q.; Cai, X.-L. OsbZIP58, a basic leucine zipper transcription factor, regulates starch biosynthesis in rice endosperm. *J. Exp. Bot.* **2013**, *64*, 3453–3466. [CrossRef]
8. Sun, C. A Novel WRKY Transcription Factor, SUSIBA2, Participates in Sugar Signaling in Barley by Binding to the Sugar-Responsive Elements of the iso1 Promoter. *Plant Cell Online* **2003**, *15*, 2076–2092. [CrossRef]
9. Schmidt, R.; Schippers, J.H.M.; Mieulet, D.; Watanabe, M.; Hoefgen, R.; Guiderdoni, E.; Mueller-Roeber, B. SALT-RESPONSIVE ERF1 Is a Negative Regulator of Grain Filling and Gibberellin-Mediated Seedling Establishment in Rice. *Mol. Plant* **2014**, *7*, 404–421. [CrossRef]
10. Chen, J.; Yi, Q.; Cao, Y.; Wei, B.; Zheng, L.; Xiao, Q.; Xie, Y.; Gu, Y.; Li, Y.; Huang, H.; et al. ZmbZIP91 regulates expression of starch synthesis-related genes by binding to ACTCAT elements in their promoters. *J. Exp. Bot.* **2016**, *67*, 1327–1338. [CrossRef] [PubMed]
11. Jiang, Y.; Zeng, B.; Zhao, H.; Zhang, M.; Xie, S.; Lai, J. Genome-wide Transcription Factor Gene Prediction and their Expressional Tissue-Specificities in Maize. *J. Integr. Plant Biol.* **2012**, *54*, 616–630. [CrossRef]
12. Cai, H.; Chen, Y.; Zhang, M.; Cai, R.; Cheng, B.; Ma, Q.; Zhao, Y. A novel GRAS transcription factor, ZmGRAS20, regulates starch biosynthesis in rice endosperm. *Physiol. Mol. Biol. Plants* **2017**, *23*, 143–154. [CrossRef] [PubMed]
13. Gramzow, L.; Ritz, M.S.; Theißen, G. On the origin of MADS-domain transcription factors. *Trends Genet.* **2010**, *26*, 149–153. [CrossRef]
14. Ma, H.; Yanofsky, M.F.; Meyerowitz, E.M. AGL1-AGL6, an Arabidopsis gene family with similarity to floral homeotic and transcription factor genes. *Genes Dev.* **1991**, *5*, 484–495. [CrossRef] [PubMed]
15. Takeda, Y.; Hizukuri, S.; Juliano, B.O. Purification and structure of amylose from rice starch. *Carbohydr. Res.* **1986**, *148*, 299–308. [CrossRef]
16. Chen, X.; Guo, L.; Du, X.; Chen, P.; Ji, Y.; Hao, H.; Xu, X. Investigation of glycerol concentration on corn starch morphologies and gelatinization behaviours during heat treatment. *Carbohydr. Polym.* **2017**, *176*, 56–64. [CrossRef]
17. Lescot, M.; Déhais, P.; Thijs, G.; Marchal, K.; Moreau, Y.; de Peer, Y.V.; Rouzé, P.; Rombauts, S. PlantCARE, a database of plant cis-acting regulatory elements and a portal to tools for in silico analysis of promoter sequences. *Nucleic Acids Res.* **2001**, *30*, 325–327. [CrossRef]
18. Qiao, Z.; Qi, W.; Wang, Q.; Feng, Y.; Yang, Q.; Zhang, N.; Wang, S.; Tang, Y.; Song, R. ZmMADS47 Regulates Zein Gene Transcription through Interaction with Opaque2. *PLoS Genet.* **2016**, *12*, e1005991. [CrossRef]
19. Deschamps, P.; Colleoni, C.; Nakamura, Y.; Suzuki, E.; Putaux, J.-L.; Buleon, A.; Haebel, S.; Ritte, G.; Steup, M.; Falcon, L.I.; et al. Metabolic Symbiosis and the Birth of the Plant Kingdom. *Mol. Biol. Evol.* **2008**, *25*, 536–548. [CrossRef]
20. Qu, J.; Xu, S.; Zhang, Z.; Chen, G.; Zhong, Y.; Liu, L.; Zhang, R.; Xue, J.; Guo, D. Evolutionary, structural and expression analysis of core genes involved in starch synthesis. *Sci. Rep.* **2018**, *8*, 12736. [CrossRef]

21. Nelson, O.; Pan, D. Starch Synthesis in Maize Endosperms. *Annu. Rev. Plant Physiol. Plant Mol. Biol.* **1995**, *46*, 475–496. [CrossRef]

22. Zhai, R.; Feng, Y.; Wang, H.; Zhan, X.; Shen, X.; Wu, W.; Zhang, Y.; Chen, D.; Dai, G.; Yang, Z.; et al. Transcriptome analysis of rice root heterosis by RNA-Seq. *BMC Genom.* **2013**, *14*, 19. [CrossRef] [PubMed]

23. Livak, K.J.; Schmittgen, T.D. Analysis of Relative Gene Expression Data Using Real-Time Quantitative PCR and the $2-\Delta\Delta CT$ Method. *Methods* **2001**, *25*, 402–408. [CrossRef] [PubMed]

24. Lin, Y.J.; Zhang, Q. Optimising the tissue culture conditions for high efficiency transformation of indica rice. *Plant Cell Rep.* **2005**, *23*, 540–547. [CrossRef]

25. Fu, F.-F.; Xue, H.-W. Coexpression Analysis Identifies Rice Starch Regulator1, a Rice AP2/EREBP Family Transcription Factor, as a Novel Rice Starch Biosynthesis Regulator. *Plant Physiol.* **2010**, *154*, 927–938. [CrossRef]

26. Bolger, A.M.; Lohse, M.; Usadel, B. Trimmomatic: A flexible trimmer for Illumina sequence data. *Bioinformatics* **2014**, *30*, 2114–2120. [CrossRef]

27. Sakai, H.; Lee, S.S.; Tanaka, T.; Numa, H.; Kim, J.; Kawahara, Y.; Wakimoto, H.; Yang, C.; Iwamoto, M.; Abe, T.; et al. Rice Annotation Project Database (RAP-DB): An Integrative and Interactive Database for Rice Genomics. *Plant Cell Physiol.* **2013**, *54*, e6. [CrossRef]

28. Kim, D.; Pertea, G.; Trapnell, C.; Pimentel, H.; Kelley, R.; Salzberg, S.L. TopHat2: Accurate alignment of transcriptomes in the presence of insertions, deletions and gene fusions. *Genome Biol.* **2013**, *14*, R36. [CrossRef]

29. Trapnell, C.; Hendrickson, D.G.; Sauvageau, M.; Goff, L.; Rinn, J.L.; Pachter, L. Differential analysis of gene regulation at transcript resolution with RNA-seq. *Nat. Biotechnol.* **2013**, *31*, 46–53. [CrossRef]

30. Trapnell, C.; Roberts, A.; Goff, L.; Pertea, G.; Kim, D.; Kelley, D.R.; Pimentel, H.; Salzberg, S.L.; Rinn, J.L.; Pachter, L. Differential gene and transcript expression analysis of RNA-seq experiments with TopHat and Cufflinks. *Nat. Protoc.* **2012**, *7*, 562–578. [CrossRef]

31. Yu, G.; Wang, L.-G.; Han, Y.; He, Q.-Y. clusterProfiler: An R Package for Comparing Biological Themes Among Gene Clusters. *OMICS J. Integr. Biol.* **2012**, *16*, 284–287. [CrossRef] [PubMed]

PERMISSIONS

All chapters in this book were first published in MDPI; hereby published with permission under the Creative Commons Attribution License or equivalent. Every chapter published in this book has been scrutinized by our experts. Their significance has been extensively debated. The topics covered herein carry significant findings which will fuel the growth of the discipline. They may even be implemented as practical applications or may be referred to as a beginning point for another development.

The contributors of this book come from diverse backgrounds, making this book a truly international effort. This book will bring forth new frontiers with its revolutionizing research information and detailed analysis of the nascent developments around the world.

We would like to thank all the contributing authors for lending their expertise to make the book truly unique. They have played a crucial role in the development of this book. Without their invaluable contributions this book wouldn't have been possible. They have made vital efforts to compile up to date information on the varied aspects of this subject to make this book a valuable addition to the collection of many professionals and students.

This book was conceptualized with the vision of imparting up-to-date information and advanced data in this field. To ensure the same, a matchless editorial board was set up. Every individual on the board went through rigorous rounds of assessment to prove their worth. After which they invested a large part of their time researching and compiling the most relevant data for our readers.

The editorial board has been involved in producing this book since its inception. They have spent rigorous hours researching and exploring the diverse topics which have resulted in the successful publishing of this book. They have passed on their knowledge of decades through this book. To expedite this challenging task, the publisher supported the team at every step. A small team of assistant editors was also appointed to further simplify the editing procedure and attain best results for the readers.

Apart from the editorial board, the designing team has also invested a significant amount of their time in understanding the subject and creating the most relevant covers. They scrutinized every image to scout for the most suitable representation of the subject and create an appropriate cover for the book.

The publishing team has been an ardent support to the editorial, designing and production team. Their endless efforts to recruit the best for this project, has resulted in the accomplishment of this book. They are a veteran in the field of academics and their pool of knowledge is as vast as their experience in printing. Their expertise and guidance has proved useful at every step. Their uncompromising quality standards have made this book an exceptional effort. Their encouragement from time to time has been an inspiration for everyone.

The publisher and the editorial board hope that this book will prove to be a valuable piece of knowledge for researchers, students, practitioners and scholars across the globe.

LIST OF CONTRIBUTORS

Sunny Ahmar and Muhammad Uzair Qasim
National Key Laboratory of Crop Genetic Improvement, College of Plant Science and Technology, Huazhong Agricultural University, Wuhan 430070, Hubei, China

Rafaqat Ali Gill
Oil Crops Research Institute, Chinese Academy of Agriculture Sciences, Wuhan 430070, China

Ki-Hong Jung
Graduate School of Biotechnology & Crop Biotech Institute, Kyung Hee University, Yongin 17104, Korea

Aroosha Faheem
State Key Laboratory of Agricultural Microbiology and State Key Laboratory of Microbial Biosensor, College of Life Sciences Huazhong Agriculture University, Wuhan 430070, China

Mustansar Mubeen
State Key Laboratory of Agricultural Microbiology and Provincial Key Laboratory of Plant Pathology of Hubei Province, College of Plant Science and Technology, Huazhong Agricultural University, Wuhan 430070, China

Weijun Zhou
Institute of Crop Science and Zhejiang Key Laboratory of Crop Germplasm, Zhejiang University, Hangzhou 310058, China

Hong Wang, Yingxin Zhang, Lianping Sun, Peng Xu, Ranran Tu, Shuai Meng, Weixun Wu, Kashif Hussain, Aamiar Riaz, Daibo Chen, Liyong Cao, Shihua Cheng and Xihong Shen
Key Laboratory for Zhejiang Super Rice Research, State Key Laboratory of Rice Biology, China National Rice Research Institute, Hangzhou 311400, Zhejiang, China

Galal Bakr Anis
Key Laboratory for Zhejiang Super Rice Research, State Key Laboratory of Rice Biology, China National Rice Research Institute, Hangzhou 311400, Zhejiang, China Rice Research and Training Center, Field Crops Research Institute, Agriculture Research Center, Kafr Elsheikh 33717, Egypt

Shiwu Gao, Yingying Yang, Liping Xu, Jinlong Guo, Yachun Su, Qibin Wu, Chunfeng Wang and Youxiong Que
Key Laboratory of Sugarcane Biology and Genetic Breeding, Ministry of Agriculture and Key Laboratory of Crop Genetics and Breeding and Comprehensive Utilization, College of Crop Science, Fujian Agriculture and Forestry University, Ministry of Education, Fuzhou 350002, China

Xiaoping Liu, Hailong Yu, Fengqing Han, Zhiyuan Li, Zhiyuan Fang, Limei Yang, Mu Zhuang, Honghao Lv, Yumei Liu, Zhansheng Li, Xing Li and Yangyong Zhang
Institute of Vegetables and Flowers, Chinese Academy of Agricultural Sciences, Key Laboratory of Biology and Genetic Improvement of Horticultural Crops, Ministry of Agriculture, Beijing 100081, China

Ling He, Yin-Huan Wu, Qian Zhao, Bei Wang, Qing-Lin Liu and Lei Zhang
Department of Ornamental Horticulture, Sichuan Agricultural University, 211 Huimin Road, Wenjiang District, Chengdu 611130, Sichuan, China

Jauhar Ali, Zilhas Ahmed Jewel and Anumalla Mahender
Rice Breeding Platform, International Rice Research Institute (IRRI), Los Baños, Laguna 4031, Philippines

Annamalai Anandan
ICAR-National Rice Research Institute, Cuttack, Odisha 753006, India

Jose Hernandez
Institute of Crop Science, College of Agriculture and Food Science, University of the Philippines Los Baños, Laguna 4031, Philippines

Zhikang Li
Institute of Crop Sciences, Chinese Academy of Agricultural Science, Beijing 100081, China

Pronob J. Paul
International Crops Research Institute for the Semi-Arid Tropics (ICRISAT), Patancheru Hyderabad 502324, India Department of Genetics and Plant Breeding, Sam Higginbottom University of Agriculture, Technology and Sciences (SHUATS), Allahabad 211007, India

Srinivasan Samineni, Mahendar Thudi, Sobhan B. Sajja, Abhishek Rathore, Roma R. Das, Rajeev. K. Varshney and Aamir W. Khan
International Crops Research Institute for the Semi-Arid Tropics (ICRISAT), Patancheru Hyderabad 502324, India

Gera Roopa Lavanya
Department of Genetics and Plant Breeding, Sam Higginbottom University of Agriculture, Technology and Sciences (SHUATS), Allahabad 211007, India

Sushil K. Chaturvedi
ICAR-Indian Institute of Pulses Research (ICAR-IIPR), Kanpur 208024, India

Pooran M. Gaur
International Crops Research Institute for the Semi-Arid Tropics (ICRISAT), Patancheru Hyderabad 502324, India
The UWA Institute of Agriculture, University of Western Australia, Perth, WA 6009, Australia

Xiaoshuang Li and Daoyuan Zhang
Key Laboratory of Biogeography and Bioresource in Arid Land, Xinjiang Institute of Ecology and Geography, Chinese Academy of Sciences, Urumqi 830011, China

Bei Gao
School of Life Sciences and State Key Laboratory of Agrobiotechnology, The Chinese University of Hong Kong, Hong Kong, China

Yuqing Liang and Xiaojie Liu
Key Laboratory of Biogeography and Bioresource in Arid Land, Xinjiang Institute of Ecology and Geography, Chinese Academy of Sciences, Urumqi 830011, China
University of Chinese Academy of Sciences, Beijing 100049, China

Jinyi Zhao
School of Life Science, University of Liverpool, Liverpool L169 3BX, UK

Jianhua Zhang
Department of Biology, Hong Kong Baptist University, Hong Kong, China

Andrew J. Wood
Department of Plant Biology, Southern Illinois University, Carbondale, IL 62901-6899, USA

Zhen-Hua Zhang, Yu-Jun Zhu, Shi-Lin Wang, Ye-Yang Fan and Jie-Yun Zhuang
State Key Laboratory of Rice Biology and Chinese National Center for Rice Improvement, China National Rice Research Institute, Hangzhou 310006, China

Rong Zhou, Donghua Li, Jingyin Yu, Jun You and Xiurong Zhang
Key Laboratory of Biology and Genetic Improvement of Oil Crops, Oil Crops Research Institute of the Chinese Academy of Agricultural Sciences, Ministry of Agriculture, No. 2 Xudong 2nd Road, Wuhan 430062, China

Komivi Dossa
Key Laboratory of Biology and Genetic Improvement of Oil Crops, Oil Crops Research Institute of the Chinese Academy of Agricultural Sciences, Ministry of Agriculture, No. 2 Xudong 2nd Road, Wuhan 430062, China
Centre d'Etude Régional Pour l'Amélioration de l'Adaptation à la Sécheresse (CERAAS), Route de Khombole, Thiès, Thiès Escale Thiès BP3320, Senegal

XinWei
Key Laboratory of Biology and Genetic Improvement of Oil Crops, Oil Crops Research Institute of the Chinese Academy of Agricultural Sciences, Ministry of Agriculture, No. 2 Xudong 2nd Road, Wuhan 430062, China
College of Life and Environmental Sciences, Shanghai Normal University, Shanghai 200234, China

Bingli Ding, Mengyu Hao, Desheng Mei, Qamar U Zaman, Shifei Sang, Hui Wang, Wenxiang Wang, Li Fu, Hongtao Cheng and Qiong Hu
Key Laboratory for Biological Sciences and Genetic Improvement of Oil Crops, Ministry of Agriculture, Oil Crops Research Institute, Chinese Academy of Agricultural Sciences, Wuhan 430062, China

Ming Lei, Zhi-ying Li, Jia-bin Wang, Yun-liu Fu, Meng-fei Ao and Li Xu
Institute of Tropical Crops Genetic Resources, Chinese Academy of Tropical Agricultural Sciences, Danzhou 571737, China
Key Laboratory of Crop Gene Resources and Germplasm Enhancement in Southern China, Danzhou 571737, China
Key Laboratory of Tropical Crops Germplasm Resources Genetic Improvement and Innovation, Danzhou 571737, China
Mid Tropical Crop Gene Bank of National Crop Resources, Danzhou 571737, China

Kangyong Zha, Haoxun Xie, Min Ge, Zimeng Wang, Yu Wang, Weina Si and Longjiang Gu
National Engineering Laboratory of Crop Stress Resistance breeding, Anhui Agricultural University, Hefei 230036, China

Index

Printed in the USA
CPSIA information can be obtained
at www.ICGtesting.com
JSHW051401091023
49903JS00006B/232